Soil Biology

Volume 43

Series Editor
Ajit Varma, Amity Institute of Microbial Technology,
Amity University Uttar Pradesh, Noida, UP, India

More information about this series at
http://www.springer.com/series/5138

K. Sowjanya Sree • Ajit Varma

Editors

Biocontrol of Lepidopteran Pests

Use of Soil Microbes and their Metabolites

 Springer

Editors
K. Sowjanya Sree
Amity Institute of Microbial Technology
Amity University Uttar Pradesh
Noida
Uttar Pradesh
India

Ajit Varma
Amity Institute of Microbial Technology
Amity University Uttar Pradesh
Noida
Uttar Pradesh
India

ISSN 1613-3382   ISSN 2196-4831   (electronic)
Soil Biology
ISBN 978-3-319-14498-6   ISBN 978-3-319-14499-3   (eBook)
DOI 10.1007/978-3-319-14499-3

Library of Congress Control Number: 2015933012

Springer Cham Heidelberg New York Dordrecht London

Printed on acid-free paper

Springer International Publishing AG Switzerland is part of Springer Science+Business Media (www.springer.com)

Foreword

Insect pest management has always been a major concern to farmers around the globe. Insect pest attack accounts to as high as 42 % loss of crop productivity, as estimated already a decade back. Lepidopteran pests contribute a major share to this crop loss. Larvae of these pests are very voracious especially at the younger instars causing a drastic damage to the crop foliage and produce. Although their natural enemies do exist, the man-made ecological imbalances do not allow them to substantially control the huge pest populations. For a long time, the effective management of these pests was relied upon the chemical and synthetic pesticides. But for over half a century now, it has been realised that the indiscriminate use of these chemicals cause dramatic ill-effects concerning health and environmental safety. This led to a paradigm shift in the pest management strategies being practiced.

The development and use of natural enemies including parasites, predators and pathogens of insect pests as biopesticides have gained impetus in the last few decades, and biopesticide has become one of the key components of Integrated Pest Management (IPM). In fact, there are some very successful cases of the use of biopesticides for insect control. A few instances are the control of sugarcane stem borer in India with its parasite *Trichogramma* spp., and saving of the staple food crop Cassava in Africa from mealybugs by aerial spray of a tiny parasitic wasp, *Anagyrus lopezi*. Amongst these biocontrol agents, the entomopathogenic micro-organisms have had an edge over the others, because of their production and application feasibility, specificity, economic viability and environmental safety. Therefore, the promotion of biopesticides for insect control holds great promise for the future.

The editors of this volume together with the authors of the individual chapters have made a remarkable contribution in collating the up-to-date information on the successful development and use of insect pathogenic viruses, bacteria, fungi and nematodes as biopesticides against the lepidopteran pests. Apart from the use of these microorganisms, their metabolites have also been screened for use as biopesticides. The use of transgenic technology in developing pest resistant crop varieties has led to a completely new era of pest management science. The overwhelming response of farmers towards the cultivation of Bt-cotton crop is inspirational in looking forward to more such transgenic resistant varieties in the near future.

I wish to congratulate the editors for bringing out a unique volume and insight into biological control of insect pests. This book will be indispensible for scientists and professionals working in the field of crop protection and can also serve as a reference book for graduate students.

February 2015   M.V. Rajam
                                                                    Department of Genetics
                                                         University of Delhi South Campus
                                                                             New Delhi
                                                                                 India

Preface

Lepidopteran pest infestation of economically important plants is a menace to farmers all over the world. These pests are in general polyphagous and voracious feeders. The larval stages cause huge economic losses to the agricultural yield. Although a number of chemical pesticides have been discovered and are in use, the increasing pest resistance to these chemicals is an alarming problem in addition to the environmental pollution caused by these chemical pesticides. The use of biocontrol agents for the control of lepidopteran pests is increasing over time and could be looked upon as an upcoming solution to resolve the environmental issues.

The entomopathogenic viruses, bacteria, fungi, nematodes and their metabolites show active insecticidal properties against these pests. This volume, comprising of 16 chapters, is focused on the characterization, mechanism of action and application of some of these entomopathogenic viruses, bacteria, fungi, nematodes and the metabolites of these microbes against the devastating lepidopteran pest species around the world. The larva of *Helicoverpa armigera*, one of the economically important lepidopteran crop pests, infected by NucleoPolyhedro Virus (NPV) is depicted in Fig. 2.4 (see Chap. 2 of this volume).

Having a focus on different aspects of entomopathogenic soil microorganisms and their metabolites for their use in biocontrol of lepidopteran pests, this volume will attract the attention of life science researchers in general and microbiologists, agricultural researchers, entomologists and applied chemists in specific. It will also serve to motivate the researchers focused on integrated pest management studies.

It was our pleasure to interact with all the authors, and we thank them for their stimulating contributions. We sincerely wish to acknowledge Hanna Hensler-Fritton and Jutta Lindenborn at Springer, Heidelberg, for their generous assistance and patience in shaping this volume. K. Sowjanya Sree acknowledges the financial support from Science and Engineering Research Board, Govt. of India, through the Fast track Young Scientist scheme.

We are thankful to Dr. Ashok K. Chauhan, Founder President of the Ritnand Balved Education Foundation (an umbrella organization of Amity Institution), New Delhi, for his kind support and constant encouragement.

Noida, India   K. Sowjanya Sree
                                                                                Ajit Varma

Contents

Chapter 1
An Introduction to Entomopathogenic Microorganisms

K. Sowjanya Sree and Ajit Varma

1.1 Introduction

The quality and quantity of the agricultural produce is influenced by a variety of abiotic and biotic factors. The abiotic factors including drought, temperature extremes and salinity partner with the biotic factors like weed and pest attack to enhance their negative impact on the crop yield. In addition, in the recent years, drastic changes in agricultural landscapes have occurred because of the intensification of agricultural systems (Robinson and Sutherland 2002; Bianchi et al. 2006). The simplified agricultural landscapes because of clearing of lands, up-scaling and expansion of crop fields have eventually led to a reduction in the occurrence of natural habitats (Robinson and Sutherland 2002; Benton et al. 2003). This has also led to the loss of biodiversity with respect to natural pest control (Ives et al. 2000; Wilby and Thomas 2002; Gurr et al. 2003). Apart from this, the changing climate in the years to come is predicted to have heavy impact on agriculture ranging from reduction in availability of fertile land (Easterling et al. 2000) to changes in the life cycle of plant pests and pathogens (Lonsdale and Gibbs 2002; Turner 2008; Collier et al. 2008).

Crop pests are one of the major contributors to economic losses in the agriculture sector (Oerke et al. 1994). Considering the pest population, it is biologically a very diverse group including pathogenic microbes such as viruses, bacteria or fungi and parasitic nematodes; invertebrates like insects and molluscs; and some vertebrates. In the present volume, we will focus on the insect pest species. Insects attack the crop plants by either herbivory, feeding on their sap or by causing or transmitting diseases, thereby reducing the crop yields globally. They can also attack the stored agricultural produce further increasing the damage. Order Lepidoptera of the class:

K.S. Sree (✉) • A. Varma
Amity Institute of Microbial Technology, Amity University Uttar Pradesh, Noida 201303,
Uttar Pradesh, India
e-mail: ksowsree@gmail.com

© Springer International Publishing Switzerland 2015
K.S. Sree, A. Varma (eds.), *Biocontrol of Lepidopteran Pests*, Soil Biology 43,
DOI 10.1007/978-3-319-14499-3_1

1

Insecta which includes butterflies and moths is one of the most diverse ones (Krenn 2010) and is the second largest order of insects. Many of the lepidopteran insect species, both in adult and larval stages, are associated with the plants. Infestation of a wide range of crop plants like tomato, cotton, maize, cauliflower, cabbage, chickpea, lentils and so on by the lepidopteran insect species causes huge economic losses globally. The lepidopteran larvae have biting and chewing type of mouth parts with well-developed mandibles facilitating them to feed voraciously on the foliage of the crop plants (Krenn 2010). These are considered to be the most damaging stages of the lepidopteran insect pests. The adults, although the females have a high fecundity rate, are floral nectar feeders and have been long established as pollinators of flowering plants (Kevan and Baker 1983; Fenster et al. 2004).

Chemical insecticides have been used now for years for the control of the damaging lepidopteran populations in the agricultural fields. However, after the ill-effects of long-term, indiscriminate use of chemical pesticides and insecticides on environment, human and animal health have been revealed in the past few decades (Carson 1962; Klaassen et al. 1996; Meiners and Morriss 2001), the use and development of alternate methods for the control of insect pest populations became a necessity. Additionally, the number of insect species gaining resistance to a variety of chemical insecticides has increased tremendously (Mallet 1989). In this context, the use of biological control agents (BCAs) is being looked upon as an environmentally safe insect pest control strategy for sustainable agriculture worldwide.

1.2 Biological Control

Biological control makes use of living organisms or their products which can manage the insect pest populations thereby minimising the damage to crop yield, both in terms of quality and quantity (Bale et al. 2008). The biological control strategies, in many cases, have been developed by researchers taking examples from nature. The natural enemies in a given agro-ecosystem play a major role in suppressing the pest populations in that region (Hajek 2004). In this context, it might be interesting to describe an instance which the author, KSS, came across. A progressive farmer, Kongara Ramesh, from the Visakhapatnam district of Andhra Pradesh, India, made a careful observation of the natural death of lepidopteran larvae each year with a cottony mass over them. He was curious to know the reason of their death and if this subject could be used as a biopesticide. With the help of researchers in this field, they found out that it was a fungal pathogen, *Beauveria bassiana*. He later went on to culture this entomopathogenic fungus for use in his agricultural fields. There exist, for sure, many other such farmers around the world. Biological control programmes operate worldwide in agriculture and forestry.

The BCAs used in pest management strategies include natural enemies of insect pests such as predatory insects, parasitoids and parasites. In order to have maximum returns using environmentally sustainable practices, integrated pest management

(IPM) strategies are being taken forward which include effective mixtures of different biological, chemical and traditional pest control measures (Flint and Van den Bosch 1981). BCAs have the potential to be self-sustainable. They display host specificity, non-phytotoxicity and human safety. However, a pre-condition for successful use of BCAs is the basic knowledge of ecology of both pest and natural enemy. Under a set circumstance, sometimes, the use of BCAs can prove to be more cost-effective than the conventional chemical control measures. On the other hand, BCAs also have some serious disadvantages because of which they are not able to cope with the competitive markets of the less expensive and more effective chemical pesticides. Most BCAs are niche products. And unlike the synthetic insecticides, these do not show immediate effects on the pest upon application. Additionally, the stringent and expensive approval and registration processes in case of the microbial BCAs significantly increase their cost of production (Mallet 1989). Lack of environmental persistence and low, unpredictable efficacy under field conditions are a few other limitations for the success of BCAs.

In the present chapter, an overview of the soil microorganisms including viruses, bacteria, fungi and nematodes pathogenic to lepidopteran insect pests and their use as effective BCAs have been outlined.

1.3 Entomopathogenic Soil Microorganisms

The entomopathogens have a wide distribution in the natural environment and are reported to cause diseases in many insect pest species, thereby emerging as potential BCAs. As many as 3,000 entomopathogens have been reported. A number of these entomopathogenic microorganisms can be mass produced, formulated appropriately and applied in the pest-infested crop fields with ease. The different categories of entomopathogens are detailed as below.

1.3.1 Entomopathogenic Viruses

Although many families of viruses have been shown to be infective to insects and invertebrates in general (Fauquet et al. 2005), much of the research on use and development of virus-based bioinsecticides has been concentrated towards baculoviruses. This is based on the fact that baculoviruses act as potent natural enemies of insect pests, lepidopteran pests to be more specific (Caballero et al. 1992). Although not very related, it is worth a mention that much of the advancements in the field of baculoviruses is also due to their use as eukaryotic expression vector systems. They have rod-shaped enveloped virions with single, covalently closed circular double-stranded DNA as their genome (Theilmann et al. 2005). Occlusion-derived virions (ODVs) and the budded virions (BVs) represent the two typical phenotypes of virions which occur in baculovirus infections. The ODVs are occluded, towards the end of the virus replication in the infected cell nucleus, in a crystalline protein matrix. Horizontal transmission of baculoviruses occurs through ODVs. They initiate virus infection in the insect midgut upon ingestion of the virus occlusion body (OB). BVs are produced when nucleocapsids bud through the plasma membrane of virus-infected cells. These are responsible for causing a systemic infection by spreading the virus to tissues throughout the host (Erlandson and Theilmann 2009).

The baculovirus-based bioinsecticides are host specific and safe to environment, human and animal health with no toxic residues and can be combined with other strategies for IPM (Monobrullah 2003; Ashour et al. 2007). Detailed classification, morphology and use of baculoviruses for control of lepidopteran pests are given under Chap. 2 of this volume by Ranga Rao et al.

1.3.2 Entomopathogenic Bacteria

Over a century back, a moth infected with a bacterium was found in Thuringia, Germany, by Ernst Berliner (Siegel 2000). This bacterium was isolated, identified and named as *Bacillus thuringiensis*. In the past few decades, research on *B. thuringiensis* has exploded into diverse innovative directions for its use as BCA and has also triggered research on other potential entomopathogenic bacteria.

Entomopathogenic bacteria fall into two categories, obligate and facultative pathogens (Bucher 1960; Krieg 1981). Obligate entomopathogenic bacteria are capable of damaging a healthy intact midgut wall of the insect larvae. A gram-positive spore-forming bacterium, *B. thuringiensis*, belongs to this class (Schnepf et al. 1998). In contrast, the facultative entomopathogens like *Serratia marcescens* can disrupt the gut wall only when the insect is under some kind of stress (Matsumoto et al. 1998) and need to be injected directly into the haemocoel for the bacterium to cause septicaemia.

The entomopathogenic bacteria have developed an array of strategies to infect and kill the host larvae. These pathogens make use of diverse virulence factors such as insecticidal toxins, exoenzymes and haemolysins. *B. thuringiensis* produces different kinds of toxin proteins at different phases which show potent insecticidal activities. Cry and Cyt toxins are produced as parasporal crystals during the sporulation phase. After ingestion, the Cry toxin binds to specific receptors in the insect midgut epithelium. By the action of proteases, the protoxin is cleaved releasing the active toxin. The activated toxin induces pore formation in the epithelial cells of the larval midgut leading to their lysis and subsequent kill of the insect larva (Daniel et al. 2000). Over 400 genes coding for these toxin variants have been identified from different strains of *B. thuringiensis* (Crickmore et al. 2007). The insecticidal spectrum of each of these toxin variants is distinct. Vegetative insecticidal proteins (Vip) are a set of toxins produced both during the vegetative and sporulation phases of the bacterium. Upon ingestion, Vip toxins cause swelling and osmotic lysis of epithelial cells of the midgut leading to host death (Estruch et al. 1996; Crickmore et al. 2007). A detailed view on this topic is given in Chaps. 3 and 4 of this volume.

Making the plants resistant to insect pests through transgenic approach by transforming them with *cry* genes (Vaeck et al. 1987; Romeis et al. 2006) has attained a great success. One of the examples which gained commercial success is '*Bt* Cotton'. However, development of resistance by the insect pests to these toxins is raising concerns in this area of research. One such case has been reported recently where it was shown that the cry1Ac toxin had a reduced binding capacity to the receptors on the brush border membrane vesicles of the field-collected cry1Ac-resistant pink bollworms (Ojha et al. 2014). Further, the safety and public acceptance of the genetically modified foods is also raising concerns for the commercialisation of transgenic crops (Shelton et al. 2002).

Research is also being focused on investigating other entomopathogenic bacteria like the species of *Photorhabdus* and *Xenorhabdus*. They are gram-negative motile bacteria living as obligate mutualists in association with the nematodes, *Heterorhabditis* and *Steinernema*, respectively (Waterfield et al. 2009; Castagnola and Stock 2014). The parasitic nematodes infect the insect pests and carry along the symbiotic bacteria from one larval host to another. These bacteria have been shown to have insecticidal properties (Silva et al. 2002; Waterfield et al. 2009) and have also displayed a plethora of toxin proteins, viz., toxin complex (Tc) proteins (Bowen and Ensign 1998), *Photorhabdus* insect-related (Pir) proteins (Waterfield et al. 2005) and insecticidal pilin proteins (Khandelwal et al. 2004). More information on nematode-associated entomopathogenic bacteria and a detailed review on their toxins can be referred from Chaps. 5 and 13 of this volume.

1.3.3 Entomopathogenic Fungi

Entomopathogenic fungi, for decades now, provide a feasible system for insect pest control and have served as model for plant-pathogen interaction studies. This group of entomopathogens are the most commonly reported natural enemies in the agricultural fields causing regular epizootics (Rios-Velasco et al. 2010). Considering entomopathogenic fungi as a whole, they infect a wide range of insect species including lepidopterans, but the individual species and strains of entomopathogenic fungus have specific host range and virulence. About 750 fungal species have been reported to be entomopathogenic although a very small number of them are used as BCAs. The species belonging to the genera *Metarhizium* (Pandey and Hasan 2009), *Beauveria* (Garcia et al. 2011), *Nomuraea* (Ingle et al. 2004), *Lecanicillium* (Anand et al. 2009) and *Isaria* (Zimmermann 2008) have been well worked out with noteworthy contributions.

These fungi naturally inhabit the soils worldwide and can be isolated from both soil- and fungus-infected insect cadavers. Fungal conidia are the infective propagules which when in contact with a susceptible host attach to its cuticle and develop a germ tube. This is however accompanied by an array of molecular responses in both the host and the pathogen. The formation of appressoria, penetration peg and the secretion of hydrolytic enzymes help the fungus to enter the insect's body. Once inside, the fungal mycelium takes up nutrients from the host for its own nourishment and causes mechanical injury to the insect leading to its death. The hyphae pierce out of the insect cadaver and sporulate on its surface. The dispersal of these conidia aids in spreading the disease and in continuing the fungal life cycle on a new host (Shah and Pell 2003; Schrank and Vainstein 2010). The advantage that the fungal BCAs have over others is that they can infect any developmental stage of the insect (Anand et al. 2009; Anand and Tiwary 2009) and are environmentally safe, also to humans (Goettel et al. 2001; Kubicek and Druzhinina 2007). More details on entomopathogenic fungi can be referred to in Chaps. 6, 7 and 10–12 of this volume.

Apart from the mechanical injury to the insect host, the fungus also produces toxins which are insecticidal in nature (Sree et al. 2008). These mycotoxins are non-ribosomal peptides and can be classified into chain peptides or cyclic peptides based on their chemical structure. A number of toxins have been reported from different entomopathogenic fungi, some of them include destruxins (Pedras et al. 2002; Sree and Padmaja 2008; Liu and Tzeng 2012), bassianolides (Suzuki et al. 1977), beauvericins (Wang and Xu 2012), enniatins (Mule et al. 1992; Hiraga et al. 2005), efrapeptins (Lardy et al. 1975) and so on. Chapters 8–11 of this volume provide more details on these mycotoxins.

1.3.4 Entomopathogenic Nematodes

Entomopathogenic nematodes are obligate parasites of insect species and are associated with symbiotic bacteria which also show potent insecticidal properties (Waterfield et al. 2009; Stock and Goodrich-Blair 2012). These nematodes belong to either of the two families, Steinernematidae and Heterorhabditidae. Their potential as insecticidal agents has been investigated against a wide range of insect species including lepidopteran pests by many researchers around the world (Begley 1990; Grewal et al. 2005; Hussain and Ahmad 2011; Lacey and Georgis 2012). The juvenile nematodes parasitise their host insects by directly penetrating the cuticle or by entry through the natural openings. The symbiotic bacteria associated with the nematodes are regurgitated in the host, which then multiply rapidly and cause host death by septicaemia, often within 48 h. The bacterial attack causes break down of the host tissue, thereby providing nourishment to the nematodes. Subsequent to host insect death, the juvenile nematodes develop into adults and reproduce, raising its next generation (Nguyen and Smart 1990; Jackson and Brooks 1995).

Use of entomopathogenic nematodes as BCAs has undergone a lot of development over the past few decades (Shapiro-Ilan et al. 2002). Foliar (Arthurs et al. 2004) and epigeal applications of these nematodes have been investigated (Shapiro-Ilan et al. 2006). Epigeal or soil surface application has yielded better results than foliar application owing to the effect of abiotic stresses on the nematodes sprayed on the foliage (Begley 1990). More elaborate details on entomopathogenic nematodes are available in Chaps. 13–15 of this volume.

1.4 Conclusions and Future Prospects

The insect order Lepidoptera comprises of a large number of economically important pests. The control of these pests in an environmentally sustainable manner is very important in the current scenario of environment protection. The use of BCAs like viruses, bacteria, fungi and nematodes plays a major role in this perspective. Also, the secondary metabolites from these microorganisms are being exploited for the development of eco-friendly bioinsecticides. Research is progressing towards the efficient use of these microorganisms and their metabolites in IPM. The rest of the 15 book chapters of the present book volume describe in detail the use of these BCAs.

Acknowledgements KSS is thankful to the SERB, Govt. of India, for financial assistance through the Fast Track Young Scientist scheme.

References

Anand R, Tiwary BN (2009) Pathogenicity of entomopathogenic fungi to eggs and larvae of *Spodoptera litura*, the common cutworm. Biocontrol Sci Technol 19:919–929

Anand R, Prasad B, Tiwary BN (2009) Relative susceptibility of *Spodoptera litura* pupae to selected entomopathogenic fungi. Biocontrol 54:85–92

Arthurs S, Heinz KM, Prasifka JR (2004) An analysis of using entomopathogenic nematodes against above-ground pests. Bull Entomol Res 94:297–306

Ashour MB, Ragheb DA, El-Sheikh EI-SA, Gomaa EI-AA, Kamita SG, Hammock BD (2007) Biosafety of recombinant and wild type nucleopolyhedrovirus as bioinsecticides. Int. J. Environ Res Public Health 4:111–125

Bale JS, van Lenteren JC, Bigler F (2008) Biological control and sustainable food production. Philos Trans R Soc B 363:761–776

Begley JW (1990) Efficacy against insects in habitats other than soil. In: Gaugler R, Kaya HK (eds) Entomopathogenic nematodes in biological control. CRC, Boca Raton, pp 215–231

Benton TG, Vickery JA, Wilson JD (2003) Farmland biodiversity: is habitat heterogeneity the key? Trends Ecol Evol 18:182–188

Bianchi FJJA, Booij CJH, Tscharntke T (2006) Sustainable pest regulation in agricultural landscapes: a review on landscape composition, biodiversity and natural pest control. Proc Biol Sci 273:1715–1727

Bowen DJ, Ensign JC (1998) Purification and characterization of a high-molecular-weight insecticidal protein complex produced by the entomopathogenic bacterium *Photorhabdus luminescens*. Appl Environ Microbiol 64:3029–3035

Bucher GE (1960) Potential bacterial pathogens of insects and their characteristics. J Insect Pathol 2:172–195

Caballero P, Aldebis HK, Vargas-Osuna E, Santiago-Alvarez C (1992) Epizootics caused by a nuclear polyhedrosis virus (NPV) in populations of *Spodoptera exigua* in southern Spain. Biocontrol Sci Technol 2:45–149

Carson R (1962) Silent spring. Houghton Mifflin Publications, Boston, pp 15–39

Castagnola A, Stock P (2014) Common virulence factors and tissue targets of entomopathogenic bacteria for biological control of lepidopteran pests. Insects 5:139–166

Collier R, Fellows J, Adams S, Semenov M, Thomas B (2008) Vulnerability of horticultural crop production to extreme weather events. In: Halford N, Jones HD, Lawlor D (eds) Effects of climate change on plants: implications for agriculture. Association of Applied Biologists, Warwick, pp 3–13

Crickmore N, Zeigler DR, Schnepf E, Van Rie J et al (2007) *Bacillus thuringiensis* toxin nomenclature. http://www.lifesci.sussex.ac.uk/Home/Neil_Crickmore/Bt/

Daniel JL, David JE, Paul J (2000) Role of proteolysis in determining potency of *Bacillus thuringiensis* Cry1Ac-endotoxin. Appl Environ Microbiol 66:5174–5181

Easterling D, Meehl GA, Parmesan C, Changnon SA, Karl TR, Mearns LO (2000) Climate extremes: observations, modeling and impacts. Science 289:2068–2074

Erlandson MA, Theilmann DA (2009) Molecular approaches to virus characterisation and detection. In: Stock SP et al (eds) Insect pathogens: molecular approaches and techniques. CAB International, Wallingford, pp 3–31

Estruch JJ, Waren GW, Mullis MA, Nye GJ et al (1996) Vip 3A, a novel *Bacillus thuringiensis* vegetative insecticidal protein with a wide spectrum of activities against lepidopteran insects. Proc Natl Acad Sci U S A 93:5389–5394

Fauquet CM, Mayo MA, Maniloff J, Desselberger U, Ball LA (eds) (2005) Virus taxonomy, Eighth Report of the International Committee on taxonomy of viruses. Elsevier/Academic, London

Fenster CB, Armbruster WS, Wilson P, Thomson JD, Dudash MR (2004) Pollination syndromes and floral specialization. Annu Rev Ecol Evol Syst 35:375–403

Flint ML, Van den Bosch R (1981) Introduction to integrated pest management. Plenum, New York

Garcia GC, Berenice GMM, Nestor BM (2011) Pathogenicity of isolates of entomopathogenic fungi against *Spodoptera frugiperda* (Lepidoptera: Noctuidae) and *Epilachna varivestis* (Coleoptera: Coccinellidae). Rev Colomb Entomol 37:217–222

Goettel MS, Hajek AE, Siegel JP, Evans HC (2001) Safety of fungal biocontrol agents. In: Butt TM, Jackson CW, Magan N (eds) Fungi as biocontrol agents: progress, problems and potential. CABI Publishing, Wallingford, pp 347–376

Grewal PS, Ehlers R-U, Shapiro-Ilan DI (eds) (2005) Nematodes as biocontrol agents. CABI Publishing, Wallingford, p 505

Gurr GM, Wratten SD, Luna JM (2003) Multi-function agricultural biodiversity: pest management and other benefits. Basic Appl Ecol 4:107–116

Hajek A (2004) Natural enemies: an introduction to biological control. Cambridge University Press, Cambridge

Hiraga K, Yamamoto S, Fukuda H, Hamanaka N, Oda K (2005) Enniatin has a new function as an inhibitor of Pdr5p, one of the ABC transporters in *Saccharomyces cerevisiae*. Biochem Biophys Res Commun 328:1119–1125

Hussain MA, Ahmad W (2011) Management of *Helicoverpa armigera* by entomopathogenic nematodes. Lambert Academic Publishing GmbH & Co. KG, Germany, p 168

Ingle YV, Lande SK, Burgoni GK, Autkar SS (2004) Natural epizootic of *Nomuraea rileyi* on lepidopterous pests of soybean and green gram. J Appl Zoolog Res 15:160–162

Ives AR, Klug JL, Gross K (2000) Stability and species richness in complex communities. Ecol Lett 3:399–411

Jackson JJ, Brooks MA (1995) Parasitism of western corn rootworm larvae and pupae by *Steinernema carpocapsae*. J Nematol 27:15–20

Kevan PG, Baker HG (1983) Insects as flower visitors and pollinators. Annu Rev Entomol 28:407–453

Khandelwal P, Choudhury D, Birah A, Reddy MK, Gupta GP, Banerjee N (2004) Insecticidal pilin subunit from the insect pathogen *Xenorhabdus nematophila*. J Bacteriol 186:6465–6476

Klaassen CD, Amdur MO, Doull J (eds) (1996) Casarett and Doull's toxicology. The basic science of poisons, 5th edn. McGraw-Hill Companies, Toronto

Krenn HW (2010) Feeding mechanisms of adult Lepidoptera: structure, function, and evolution of the mouthparts. Annu Rev Entomol 55:307–327

Krieg A (1981) The genus *Bacillus*: insect pathogens. In: Starr MP, Stolp H, Trueper HG, Balows A, Schlegel HG (eds) The prokaryotes. Springer, Berlin, pp 1743–1755

Kubicek CP, Druzhinina IS (2007) Environmental and microbial relationships, 2nd edn, The Mycota IV. Springer, Berlin, pp 159–187

Lacey LA, Georgis R (2012) Entomopathogenic nematodes for control of insect pests above and below ground with comments on commercial production. J Nematol 44:218–225

Lardy H, Reed P, Lin C-HC (1975) Antibiotic inhibitors of mitochondrial ATP synthesis. Fed Proc Fed Am Soc Exp Biol 34:1707–1710

Liu BL, Tzeng YM (2012) Development and applications of destruxins: a review. Biotechnol Adv 30:1242–1254

Lonsdale D, Gibbs JN (2002) Effects of climate change on fungal diseases of trees. In: Broadmeadow MSJ (ed) Climate change: impacts on UK forests, Forestry Commission Bulletin No 125. Forestry Commission, Edinburgh, pp 83–97

Mallet J (1989) The evolution of insecticide resistance: have the insect won? Trends Ecol Evol 4:336–340

Matsumoto H, Noguchi H, Hayakawa Y (1998) Primary cause of mortality in the armyworm larvae simultaneously parasitized by parasitic wasp and infected with bacteria. Eur J Biochem 252:299–304

Meiners RE, Morriss AP (2001) The legacy of the DDT Ban, DDT's legacy: malaria's return. Property and Environment Research Center Reports 19

Monobrullah M (2003) Optical brighteners - pathogenecity enhancers of entomopathogenic viruses. Curr Sci 84:640–645

Mule G, D'Ambrosio A, Logrieco A, Bottalico A (1992) Toxicity of mycotoxins of *Fusarium sambucinum* for feeding in *Galleria mellonella*. Entomol Exp Appl 62:17–22

Nguyen KB, Smart GC (1990) *Steinernema scapterisci* n. sp. (Steinernematidae: Nematoda). J Nematol 22:187–199

Oerke EC, Dehne HW, Schnbeck F, Weber A (1994) Crop production and crop protection: estimated losses in major food and cash crops. Elsevier, Amsterdam

Ojha A, Sree KS, Sachdev B, Rashmi MA, Ravi KC, Suresh PJ, Mohan KS, Bhatnagar RK (2014) Analysis of resistance to Cry1Ac in field-collected pink bollworm, *Pectinophora gossypiella* (Lepidoptera. Gelechiidae), populations. GM Crops Food 5:280–286. doi·10 4161/21645698. 2014.947800

Pandey R, Hasan W (2009) Pathogenicity of entomopathogenic fungi, *Metarhizium anisopliae* against tobacco caterpillar, *Spodoptera litura* (Fabricius). Trends Biosci 2:29–30

Pedras MSC, Irina ZL, Ward DE (2002) The destruxins: synthesis, biosynthesis, biotransformation and biological activity. Phytochemistry 59:579–596

Rios-Velasco C, Cerna-Chavez E, Pena SS, Morales GG (2010) Natural epizootic of the entomopathogenic fungus *Nomuraea rileyi* (Farlow) Samson infecting *Spodoptera frugiperda* (Lepidoptera: Noctuidae) in Coahuila Mexico. J Res Lepid 43:7–8

Robinson RA, Sutherland WJ (2002) Post-war changes in arable farming and biodiversity in Great Britain. J Appl Ecol 39:157–176

Romeis J, Meissle M, Bigler F (2006) Transgenic crops expressing *Bacillus thuringiensis* toxins and biological control. Nat Biotech 24:63–71

Schnepf HE, Cricmore N, Vanrie J, Lereclus D, Baum J, Feitelson J, Zfider DR, Dean DH (1998) *Bacillus thuringiensis* and its pesticidal crystal proteins. Microbiol Mol Biol Rev 62:775–806

Schrank A, Vainstein MH (2010) *Metarhizium anisopliae* enzymes and toxins. Toxicon 56:1267–1274

Shah PA, Pell JK (2003) Entomopathogenic fungi as biological control agents. Appl Microbiol Biotechnol 61:413–423

Shapiro-Ilan DI, Gouge DH, Koppenhöfer AM (2002) Factors affecting commercial success: case studies in cotton, turf and citrus. In: Gaugler R (ed) Entomopathogenic nematology. CABI Publishing, Wallingford, pp 333–356

Shapiro-Ilan DI, Gouge DH, Piggott SJ, Fife JP (2006) Application technology and environmental considerations for use of entomopathogenic nematodes in biological control. Biol Control 38:124–133

Shelton AM, Zhao J-Z, Roush RT (2002) Economic, ecological, food safety and social consequences of the deployment of Bt transgenic plants. Annu Rev Entomol 47:845–881

Siegel JP (2000) Bacteria. In: Lacey LL, Kaya HK (eds) Field manual of techniques in invertebrate pathology. Kluwer Academic, Boston, pp 209–230

Silva CP, Waterfield NR, Daborn PJ, Dean P, Chilver T, Au CP, Sharma S, Potter U, Reynolds SE, ffrench-Constant R (2002) Bacterial infection of a model insect: *Photorhabdus luminescens* and *Manduca sexta*. Cell Microbiol 4:329–339

Sree KS, Padmaja V (2008) Oxidative stress induced by destruxin from *Metarhizium anisopliae* (Metch.) involves changes in glutathione and ascorbate metabolism and instigates ultrastructural changes in the salivary glands of *Spodoptera litura* (Fab.) larvae. Toxicon 51:1140–1150

Sree KS, Padmaja V, Murthy LNY (2008) Insecticidal activity of destruxin, a mycotoxin from *Metarhizium anisopliae* (Hypocreales), against *Spodoptera litura* (Lepidoptera: Noctuidae) larval stages. Pest Manag Sci 64:119–125

Stock SP, Goodrich-Blair H (2012) Nematode parasites, pathogens and associates of insects and invertebrates of economic importance. In: Lacey LA (ed) Manual of techniques in invertebrate pathology, 2nd edn. Academic, San Diego, pp 373–426

Suzuki A, Kanaoka M, Isogai A, Murakoshi S, Ichinoe M, Tamura S (1977) Bassianolide, a new insecticidal cyclodepsipeptide from *Beauveria bassiana* and *Verticillium lecanii*. Tetrahedron Lett 18:2167–2170

Theilmann DA, Blissard GW, Bonning B, Jehle J, O'Reilly DR, Rohrmann GF, Theim S, Vlak J (2005) Family *Baculoviridae*. In: Fauquet CM, Mayo MA, Maniloff M, Desselberger U, Ball LA (eds) Virus taxonomy. Eighth report of the International committee on virus taxonomy. Elsevier, San Diego, pp 177–185

Turner JA (2008) Tracking changes in the importance and distribution of diseases under climate change. In: Proceedings HGCA R&D conference: arable cropping in a changing climate 2008, pp 68–77

Vaeck M, Reynaerts A, Hofte H, Jansens S et al (1987) Transgenic plants protected from insect attack. Nature 328:33–37

Wang Q, Xu L (2012) Beauvericin, a bioactive compound produced by fungi: a short review. Molecules 17:2367–2377

Waterfield N, Kamita SG, Hammock BD, ffrench-Constant R (2005) The *Photorhabdus* Pir toxins are similar to a developmentally regulated insect protein but show no juvenile hormone esterase activity. FEMS Microbiol Lett 245:47–52

Waterfield NR, Ciche T, Clarke D (2009) *Photorhabdus* and a host of hosts. Annu Rev Microbiol 63:557–574

Wilby A, Thomas MB (2002) Natural enemy diversity and pest control: patterns of pest emergence with agricultural intensification. Ecol Lett 5:353–360

Zimmermann G (2008) The entomopathogenic fungi *Isaria farinosa* (formerly *Paecilomyces farinosus*) and the *Isaria fumosorosea* species complex (formerly *Paecilomyces fumosoroseus*): biology, ecology and use in biological control. Biocontrol Sci Technol 18:865–901

Chapter 2
Role of Nucleopolyhedroviruses (NPVs) in the Management of Lepidopteran Pests in Asia

G.V. Ranga Rao, Ch. Sridhar Kumar, K. Sireesha, and P. Lava Kumar

2.1 Introduction

The use of synthetic insecticides has been the major approach in modern agriculture for controlling insect pests on different crops in most of the developing countries. Chemical control is one of the effective and quicker methods in reducing pest population, where farmer obtains spectacular results within a short period. However, overreliance and indiscriminate unscientific use of pesticides for longer periods resulted in a series of problems, mainly risk of environmental contamination, loss of biodiversity, development of insecticide-resistant pest populations, resurgence, outbreaks of the secondary pests, increase in inputs on chemicals and toxicological hazards due to accumulation of pesticide residues in the food chain, etc., ultimately contributing not only to inefficient insect control but also environmental and health hazards (Armes et al. 1992; Kranthi et al. 2002). Therefore, there is an urgent need to rationalize the use of chemical pesticides for the management of insect pests. In recent years, the growing public concern over potential health hazards of synthetic pesticides has led to the exploration of alternative pest management options, such as adoption of integrated pest management (IPM). IPM combines cultural, biological, and chemical measures in the most effective, environmentally sound, and socially acceptable way of managing pests, diseases, and weeds. IPM aims at suppressing the pest population by combining available methods in a harmonious way with emphasis on farm health and net returns. In an attempt to overcome the present crisis and to find alternatives to synthetic

G.V.R. Rao (✉) • Ch.S. Kumar • K. Sireesha
International Crops Research Institute for the Semi-Arid Tropics (ICRISAT), Patancheru, Telangana 502324, India
e-mail: g.rangarao@cgiar.org

P.L. Kumar
International Institute of Tropical Agriculture (IITA), Ibadan, Oyo State, PMB 5320, Nigeria
e-mail: L.kumar@cgiar.org

© Springer International Publishing Switzerland 2015
K.S. Sree, A. Varma (eds.), *Biocontrol of Lepidopteran Pests*, Soil Biology 43,
DOI 10.1007/978-3-319-14499-3_2

insecticides, the application of "bio-pesticides" as an eco-friendly measure for pest suppression has come up as one of the effective tools in IPM approach.

Bio-pesticides are developed from natural plant or animal origin, which can intervene in the life cycle of insect pests in such a way that the crop damage is minimized. The biological agents employed for this purpose include parasites, predators, and disease-causing fungi, bacteria, nematodes, and viruses, which are the natural enemies/pathogens of pests. More than three thousand microorganisms, comprising viruses, bacteria, fungi, protozoa, and nematodes, have been reported as insect pathogens. Of these, microbial pathogens gained significance for use as bio-pesticides primarily due to ease in production, application, wider adoptability, persistence, economic feasibility, and environmental compatibility. Many species of insect pathogenic microorganisms have been exploited as bio-pesticides, and some species have been developed into commercial formulations that are being used in many countries. Though farmers in Asia are aware of the importance of IPM and its impact on health and environment, the adoption level was not up to the expected levels. However, recent estimates are quite encouraging with reduction in chemical use to $25.3 billion in 2010 compared to $26.7 billion in 2005. On the other hand, interestingly, the bio-pesticides market is growing rapidly from $672 million in 2005 to over $1 billion in 2010 (Anon 2009). Several viruses belonging to 18 different families are known to infect invertebrates and insects (Fauquet et al. 2004). However, bio-pesticide development is concerned almost exclusively with members of one family, the Baculoviridae, because of their common occurrence in most important insect pests primarily in the order of Lepidoptera and their action as natural regulators of pest populations (Weiser 1987; Gelernter and Federici 1990; Caballero et al. 1992; Blissard et al. 2000). The potential of baculoviruses to be employed as insecticides is known since more than 75 years ago (Benz 1986). To date, over 30 different baculoviruses are used to control several insect pests in agriculture and forestry (Moscardi 1999). The use of baculovirus as insecticides is based on a set of useful properties, such as pathogenicity, specificity, narrow host range, environmental persistence, suitability to add to other bio-agents with synergism, and ability to induce epizootics. There are several advantages of using insect viruses in pest management over traditional synthetic chemical insecticides: these are highly host specific and are known to be completely safe to humans, animals, and non-target beneficial insects such as bees, predatory insects, and parasitoids (Groner 1986; Monobrullah and Nagata 1999; Nakai et al. 2003; Ashour et al. 2007); lack of toxic residues allowing growers to treat their crops even shortly before harvest, with low probability to develop stable resistance (Monobrullah 2003). These are highly compatible with other methods of pest control and are well suited for use in IPM programs. Another important reason for the interest in baculoviruses as potential insect control agents is that they are relatively easy to visualize and monitor using a light microscope. In vivo and in vitro tests with several vertebrate, invertebrate, and plant species have not demonstrated any pathogenic, toxic, carcinogenic, or teratogenetic effects after exposure to these viruses (Ignoffo and Rafajko 1972; Ignoffo 1973; Banowetz et al. 1976; Lautenschlager et al. 1977; Roder and Punter 1977; Huber and Krieg

1978). Baculoviruses are stable and can be stored as aqueous suspensions or dried powders for long periods without any loss of activity (David and Gardiner 1967a). They are resistant to many chemicals and persist in the soil for many years (David and Gardiner 1967b), and their activity is not altered significantly by relative humidity (David et al. 1971), precipitation (David and Gardiner 1966), or prolonged exposure to normal field temperatures (Yendol and Hamlen 1973). They can be used concurrently with most chemical insecticides and permit the reduction of the number of applications needed to keep the insect plague under control in crops, thus contributing to the reduction of the costs of protection. Finally, its use in replacement of synthetic insecticides helps to reduce the overall levels of chemical pollution (Falcon 1971; Hunter et al. 1975; Jacques and Long 1978). Baculoviruses differ significantly from chemical insecticides in that they are components of nature. Large quantities of virus are released into the environment during natural epizootics, which are common, widespread, and often important in regulating insect population levels (Injac 1973; Federici 1978). There is evidence that the amount of virus which is artificially placed into the environment as bio-pesticide is minimal compared with the amount produced during such epizo-otics (Thomas 1975).

2.2 Classification of Baculoviruses

Baculoviruses are occluded, double-stranded DNA (dsDNA) viruses and charac-terized by the presence of occlusion or inclusion bodies (OBs). The nature and significance of these OBs remained a mystery for a long time until the electron microscope (EM) was available that the virus particle could be isolated and identified as the infectious viral agent. Based on the size, shape, and occluded virion phenotype, the baculoviruses are classified into two genera, nucleopolyhe-droviruses (NPVs) and granuloviruses (GVs) (Rohrmann 1999; Winstanley and O'Reilly 1999; Blissard et al. 2000; Fauquet et al. 2004). The EM observation of NPVs reveals polyhedral to irregular shaped OBs with size 0.15–15 μm in diameter (Figs. 2.1 and 2.2) composed of matrix protein (30–40 % of total viral protein) called polyhedrin, which crystallizes around many enveloped nucleocapsids (Hooft van Iddekinge et al. 1983). Different NPVs are characterized by their occluded virions being present either as single (SNPV) or multiple (MNPV) nucleocapsids within the envelope (Figs. 2.1 and 2.3). Both SNPVs and MNPVs may contain 20–200 virions depending upon species (Rohrmann 1999). The GVs have small OBs (0.25–0.5 μm in cross section), are ellipsoidal in shape, and normally contain a single nucleocapsid, which is enveloped and is composed of a major matrix protein called granulin (Funk et al. 1997; Winstanley and O'Reilly 1999). NPVs are found mostly in the order Lepidoptera, Hymenoptera (31 species), Diptera (27 species), and Coleoptera (5 species) as well as from the crustacean class Decapoda (shrimp), whereas GVs are only found within the order Lepidoptera (Federici 1997; Blissard et al. 2000). Virions consist of one or more nucleocapsids embedded in a

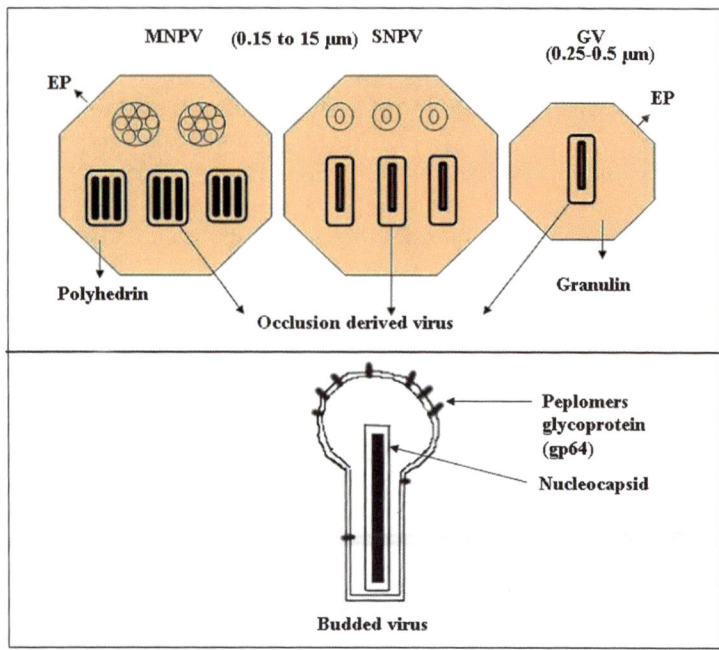

Fig. 2.1 Morphological characteristics of nucleopolyhedroviruses (NPVs) and granuloviruses (GVs)

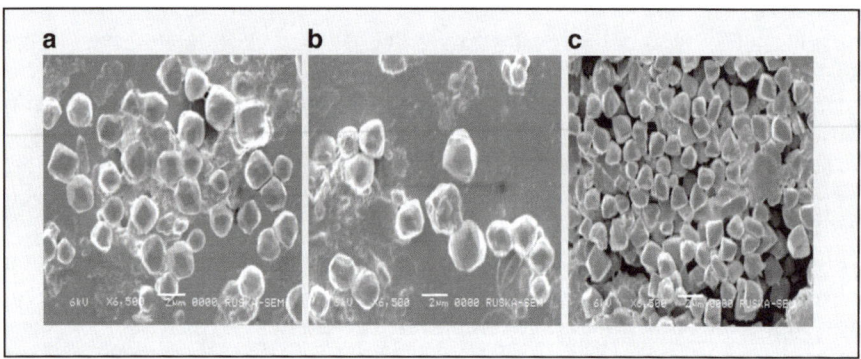

Fig. 2.2 Scanning electron micrograph of POBs: scanning electron micrograph (SEM) images of occlusion bodies (OBs) of baculoviruses. The purified aqueous OBs of baculoviruses isolated from (**a**) *H. armigera*, (**b**) *S. litura*, and (**c**) *A. albistriga* were dehydrated, mounted over the stubs, applied with a thin layer of gold metal over the sample using sputter coater, and then scanned under EM. Magnification = 6,500×

membranous envelope. Two morphologically distinct but genetically identical viral forms are produced during postinfection: (a) budded virus particles (BV) which serve for the transmission of the virus to other tissues of the infected pest and

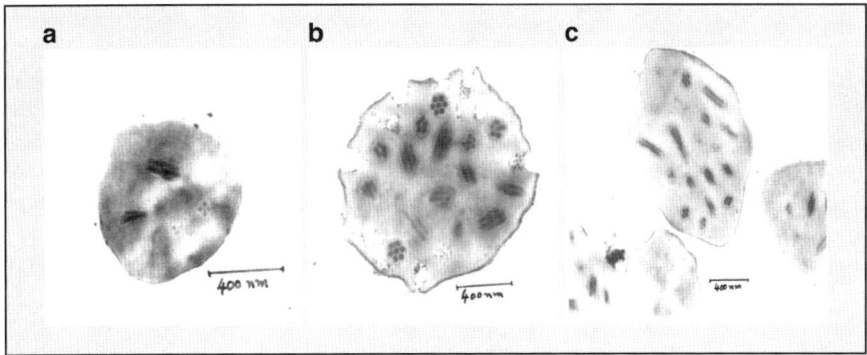

Fig. 2.3 Transmission electron micrographs of the cross section of poly-occlusion bodies (POBs): Pellets of purified OBs of baculoviruses isolated from (**a**) *H. armigera*, (**b**) *S. litura*, and (**c**) *A. albistriga* were subjected to ultrathin sections, mounted on copper grids, and stained with saturated aqueous uranyl acetate and counterstained with 4 % lead citrate and observed under TEM. Magnification = 25,000×

(b) OBs which are responsible for the survival of the virus and spread of the disease. The OB (polyhedra) of NPV contains many occlusion-derived virions (ODV) surrounded by a matrix composed of mainly polyhedrin, a major structural protein (Braunagel et al. 2003). Polyhedrin is produced in large quantities (around 30 % of total protein mass at the time of host death). Polyhedra are relatively stable and under favorable conditions virions can survive in the environment for more than 20 years. Under magnification of around 1,000 ×, polyhedra resemble clear, irregular crystals of salt; so they are big enough to be seen in a light microscope. Some common symptoms of the virus attack are sluggishness, discoloration of skin, wet or extremely moist droppings, and regurgitation of fluids (a sign of stress).

The size and shape of occlusion bodies in NPVs varies considerably not only between the polyhedral occlusion bodies (POBs) from different insects but often also within the same species. For example, majority of the POBs of *Helicoverpa armigera* NPV are spherical, while some of them are irregular in shape and the size ranges from 0.6 to 2.3 μm, averaging to 1.35 μm (Fig. 2.2). The diameter of polyhedra ranges about 0.5–1.5 μm, depending on the insect species (Fig. 2.3). The differentiation in the cross section of typical baculovirus (NPV and GV) occlusion bodies (OBs) is clearly represented in Fig. 2.1. In the boundary of OB, the protein envelope (PE) appears as an electron-dense layer made up of PE protein or envelope protein and shown to be very sensitive to alkaline proteases (Russell and Rohrmann 1990; Van Lent et al. 1990). The distance between envelope and crystalline matrix (polyhedrin or granulin) is not uniform around the occlusion body. The fine structure of occlusion body reveals crystalline lattice of the occlusion body protein molecules, which are arranged in cubic system. Although there is no true membrane covering the OB, difficulties in staining OB, the retention of their shape, and the presence of a membrane-like coat following chemical and physical treatment indicate that the exterior portion of OB is different from the interior portion. On the whole, they are very stable and can persist indefinitely in the environment (Bergold 1982). The infectious, rod-shaped virions are randomly occluded in OBs without any apparent disruption of the lattice; an 8-nm layer separates virion from the protein matrix. The size of the virions is with dimensions in the range from 4.0 to 140×250 to 400 nm. Alkaline-liberated virions readily lose their envelopes to reveal nucleocapsids each made up of a capsid surrounding a DNA core. The capsid, in turn, consists of protein subunits arranged along its long axis. The virions contain large circular, covalently closed, dsDNA genome with size in range of 80–180 kbp packed in the nucleocapsid (Blissard and Rohrmann 1990; Volkman et al. 1995).

2.3 Examples of Some Commercial Baculovirus-Based Products Registered in Different Countries for Pest Management

Over 20 species of baculoviruses have been developed or registered as commercially available insecticides, and over 30 different products have been registered as commercial insecticides based upon NPV or GV. *Autographa californica* and *Anagrapha falcifera* NPVs were registered in various countries and have relatively broad host spectrum and potentially can be used on a variety of crops infested with pests including *Spodoptera* and *Helicoverpa*. GV is the active component of a number of bio-pesticides used for protection of apple and pear orchards against the codling moth, *Cydia pomonella*. Some of the trademarks of GV-based products are the following: Granusal™ in Germany, Carpovirusine™ in France, Madex™ and Granupom™ in Switzerland, and Virin-CyAP in Russia. Annually, up to 250,000 ha of orchards have been protected with Madex™ in different European countries (Vincent et al. 2007). Another GV infecting *Erinnyis ello* (cassava hornworm) was found to be very efficient in protection of cassava plantations (Bellotti 1999). This GV has been used for spraying cassava crops in South American countries. Two commercial preparations based on *Spodoptera* NPV have been available. These are SPOD-X™ containing *Spodoptera exigua* NPV to control insects on vegetable crops and Spodopterin™ containing *Spodoptera littoralis* NPV which is used to protect cotton, corn, and tomatoes (Szewczyk et al. 2006). In China, twelve baculoviruses have been authorized as commercial insecticides (Sun and Peng 2007), including *H. armigera* NPV (the most widely used virus in China for cotton, pepper, and tobacco protection), *Spodoptera litura* NPV (vegetables), *S. exigua* NPV (vegetables), *Buzura suppressaria* NPV (tea), *Pieris rapae* GV, and *Plutella xylostella* GV (vegetables). China is the largest user of baculoviruses worldwide, with maximum number of viruses being registered for insect control. The well-known success of employing baculovirus as a bio-pesticide is the case of *Anticarsia gemmatalis* nucleopolyhedrovirus (AgMNPV) used to control the velvetbean caterpillar in soybean (Moscardi 1999). This program was implemented in Brazil in the early 1980s and came up to over 2,000,000 ha of soybean treated annually with the virus. The use of AgMNPV in Brazil brought about many economical, ecological, and social benefits. The protection of soybean fields in Brazil has proven that baculoviral control agents can be effectively produced on a large scale and they may be an alternative to broad-spectrum chemical insecticides. The forests of temperate regions are very often attacked and defoliated by larvae of Lepidoptera (the most common pest species are *Lymantria dispar*, *Lymantria monacha*, *Orgiya pseudotsugata*, and *Panolis flammea*) and some hymenopteran species (mainly *Neodiprion sertifer* and *Diprion pini*). *L. dispar* MNPV formulations marketed under trade names Gypchek, Disparivirus, and Virin-ENSH and *O. pseudotsugata* MNPV under trade names BioControl-1 and Virtuss (Reardon et al. 1996) are sometimes used for forest protection. Forest ecosystems tend to be more stable than agricultural systems, allowing for natural or applied baculoviruses to remain in the environment for long periods of time increasing the chance of natural epizootics.

2.4 Isolation and Characterization

Among the baculoviruses, NPVs attracted more attention of plant protection scientists who were looking for an alternative to pesticides because they cause a highly infectious disease that kills the pest within 5–7 days. These viruses attack some of the most important Lepidopteran crop pests including the species of *Helicoverpa*, *Spodoptera*, and *Amsacta*. Some of the related GV species are also highly infectious, e.g., *Cydia pomonella* (apple codling moth) GV and *P. xylostella* (diamond back moth) GV. However, not all GVs are as fast acting as NPV because morphologically they had single envelope with single nucleocapsid per occlusion body (Winstanley and O'Reilly 1999) (Fig. 2.1). In general, the host range of most NPV is restricted to one or a few species of the genus or family of the host where they were originally isolated. However, it also addresses an important commercial drawback, restricting the use of these products to specific pest or closely related pest complexes, such as *Helicoverpa* species (Chakraborthy et al. 1999). Some of the few exceptions having a broader host range are (1) *A. californica* MNPV infecting more than 30 species from about 10 insect families, all within the order Lepidoptera; (2) *A. falcifera* NPV infecting more than 31 species of Lepidoptera from ten insect families; and (3) *Mamestra brassicae* MNPV which was found to infect 32 out of 66 tested Lepidopteran species from four different families (Groner 1986; Doyle et al. 1990; Hostetter and Puttler 1991). In contrast to NPV, the host range of GV appears to be even narrower and mostly restricted to a single species. In India, about 35 insect viruses have been recorded from the baculovirus group, the most important being the NPVs of *H. armigera*, *S. litura*, *Spilosoma obliqua* (Walker), *Achaea janata* (L.), and *Amsacta albistriga* (Walker) and the GVs of *A. janata*, *S. litura*, *H. armigera*, and *Chilo* spp. (Pawar and Thombre 1992).

2.4.1 Morphological Characterization

During natural epizootic conditions, baculoviruses were isolated from larvae of
H. armigera (Hübner) (Lepidoptera: Noctuidae), *S. litura* (Fabricius) (Lepidoptera:
Noctuidae), and *A. albistriga* (Walker) (Lepidoptera: Arctiidae) at ICRISAT farms,
and the viruses isolated from these insect pests were characterized as MNPVs by
conducting morphological and biological studies (Sridhar Kumar et al. 2011). The
diseased larvae showed the typical baculovirus infection symptoms. The infected
larvae showed pale swollen bodies and are moribund. The larvae of *H. armigera*
and *A. albistriga* crawled to the top of the twigs (negative geotropism) on which
they fed (Fig. 2.4). But the larva of *S. litura* did not show this feature due to its soil
inhabiting nature and nocturnal habitat. The initial signs of baculoviral infection are
gradual changes in the color and luster of the integument. Infection of the epidermis
caused the host to appear soft and in some larvae the cuticle was ruptured and
discharging of body fluid onto plant parts was observed. Earlier, these symptoms
were also reported by others (Aizawa 1963; Tanada and Kaya 1993; Federici 1997).
Observation of discharged body fluid under phase contrast microscope revealed that
it consists of OBs (Fig. 2.5). Electron microscopic studies of OBs indicated that the
viruses isolated were NPVs rather than GVs. Under scanning electron microscope,
the OBs appeared as irregular shaped structures with sizes ranging from 0.5 to
2.5 μm (*Ha*NPV), 0.9 to 2.92 μm (*Sl*NPV), and 1.0 to 2.0 μm (AmalNPV) in
diameter (Fig. 2.2). Transmission electron microscopic (TEM) studies on cross
sections of purified POBs of these viruses showed that each occlusion body contains
2–7 (multiple) nucleocapsids packaged within a single viral envelope. The nucleocapsids are elongated with parallel sides and two straight ends, measuring
277.7×41.6 nm (*Ha*NPV), 285.7×34.2 nm (*Sl*NPV), and 228.5×22.8 nm
(AmalNPV) (Fig. 2.3). Before characterization of any baculovirus from an insect
host, initially, it is necessary to conduct electron microscopic study (SEM and
TEM) to determine whether it is NPV or GV or SNPV or MNPV. Similarly, Tuan
et al. (1999) reported that the occlusion bodies of *Ha*NPV and *Sl*NPV isolated in
Taiwan were irregular shaped with sizes ranging from 0.79 ± 0.22 μm (*Ha*NPV)

Fig. 2.4 NPV-infected larvae of *H. armigera* and *A. albistriga*: NPV-infected larvae of
H. armigera on pigeon pea (**a**) and *A. albistriga* on groundnut (**b**)

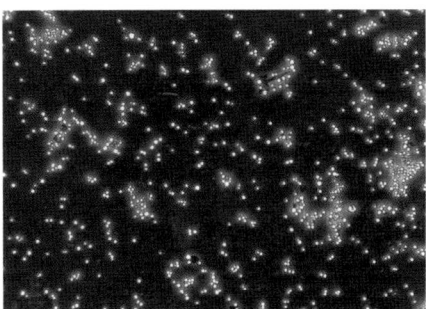

Fig. 2.5 Enumeration of poly-occlusion bodies (POBs) under phase-contrast microscope: POBs of NPVs were purified by differential centrifugation and enumerated under phase-contrast microscope at 1,000× magnification

and 1.61 ± 0.32 μm (*Sl*NPV); both viruses were MNPVs and the nucleocapsids were bacilliform to cylindrical tubular-shaped structures with dimensions of $319.80 \pm 7.80 \times 44.45 \pm 4.54$ nm (*Ha*NPV) and $332.26 \pm 13.55 \times 47.16 \pm 1.42$ nm (*Sl*NPV). In another study, the polyhedra of the *L. dispar* MNPV-NM isolate were irregularly shaped with an average diameter of 1.62 ± 0.33 μm. TEM revealed that LdMNPV-NM had bundles of virions in the nucleocapsid, which belonged to MNPV (Shim et al. 2003). Wolff et al. (2002) reported the morphology of an MNPV isolated from *Lonomia obliqua* (Lepidoptera: Saturniidae) with size ranging from 1 to 1.4 μm and nucleocapsid dimensions of 270×36 nm. Ma et al. (2006) observed the occlusion bodies in the midgut tissues of the tea looper (*Ectropis obliqua*) under TEM, the micrograph showed that the EcobSNPVs were irregular shaped with size ranging from 0.7 to 1.7 μm in diameter, and multiple rod-shaped virions measuring about 250×40 nm were embedded in each OB with a single nucleocapsid packaged within the envelope of the virion.

2.4.2 Biological Characterization

Biological assessment studies of the MNPVs isolated at ICRISAT farms from *H. armigera*, *S. litura*, and *A. albistriga* indicated (Tables 2.1 and 2.2) that they are highly virulent strains and have good potential for use as bio-control agents against these important pests (Sridhar Kumar et al. 2011). The efficacy of *Ha*NPV isolates collected from six geographical locations of India was tested by conducting bioassay experiments at ICRISAT with the second and third instar larvae of *H. armigera*, and it was found that ICRISAT *Ha*NPV was superior amongst the isolates tested (Sireesha 2006). Similar differences in virulence among NPV isolates have been established in previous studies conducted by Shapiro and Ignoffo (1970) on the variations in virulence of 34 isolates of *Ha*NPV. A 56-fold difference in the activity of the isolates was recorded, and it was opined that the difference in activity must be due to some characteristics of the occlusion body and/or its occluded virions. They also did not exclude the possibility that some other factors such as solubility of occlusions and availability of occluded virions may account for the differences observed. Abul-Nasr and Elnagar (1980) reported the differential biological activity of two *Spodoptera littoralis* isolates both at laboratory- and field-level studies. Hughes et al. (1983) compared the time mortality response of *Heliothis zea* to 14 isolates of HzNPV and identified six activity classes. Shapiro et al. (1984) tested 19 NPV isolates of *L. dispar* and reported nearly 1,000-fold difference in activity. Rabindra (1992) demonstrated the tremendous variation in virulence among the three *Ha*NPV isolates and recorded the lowest LC_{50} value of 3.467×10^4 POBs/ml for the *Ha*NPV isolate from Nilgiris. Somasekhar et al. (1993) on characterizing five Indian isolates of *Ha*NPV found that the most virulent isolate was that from Ooty in Tamil Nadu, India, with the lowest LC_{50} value of 2.538×10^3 POBs/ml, followed by the isolate from Coimbatore in Tamil Nadu, India (2.965×10^3 POBs/ml), and the Rajasthan isolate with LC_{50} value of 13.08×10^3 POBs/ml was the least effective. Geetha and Rabindra (1999) found that among the 11 *Ha*NPV isolates collected from different regions in India, Negamam and Ooty isolates from Tamil Nadu were significantly more virulent with LC_{50} values of 83.807 and 93.926 POBs/cm^2, respectively. The Rajasthan isolate was the least potent with LC_{50} value of 111.778 POBs/cm^2.

All these studies indicate that there is a significant variation in LC_{50} values with overlapping fiducial limits and the use of locally produced NPV appeared to be more useful for managing the respective insect pests than the commercially available NPV from other parts of the country. Geetha and Rabindra (1999) also reported overlapping fiducial limits of LC_{50} values of eight *Ha*NPV isolates among the 11 isolates evaluated. The variation in the activity of different isolates may be due to different reasons. Inherent genetically controlled factors may logically be an important reason. The other reason may be that the different isolates had different number of passages in the host either under natural conditions or in the laboratory (Geetha and Rabindra 1999). In Log concentration–probit mortality relationship, the lower the slope value, the greater is the variability. Normally, the slope values were very low in bioassay studies with insect pathogens (Burges and Thompson 1971). Battu and Ramakrishna (1987) reported the respective slope values of 0.1674, 0.4078, and 0.1215 for 6-, 9-, and 11-day-old *S. obliqua* larvae, respectively, when inoculated with its own NPV. Arora et al. (1997) reported slope values varying from 0.58 to 0.96 for the five *Ha*NPV isolates evaluated against the second instar larvae of *H. armigera*. The low slope of dosage–mortality curves for insect pathogens often indicates a more stable host–pathogen relationship.

2.4.3 Comparative Analysis of Viral Proteins of Different NPV Isolates

The crystalline matrix of the occlusion body mainly consists of a single protein, called polyhedrin or granulin. These proteins have about 245 amino acids (29 kDa)

Table 2.1 LC$_{50}$ values of NPV isolates against the second and third instar larvae on the 7th day

NPV isolate	Regression equation		Heterogeneity	LC$_{50}$ (POB/ml)	Fiducial limits		Chi-square
	Intercept	Slope			Lower	Upper	
Second instar larvae							
HaNPV	3.84	0.26	0.220	2.3×10^4	5.9×10^3	7.8×10^4	0.88
SlNPV	3.87	0.24	0.140	3.5×10^4	1.19×10^4	9.9×10^4	0.56
AmalNPV	3.87	0.23	0.060	5.6×10^4	2.7×10^4	1.15×10^5	0.25
Third instar larvae							
HaNPV	3.66	0.26	0.09	1.5×10^5	6.1×10^4	3.3×10^5	0.36
SlNPV	3.14	0.34	0.2	2.4×10^5	8.3×10^4	6.1×10^5	0.82
AmalNPV	3.026	0.352	0.292	3.96×10^5	1.16×10^5	1.1×10^6	1.17

Table 2.2 LT_{50} values of NPV isolates against the second and third instar larvae

Virus concentration (POB/ml)	LT_{50} (h) values					
	Second instar larvae			Third instar larvae		
	HaNPV	SlNPV	AmalNPV	HaNPV	SlNPV	AmalNPV
1.8×10^8	–	–	–	123.60	132.72	128.64
1.8×10^7	122.64	128.58	132.52	134.25	140.4	144.0
1.8×10^6	131.28	133.62	136.64	136.42	143.0	148.64
1.8×10^5	142.32	146.76	149.72	150.0	156.12	162.42
1.8×10^4	153.30	158.60	162.72	161.22	176.08	182.06
1.8×10^3	191.18	195.60	199.20	216.07	228.96	236.16
1.8×10^2	230.68	234.60	238.06	–	–	–

and are hyper-expressed during the very late phase of virus infection and are not required for virus replication (Rohrmann 1986, 1992; Funk et al. 1997) and constitute up to 18 % or more of total alkali-soluble protein late in infection (Quant et al. 1984). It is a highly stable protein, insoluble in many solvents at neutral pH values and physiological conditions, highly resistant against the action of proteolytic enzymes, and at the same time it is highly sensitive to alkaline conditions (Bergold 1947; 1948). At ICRISAT, with an aim of production of polyclonal antibodies against poly-occlusion body protein (polyhedrin) for the development of diagnostic and quality control tools during mass production of NPVs, the purification protocol for polyhedrin protein was standardized (Sridhar Kumar et al. 2007) with slight modifications to the methods suggested by Summers and Egawa (1973), Harrap et al. (1977), and Quant et al. (1984). In 12 % SDS-PAGE analysis, the denatured purified protein preparations of three viruses resolved as single band (Fig. 2.6) with estimated molecular weights of 31.65 kDa (± 0.00) (HearNPV), 31.29 kDa (± 0.00) (SpltNPV), and 31.67 kDa (± 0.295) (AmalNPV). This report is similar to that reported by Tuan et al. (1999) for three lepidopteran NPVs such as *Ha*NPV, *Sl*NPV, and SeNPV. Recently, Ashour et al. (2007) reported the molecular weight of 32 kDa for recombinant and wild-type *A. californica* NPV (AcAaIT and AcMNPV). In addition to the major polyhedrin, they are contaminated with some minor low molecular weight peptides of about 7–27 kDa and a high molecular weight peptide of about 60–70 kDa, which could be the degraded peptides or dimmers of the 31 kDa polyhedrin protein. This has revealed that these three NPVs have six to eight minor polypeptides.

To characterize NPV protein structure for the purpose of providing reliable identification methodology and developing specific and sensitive serological detection techniques, the nucleocapsids of different *Ha*NPV isolates across India were purified by alkali dissolution of POBs followed by 25–60 % sucrose gradient centrifugation (Sireesha 2006). Purified samples of *Ha*NPV from ICRISAT, University of Agricultural Sciences of Dharwad, Tamil Nadu Agricultural University, Panjabrao Deshmukh Krishi Vidyapeth in Akola, Punjab Agricultural University, and Gujarat Agricultural University (UASD, TN, AK, PAU, and GAU) were analyzed in 12 % SDS-PAGE gels for proteins. This has revealed that all the

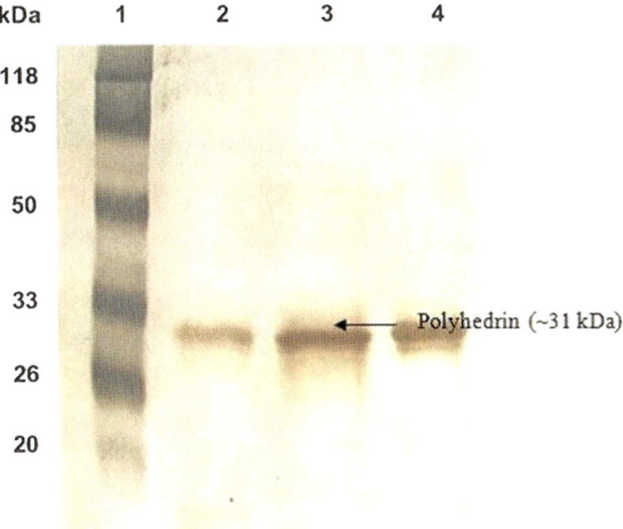

Fig. 2.6 SDS-PAGE (12 %) profile of polyhedrin protein preparations of NPVs. Purified poly-hedral protein (polyhedrin) preparations of NPVs were separated in 12 % SDS-PAGE. The polyhedrin was appeared as single protein band in silver stained gel and the protein band at ~31 kDa was indicated with arrow mark. *Lane 1*: Protein molecular weight marker; *Lane 2*: HaNPV polyhedrin; *Lane 3*: SlNPV polyhedrin; *Lane 4*: AmalNPV polyhedrin

isolates have 4 to 5 major polypeptides of 42.32 ($\pm$ 0.92) kDa, 34.74 ($\pm$ 0.27) kDa, 31.77 ($\pm$ 0.44) kDa, 30.66 ($\pm$ 0.27) kDa, and 19 ($\pm$ 1.41) kDa and several minor peptides. Three major proteins were present in all except in GAU isolate. The molecular weights of the major proteins were nearly similar, but not identical. GAU sample was unique in that it lacked the ca. 42 and ca. 34 kDa proteins. Several minor proteins were also seen in the gel (indicated with arrows). GAU isolate recorded one extra protein of 19 ($\pm$ 1.41) kDa. It was also noticed in other isolates, but it was not as conspicuous as in the case of GAU. Summers and Smith (1978) studied the structural polypeptides of eight insect baculoviruses which revealed a complex but unique composition of 15–25 bands with molecular weights ranging from 15,000 to 1,60,000 Da. *A. californica* MNPV capsids contained two major polypeptides VP18.5 and VP37; *Rachiplusia ou* MNPV capsids contained VP16, VP18, VP30, and VP36; *A. gemmatalis* MNPV contained one major capsid protein VP29; and the major capsid proteins of *H. zea* SNPV were VP16, VP28, and VP63.

Kelly et al. (1980) observed a high degree of similarity between the polypeptides of two SNPVs of *H. armigera and H. zea*. Monroe and McCarthy (1984) charac-terized the structural polypeptides of *H. armigera* NPV from India, China, and USSR. For Indian isolate, the molecular weights of polypeptides ranged from 14.2 to 90.0 kDa. Harrap et al. (1977) purified polyhedron proteins from three closely related insect pest NPVs, viz., *S. littoralis*, *Spodoptera exempta*, and *Spodoptera frugiperda*, after dissolution in 0.05 M sodium carbonate and separation on 7 and

10 % SDS-PAGE. They did not observe significant differences between the sizes of major proteins of the three viruses whose molecular weights ranged from 28 to 29 kDa. They also concluded that the smaller polypeptides of each virus preparation represented an initial breakdown product derived by proteolytic cleavage of larger molecule. Maskos and Miltenburger (1981) compared the polyhedral polypeptides of *L. dispar*, *M. brassicae*, and *A. californica* NPVs using SDS-PAGE. They observed eight distinct polypeptide bands with major polypeptides in the range of 28.0–30.0 kDa. They found characteristic differences between the species for minor polypeptides having molecular weights in the range from 12.4 to 62.0 kDa. Caballero et al. (1992) observed similar mobility profiles of the occluded virion polypeptides and polyhedrins of four *S. exigua* NPV isolates originating from the United States, Thailand, and two locations in Spain.

2.4.4   Efficacy of NPV Under Different Storage Conditions

Unlike chemical pesticides, viral pesticides often have a shorter half-life of infectivity (Shieh 1978) that requires special attention for commercial operations. Viral insecticides cannot be developed commercially until formulations of these are physically, chemically, and environmentally stable in storage and distribution. At ICRISAT, *Ha*NPV storage studies were conducted over a period of time under different set of storage conditions, and their efficacy was tested against the second instar *H. armigera* larvae at the rate of 10^6 POB/ml at an interval of 2 months up to 10 months, and the mortality data on the 5th, 7th, and 9th days were analyzed using two-way ANOVA. Variation in the efficacy was observed when stored under different conditions of storage over a period of time (Table 2.3). NPV sample, which was stored under refrigerated condition, could maintain its efficacy up to 8 months (100 %), and by the tenth month, there was a slight decline (97.50 %) but it was not significant, whereas NPV sample stored in earthen pot and at room temperature (both in amber-colored bottle and glass bottle) maintained its efficacy up to 4 months and after that virulence started to decrease. This decreased efficacy of samples stored under room temperature may be due to increased bacterial activity. When the samples were tested for the bacterial load, it was 3.47 times more in the samples stored at room temperatures after 6 months of storage. Gopali and Lingappa (2001b) also recorded decreased efficacy of NPV when stored under open house conditions, and it was opined that the change in the pH of viral suspension stored under refrigerated condition was very slow from acidic to normal (5–7) pH as against becoming excessively alkaline at ambient and earthen pot conditions. It was also reported that this change was mainly brought about by the growth of other microbes and warm conditions, which resulted in lowering of virulence of viral bodies. Attathom et al. (1990) also reported the same.

The stability of *Ha*NPV appears to be dependent on the resistance of the inclusion body protein to decomposition. Many scientists (Stairs et al. 1981) reported that the inclusion body protein is broken down by weak alkalies but it

Table 2.3 Efficacy of HaNPV under different storage conditions (on 9th day)

Storage conditions	Percent larval mortality due to HaNPV (10^6 POBs) stored for different months						Mean
	0	2	4	6	8	10	
Refrigerated	100 (71.56)	100 (71.56)	100 (71.56)	100 (71.56)	100 (71.56)	97.50 (68.11)	70.99
Earthen pot	100 (71.56)	100 (71.56)	100 (71.56)	95.00 (64.67)	92.50 (61.22)	87.50 (56.52)	66.18
RT (amb)	100 (71.56)	100 (71.56)	100 (71.56)	92.50 (61.22)	80.00 (52.79)	70.00 (48.90)	62.93
RT (glass bottle)	100 (71.56)	100 (71.56)	100 (71.56)	90.00 (57.77)	75.00 (51.82)	67.50 (48.03)	62.05
Control	5.00 (12.71)	5.00 (12.71)	7.50 (19.07)	5.00 (12.71)	7.50 (19.07)	10.00 (25.42)	16.95
Mean	59.7947	59.7947	61.0661	53.5892	51.2944	49.4025	

	Treatment	Storage	Interaction
SE±	2.39	1.004	3.147
LSD	7.203	2.827	8.991
F (prob. at 5 %)	<0.001	<0.001	<0.001

can withstand exposure to relatively strong acids and many other chemicals. Ebora et al. (1990) reported that virulence was greatest around neutral pH and reduced when subjected to high pH (12). Shapiro and Ignoffo (1969) showed that activity of virions of *Ha*NPV released from polyhedral cover lost about half of their activity when stored for 60 days at 37 °C, whereas virus particles covered with polyhedral layer retain their activity for a longer period withstanding freezing and prolonged normal field temperature than free virions (Yendol and Hamlen 1973). Many scientists reported that virus could be preserved for more than 10 years at 4 °C without loss in virulence (Narayanan 1985). Gudauskas and Cannerday (1968) found the thermal inactivation point of *Ha*NPV to be 75–80 °C for 10 min. The virulence of virus depends on the quality of the virus, storage conditions, and duration of storage, temperature, and pH of the product.

2.4.5   Effect of Chemicals on the Shelf Life of NPV

Information on screening of different chemicals or disinfectants on the shelf life of NPV products is scarce. It is known that the NPV produced in live insects may contain bacterial contamination (Podgwaite et al. 1983; Huber 1985) presenting a potential health hazard. Although these studies on *L. dispar* NPV and *C. pomonella* GV failed to detect human pathogens in the product, these viruses were produced under temperate conditions, and it might be anticipated that production in a near tropical situation would involve greater microbiological contamination problems. Grzywacz et al. (1997) quantified microbial contaminants level in *S. littoralis* NPV. They found 10^6–10^9 bacterial colony forming units/ml in virus suspension containing 2.1×10^9 POBs/ml. They concluded that none of the bacterial contaminants found were harmful to cause potential health hazard. But some bacteria such as *Bacillus cereus* might be of concern, from the point of view of standardizing the product. Therefore, the development of production procedures, which reduce these contaminants to a lower, more consistent level, would be valuable in promoting a wider acceptance of viral insecticides as safe control agents. Simple centrifugation in water does not remove many microbial contaminants, as bacterial spores, etc., tend to pellet with NPV OB. Grzywacz et al. (1997) suggested the use of bacteriostatic agents and pH buffers to stabilize the formulations by reducing the multiplication of contaminants.

In our study, the *Ha*NPV samples preserved in different chemicals over different periods of storage varied in their efficacy and bacterial contamination levels (Tables 2.4 and 2.5). The sample that was stored in distilled water maintained its virulence up to 10 months as evidenced by the 5th, 7th, and 9th day mortality. Cumulative mortality on the ninth day showed that the samples stored in 10 % Dettol, 2 % phenyl, 10 % ethyl alcohol, and 10 % methanol recorded 100 % mortality followed by 10 % acetone with 96.66 and 10 % ethyl acetate with 90 % mortality after the period of 2 months. NPV sample stored in 10 % ethyl acetate consistently reduced its efficacy as evidenced by the mortality on the 5th, 7th, and

Table 2.4 Effect of chemicals on the shelf life of HaNPV (on 5th day)

Treatment	Mortality (%)					Mean
	2	4	6	8	10	
NPV + 10 % Dettol	40.00 (39.23)	33.33 (35.22)	33.33 (35.22)	23.33 (28.78)	20.00 (26.57)	33.0025
NPV + 10 % acetone	73.33 (59.00)	63.33 (52.78)	60.00 (50.77)	56.66 (48.85)	53.33 (46.92)	51.6633
NPV + 2 % phenyl	50.00 (45.00)	43.33 (41.15)	43.33 (41.15)	40.00 (39.23)	36.66 (37.22)	40.753
NPV + 10 % ethyl acetate	40.00 (39.23)	40.00 (39.23)	40.00 (39.23)	33.33 (35.22)	30.00 (33.21)	37.2247
NPV + 10 % ethyl alcohol	53.33 (46.92)	53.33 (46.92)	53.33 (46.92)	50.00 (45.00)	50.00 (45.00)	46.1537
NPV + 10 % methanol	40.00 (39.23)	36.66 (37.22)	36.66 (37.22)	36.66 (37.22)	33.33 (35.22)	37.2247
NPV + distilled water	76.66 (61.22)	76.66 (61.22)	73.33 (59.00)	73.33 (59.00)	70.00 (56.79)	59.4474
Mean	47.12	44.82	43.22	41.90	40.13	

	Treatment	Storage	Interaction
SE±	0.710	0.60	1.588
LSD	2.00	1.69	4.48
F (prob. at 5 %)	<0.001	<0.001	0.169

Table 2.5 Effect of chemicals on the bacterial contaminants in storage

Treatment	Before storage		After a period of 6 months	
	No. of bacterial colonies	CFU/ ml $\times 10^6$	No. of bacterial colonies	CFU/ ml $\times 10^6$
NPV + 10 % acetone	20.33	1.01	19.66	0.98
NPV + 10 % ethyl alcohol	22	1.1	21.33	1.06
NPV + 10 % ethyl acetate	25	1.25	8	0.40
NPV + 10 % Dettol	23.33	1.16	25	1.25
NPV + 2 % phenyl	18	0.9	7.33	0.35
NPV + 10 % methanol	22	1.1	24.66	1.23
NPV + distilled water	50.66	2.53	250	12.5
SE±	0.433	0.02	0.349	0.01
LSD	1.335	0.06	1.075	0.050
F (prob. at 5 %)	<0.001	<0.001	<0.001	<0.001

9th day. The results clearly indicated that, though all the chemicals could effectively reduce the bad odor problem in storage, the samples stored in acetone and ethyl alcohol only recorded 73.33 and 70.00 % mortality, respectively, by the end of 10 months. However, NPV samples in all the treatments gave 90 % mortality after storing for a period of 2 months (Tables 2.3 and 2.4). Ignoffo and Shapiro (1978) suggested the use of acetone in purification of NPV POBs. Acetone, being a potential antimicrobial agent, regulates bacterial infection and, being a lipid solvent, removes the lipid (fat cells) from the larval homogenate, thereby inhibiting bacterial lipid degradation and in turn the malodor. To address the problem of bacterial contaminants, Rao and Meher (2004) used 10 % acetone solution. They could clear the lipid mass and leftover contaminating bacteria in the preparation and regulated the malodor problem. They confirmed the viability of the virus after 1 month of acetone clarification by conducting bioassay studies.

2.4.6 Molecular Characterization

Molecular-level identification, characterization, and evaluation of phylogenetic status of a particular baculovirus are also important for the establishment of purity of seed stock or master stock. Apart from the multiple or singly enveloped feature, NPV or GV cannot be identified visually from either light or electron microscopic studies. Microscopic and serological tools are unreliable for establishing the real identity of a given isolate and are not particularly helpful in providing clues about its host range and infectivity (Rovesti et al. 2000). Differences between viruses are usually reflections of intrinsic differences in their viral genomes. At one extreme, viruses may be readily distinguished by the nature of the nucleic acid (RNA or

DNA) and its strandedness (single or double stranded), while more closely related viruses may differ only by small regions of distinct base sequences which can be defined only by more sophisticated techniques such as restriction endonuclease (REN) analysis or molecular probes, or sequencing of conserved and unique gene sequences which offers a relatively simple method for the identification and differentiation of baculoviruses (Smith and Summers 1978).

The only nucleic acid type found within the enveloped nucleocapsids of these viruses is a dsDNA molecule. DNA of baculoviruses is a large circular molecule. REN analysis can provide a measure for baculovirus DNA molecular weight. It has a more useful role in virus identification and ultimately in mapping of the viral genome. Examination of the DNA using these techniques has shown that many variants of a species may exist, for example, the MNPVs from *A. californica, Trichoplusia ni, S. exempta, R. ou, A. falcifera,* and *Galleria mellonella* can be considered to be variants of the same virus (Miller and Dawes 1978; Smith and summers 1979; Summers et al. 1980; Brown et al. 1984; Harrison and Bonning 1999). Many of the known baculoviruses could be grouped together depending on their degree of genetic relationship, which does not reflect the taxonomic grouping of their host/hosts (Zanotto et al. 1993). Among the NPVs with potential as pest control agents, the MNPVs isolated from *M. brassicae* (Lepidoptera: Noctuidae) and *H. armigera* (Lepidoptera: Noctuidae) were shown to be similar in terms of both biological activity and genomic homology (Smith and Summers 1982; Figueiredo et al. 1999; Rovesti et al. 2000). For identification of a particular NPV strain, the bioassay studies and REN profiles of different NPV species have been studied and compared by several investigators (Shapiro and Ignoffo 1970; Hughes et al. 1983; William and Payne 1984; Rabindra 1992; Somasekhar et al. 1993; Arora et al. 1997; Geetha and Rabindra 1999; Sudhakar and Mathavan 1999; Figueiredo et al. 1999; Rovesti et al. 2000). The existence of genetic variants with different biological activities may have important implications for development of bio-pesticides both in the possibility to select better naturally occurring strains and as a source material for genetic manipulation (Guo et al. 2006).

In order to establish the purity of seed stock or master stock of *Ha*NPV used for commercial insecticide preparations, an attempt was made at ICRISAT by molecular characterization of *Ha*NPV done by isolation, cloning, sequencing of polyhedrin gene, and evaluation of the phylogenetic status (Sridhar Kumar 2008). Considering the sizes of previously published polyhedrin gene sequences, most amplification products were between 730 and 750 bp long. Gene sequencing analysis of selected clones resulted in 744-nucleotide-long ORF with a predicted coding capacity for a polypeptide of 247 amino acids as in the AmalNPV (AF118850). Similarly, Rivkin et al. (1998) reported a 246 amino acid polypeptide in a local strain of Israeli *Ha*NPV polyhedrin, and Bansal et al. (1997) reported same length of amino acid polypeptide in *Sl*NPV polyhedrin. The polyhedrin is the major protein of the virus OB and is the most conserved protein of NPVs (Rohrmann 1992). After the first report about localization of the polyhedrin gene in AcNPV, its nucleotide sequence was determined (Vlak and Smith 1982; Hooft van Iddekinge et al. 1983). Polyhedrin/granulin is a protein of about 245–250 amino acids and appears to be the most highly conserved baculovirus protein. These characteristics lead to the use of polyhedrin or granulin sequences as the base of baculovirus phylogenetic studies, since this is the gene from which a larger number of different examples are available for comparison (Zanotto et al. 1993). Although polyhedrin gene is still considered a reasonable marker for identification of an NPV and its neighbors, Herniou et al. (2003) and Lange et al. (2004) argued that it might not be the best baculovirus gene for phylogenetic studies because polyhedrin phylogenies often disagree with other gene phylogenies. While other phylogenetic analyses consistently group AcMNPV and BmNPV together, phylogenies based on polyhedrin have AcMNPV as a sister group to the rest of the group-I NPVs (Herniou et al. 2003). Phylogenies based on combined sequences of shared genes have been found to be more robust than those based on the sequences of individual genes (Herniou et al. 2003).

PCR, when combined with the use of REN analysis, can provide considerable resolution for use in diagnostics; it is relatively simple to use and can provide quick results. Not surprisingly, this approach is now widely used for detection and identification of a range of insect viruses (Kool et al. 1991; Williams 1993; De Moraes and Maruniak 1997; Bulach et al. 1999). Christian et al. (2001) developed a rapid method based on PCR–RFLP analysis for identification and differentiation of HaSNPV and AcMNPV groups by using a set of redundant primers to highly conserved region of polyhedrin gene. Based on REN analysis, Rovesti et al. (2000) reported that the *Ha*NPV isolate was shown to be a mixture of many closely related genotypes but individual genotypes remained unchanged on passage in either *H. armigera* or *M. brassicae*. Doyle et al. (1990) noted that when MbMNPVD was passed in *Orthosia cruda*, there were minor changes in the restriction enzyme profile, which was attributed to the selection of a different variant. However, bioassay studies of Rovesti et al. (2000) showed that the two viruses HaMNPV and MbMNPV were successfully replicated in *H. armigera*, *M. brassicae*, and *H. zea*, resulting, in each case, in progeny virus which was essentially similar to the inoculum. Therefore, a viral insecticide based on these NPVs from *H. armigera* and *M. brassicae* would be more appropriately targeted against both insects. Similarly, Murillo et al. (2001) have reported that REN profiles of two SeNPV isolates (SeUZB and SeSP3) in Uzbekistan and Spain and MbNPV (Mb-PL) in Poland were closely related to previously described Spanish isolates of SeNPVs. At the same time, the *Pst-I* and *Bgl-II* profiles of SeUZB and Mb-PL were identical and very similar to the REN profiles of MbMNPV strain which is the active component of Mamestrin[R], a commercial bio-insecticide in France (NPP, Nogueres, France). In addition to SeMNPV, *S. exigua* is susceptible to other NPVs such as *A. californica* MNPV (Smits and Vlak 1987) and MbMNPV (Wiegers and Vlak 1984). In another case, AcMNPV and BmNPV also show a high degree of genomic homology and different REN fragment profiles but do not share an overlapping host range and can be regarded as two different species. It is interesting to note that only minor changes in the virus genome, namely, one or two amino acid substitutions in the AcMNPV helicase p143, are sufficient to expand the host range of AcMNPV to *Bombyx mori* larvae (Kamita and Maeda 1997; Arguad et al. 1998).

Clearly, there are problems in naming a baculovirus simply on the basis of the host from which it was originally isolated, and therefore, greater vigilance should be taken in naming new (and old) viruses. For example, studies on viruses from members of the same or different host species show similarity or variability in morphology, virulence, and biological characteristics (Shim et al. 2003). Many of the known baculoviruses could be grouped together depending on their degree of genetic relatedness, which does not reflect the taxonomic grouping of their host/hosts (Zanotto et al. 1993). Based on the above reports, one can comment like "variants of baculoviruses with heritable similarities in virulence and variations in host range arise spontaneously in nature." In the earlier days of baculovirology, it was believed that baculoviruses could only infect a single host species and that no cross-infection occurred. This generalized the use of binomial Latin names of the insect species hosts to describe the new viral isolates. However, this practice can affect our understanding of baculovirus biology and can also lead to confusion upon studying the classification and taxonomy of baculoviruses (Federici and Hice 1997) and should be changed by more reliable methods based, for instance, on the genotypic characteristics of the viruses. A useful means for identification or description of baculoviruses is REN analysis of viral DNA, as firstly demonstrated by Lee and Miller (1978). This method has proved to be very useful not only for distinguishing distinct NPV species but also different strains of one virus or even different genotypes within the same virus isolate (Smith and Crook 1988; Munoz et al. 1998). In general, baculovirus wild-type populations, from different geographical isolates of the same virus and within a single isolate, where several genotypic variants frequently coexist, show a considerable genetic heterogeneity. This heterogeneity is due to the enormous plasticity displayed by baculoviruses genomes which can undergo deletions (Munoz et al. 1998), insertions (Jehle et al. 1995), point mutations, recombinations (Croizier and Ribeiro 1992), etc. This plasticity suggests that field isolates may be adapting to host and environmental conditions and that those isolates containing heterogeneous populations may be more valid for viral survival in the field (Possee and Rohrmann 1997). Investigations of geographic variability and the role of genotypic differences in the biology of baculoviruses are an important area of current research. Such studies may provide insight into the evolution of baculoviruses and their hosts and may also aid in the development of more effective virus strains for biological control of insects.

2.5 Production Technologies

Historically, several entomopathogenic viruses have been produced in susceptible host insects, because of the following reasons: (1) The insect host is an efficient virus producer (Ignoffo and Couch 1981). (2) Automation of in vivo rearing and in vivo production systems is feasible (Powell and Robertson 1993). Some baculovirus species may be produced in insect cell cultures, but the associated costs are relatively high (Hink 1982). Therefore, all NPVs that have been developed as commercial products thus far have been produced in host larvae.

2.5.1   In Vivo Mass Production

In vivo virus production has several advantages like (1) Successful use of viruses produced in the host to control insect pests (Ignoffo and Couch 1981; Bell 1991). (2) Research is continuing in this area to produce more efficient systems, which makes this approach an economically viable one. (3) In many areas of the world, virus production in the host is the only approach feasible (Katagiri 1981; Moscardi et al. 1981). In vivo mass production systems have changed little over the past 30 years. The development of semi-synthetic artificial diets by Vanderzant et al. (1962) resulted in rearing and virus production systems for the cotton bollworm (*H. zea*), the tobacco budworm (*Heliothis virescens*), and the cabbage looper (*T. ni*) by Ignoffo (1965). The initial rearing system was made more efficient by the introduction of disposable multi-celled plastic trays (Ignoffo and Boeing 1970), automation in rearing, and automation in virus inoculation and harvesting. Optimal virus production is the result of interrelationships of host–pathogen–environment, and each factor in this triad must be assessed for influence on quantity and quality of the product. Research in these areas has been summarized (Shapiro et al. 1986; Shapiro and Bell 1981, 1982). A broader and more complete account of some aspects on virus production and role of virus in insect pest control has been given by Burges (1981). Subsequent development and industrialization for mass rearing process, improvements in viral recovery procedures, and formulation of the viral product made it possible for commercialization of *Ha*NPV (Shieh 1978). Further, Ignoffo and Couch (1981) improved the method of mass production of baculovirus of *Helicoverpa* from the laboratory-reared *Helicoverpa* larvae through which 7–9 times more active virus and 2–5 times more POBs were obtained from dead and diseased larvae.

Field collection of diseased larvae led to contamination with adventitious agents which would pose a major problem in terms of safety and quality control, and as such, it was not desirable for *Ha*NPV production (Sherman 1985). Because of the developments in semi-synthetic diet, containerization, and automation, laboratory-reared insects have been the hosts of choice. The advantages of these insects are (1) Laboratory-reared insects tend to be larger than field-collected insects, because of the selection and adaptation to the laboratory environment (diet, temperature, humidity, and photoperiod). (2) They are normally disease free, which should result in virus product that is free from other pathogens. (3) The growth and development of laboratory-reared insects tend to be faster than field insects, because of selection. (4) Virus yield among laboratory-reared insects tends to be greater than among field insects, since virus yield is dependent on host biomass (Hedlund and Yendol 1974; Shapiro and Bell 1981). Although laboratory-colonized insects provide several advantages over field insects as virus producers, field insects have also been used successfully to produce NPV from larvae of the potato tuber moth (*Phthorimaea operculella* [Zeller]) in Australia (Matthiessen et al. 1978), the velvetbean caterpillar (*A. gemmatalis*) in Brazil (Moscardi et al. 1981), and the European pine sawfly (*N. sertifer*) in the United States (Rollinson et al. 1970) and CPV from the pine caterpillar (*Dipodomys spectabilis*) in Japan (Katagiri 1981) on natural foliage.

Different methods of mass production of baculoviruses, according to Pawar and Thombre (1992), are (1) large-scale rearing of insects in the laboratory, (2) field collection of host larvae from infested crops and infecting them in the laboratory, and (3) field collection of diseased larvae from infested fields. Large collection of insect viruses at the rate of 20,000 host larvae have been reported from different crops, viz., cotton, sunflower, pigeonpea, and chickpea (Ignoffo 1966a, b; Anderson et al. 1972; Battu 1992). Battu (1992) reported relatively lower levels of POBs obtained from field-collected, diseased, and dead insects. The number of larvae required to produce one larval equivalent (LE) (6×10^9 POBs) of virus from field-collected larvae was higher (2.97) than laboratory-reared ones (2.14) since field-collected larvae were of different sizes unlike the uniform stages in the laboratory-reared ones (Gopali and Lingappa 2001a). At ICRISAT, for effective mass multiplication of *Amal*NPV, the field-collected larvae are released into an aluminum or polythene grid/enclosure (10 cm height) to confine the larvae inside the shaded enclosure and feed with plants already inoculated with the virus. The field technique for rearing larvae is advantageous, particularly in avoiding the handling of huge larval populations, rearing, and inoculation. This would also facilitate farm-level production and access to the bio-pesticide at the village level (Rao 2006). The laboratory-level mass production technique for *Amal*NPV has been standardized by Veenakumari et al. (2006). In situ field-level mass production of *Amal*NPV in a groundnut ecosystem was developed for the first time at Project Directorate of Biological Control (PDBC), Bangalore, India (Veenakumari et al. 2007).

The virus used for the inoculation must confirm the quality control specifications of viral products as reported by Shieh and Bohmfalk (1980). The inoculation dose is expressed in units of POB/ml, and the optimal dose varies with the virulent virus and age of the host (Ignoffo and Couch 1981). Angelini and Labonne (1970) suggested that the best method to propagate the virus was to spray a suspension on larval diet. They could get the larval mortality after 7–8 days. Shapiro and Bell (1981) reported that surface treatment is an efficient system that is easily automated and requires much less virus than diet incorporation. However, Odak et al. (1984) used soaked chickpea seeds treated with *Ha*NPV to feed *Helicoverpa* larvae and found that the method was effective for mass production of virus. Bioassays were used to determine the activity of each batch of virus. Several modes of administration of virus were tried using different larval instars, viz., surface treatment, diet incorporation, and direct feeding (Ignoffo 1966a). Earlier instars were highly susceptible to the virus (Rabindra and Subramanian 1974) with LT_{50} shorter than older ones. Narayanan (1979) report that the early instars recorded 100 % mortality, whereas the late instars particularly from the fifth instars pupated and gave rise to malformed adults with short and ruffled wings. The effect of NPV was directly related to the age of the larvae at the time of infection (Battu 1990). Further, Battu

(1992) reported that increasing dosages are required to kill the older larvae. The relative resistance of 8-day-old larvae was 2,000 times more than that of 1-day-old larvae. Further, Battu observed that the fifth and sixth instar larvae could not be infected with the virus even at higher concentrations. Rabindra and Subramanian (1974) inoculated the fourth instar larvae with a dose of 10^6 POBs/ml to harvest maximum yield. The LC_{50} values for the first and third instars of *H. armigera* were 8.3×10^3 and 28.6×10^5 POBs per larva, respectively (Backwad 1979). Narayanan (1979) found that the optimum dose of inoculum required for obtaining maximum harvest of virus from the fourth instar larvae was 5×10^4 POBs/cavity/larva by the diet surface contamination method, whereas Shieh (1978) used 5×10^5–5×10^6 POBs/ml inoculum in each cavity and observed that there was significant interaction between the age of the larvae and dose of the virus with the recovery of POB.

Taun et al. (1989) described the pathogenicity of *Ha*NPV to *H. armigera* using three different inoculation methods. The LD_{50} values of the fourth instar larvae that were fed on diet containing NPV or maize kernels soaked in virus suspension were 1.85×10^6 and 2.55×10^5 POBs per larva, respectively. The inoculum-imbibing method was more sensitive and convenient for inoculating the pest with virus, whereas Jayaraj and Sathiah (1993) described the three methods of inoculation, viz., head dipping, oral feeding, and diet surface contamination, and the latter method was the most economical and convenient for easy application. Ignoffo (1966b) estimated that at least 6×10^9 virus polyhedra were produced per larva in late instars of *H. zea*, and he defined it as "one larval equivalent." The average yield of virus per larva infected after 5–7 days at 30 °C was 1.5×10^9 polyhedra (Ignoffo 1973). Teakle et al. (1985) observed that the least yield of 1.18×10^7 POBs/insect was from younger larvae of *H. armigera* compared to 3.6×10^9 POBs/insect from grown-up larvae, whereas Shieh (1978) recovered 5×10^9 POBs/larva, indicating that the yield of POBs was directly related to the age of the infected larvae. The host insect, insect diet, insect age and virus dosage, incubation, environment, and preservation of virus infectivity were some of the major factors that optimize the production of HzNPV (Carter 1984). The virus yield increased exponentially with the age of larva at dosing in the range of zero to six days, the overall increase being approximately 100-fold (Teakle and Byrne 1989). Battu (1990) reported an average yield of 1.81×10^9 POBs per larva of *H. armigera*. Similarly, Pawar and Thombre (1992) reported that *Ha*NPV yields per larva ranged from 0.95×10^9 to 3.5×10^9. Gopali and Lingappa (2001a) suggested 10^8 POBs/ml as the optimum dose required for the third and fourth instar larvae to achieve quicker and higher mortality of larvae for virus production, and among different instars of *H. armigera*, the fourth instar larva was found ideal for virus production as it yielded higher quantity of virus per larva (2.81×10^9 POBs). In vivo mass production and control efficacy studies of *S. litura* NPV (*Sl*NPV) were positively correlated with larval weight from the third instar to the fifth instar larvae; a maximum yield of 1.4×10^9 POBs/ml was obtained with the early fifth instar larvae individually infected by diet-incorporation of inocula of 3×10^6 POBs/ml for 7 days of incubation at 30 °C (Tuan et al. 1998). Similarly, a maximum yield of 5.57×10^9 POBs/larva was obtained at the inoculum dose of 1966.2 POB/mm^2 of *S. litura* NPV when exclusive harvest of cadaver was done (Senthil Kumar et al. 2005). Jun et al. (2007) reported that the volume of POBs of SlMNPV harvested on the 5th day of the postinoculation period was significantly lower than that harvested on the 7th day of the post-incubation period which was significantly lower than that harvested after larval death, and a similar trend was observed in biological activity by dosing the fifth instar larvae. To study the influence of virus inoculation method and host larval age on the productivity of the NPV of the teak defoliator, *Hyblaea puera* (Cramer) was determined by different methods of inoculation (Biji et al. 2006).

At present in India, in vivo propagation is being practiced for NPV mass production at commercial scale and even at farmer level. Healthy larvae reared in the laboratory or collected from the fields are fed with low dose of NPV, and the virus produced in the insect is harvested and its concentration is estimated by counting POBs using a light microscope fitted with a hemocytometer. Recently, local production and utilization of NPV gained momentum in India through participation of scientists, farmers, NGOs, and state agricultural and extension departments. In the fields, natural mortality of *Helicoverpa* and *Spodoptera* can be seen due to infestation of disease-causing virus particles. Such larvae can be collected and may be utilized for in vivo mass propagation and again for checking their efficacy against pest populations. ICRISAT (Rao et al. 2007) trained several national agricultural research and extension systems (NARES) scientists and farmers on bio-pesticide production and established 96 village-level NPV production units in India and Nepal to encourage their use. As the selection of virulent strain of NPV is key to the development of effective bio-pesticides, local strains are always preferred for sustainability, adaptability, and efficacy under a given set of agroecosystem and hold an ample scope for their widespread multiplication and commercial use in a particular region (Gupta et al. 2007, 2010). It is well recognized that factors such as the geographic origins of both the virus and host can affect the characteristics of the dose–response curve and the period of survival of infected hosts (Maeda et al. 1990). For production of *Ha*NPV and *Sl*NPV, a host insect larva has to be multiplied on artificial or semisynthetic diet or soaked chickpea seeds. Crude *Ha*NPV is commercially produced at Dr. Panjabrao Deshmukh Krishi Vidyapeeth, Krishi Nagar, Akola (Maharashtra), and at Agricultural Research Station, Gulbarga (Karnataka), by following the procedures: host insect multiplication, virus inoculation and harvest, extraction and purification of virus, and standardization of NPV.

2.5.2 Problems Associated with Commercialization of In Vivo-Produced NPV Products

In vivo mass production of NPVs is labor intensive and involves mass rearing and infection of insect larvae, which account for high production costs. In addition, the products have some quality and storage issues which severely affect the efficacy and quality of the products. One of the major problems involved in harvesting virus from dead larvae was that they were often heavily contaminated with bacteria. Purification methods based on centrifugation were found to be less effective in removing bacteria (Sireesha 2006; Sridhar Kumar 2008). Other problems are inconsistency in the yield and malodor during production and even in the finished product also. Another important problem associated with the commercialization of NPV products is the lack of proper diagnostic systems to quantify the virus titer, microscopic counting procedure used to screen the larvae for NPV infection, and quality control of the viral insecticide batches which has low-detection efficiency, unknown specificity, and is laborious and requires considerable skill (Wigley 1976). Because of this, many NPV products produced in India have poor efficacy and are found to be ineffective under field conditions. To overcome this problem and for effective production of viral insecticides, it is necessary to have an efficient strategy for virus production, combined with rapid and specific diagnostic and quality control system (Shieh 1989).

Microbial pesticides including NPVs and GVs have now been brought under the ambit of the Central Insecticide Act, 1968. Commercialization of microbial pesti-cides is possible only after registration with the Central Insecticide Board (CIB) in India, a body constituted under the insecticides act that regulated their manufacture and use. Section 9(3b) of the act specifies a minimum quantity of active ingredient in the formulation. Many of the viral products available in the markets in develop-ing countries were classified as weak, with poor efficacy, questionable quality control (Harris 1997), and are failing to meet acceptable standards (Kern and Vaagt 1996). Unless this matter is addressed effectively, there is serious danger in these countries that poor quality products with their inevitable failures will erode the farmer's confidence in microbial control products such as NPV and significantly retard the promotion of this potential technology. There has been a rapid growth in the production and use of NPV products in the last decade in India, but this has exacerbated quality problems (Kambrekar et al. 2007). The causes of poor quality lay in deficiencies in production technologies and poor quality control procedures. NPVs of *Helicoverpa* and *Spodoptera* have been most extensively researched and studied with regard to their efficacy, mass production, and compatibility with botanicals and other insecticides and against several nontarget organisms (Hunter-Fujita et al. 1998 and Saxene and Ahmad 2005). Pathogenicity of the isolates varies according to localities and therefore needs to be screened. Develop-ment of economical in vitro cell culture techniques for large-scale production of NPVs which will go a long way has not yet been developed for agricultural use.

2.5.3   Is In Vitro Production the Only Solution to Address Quality Control Issues?

While several improvements in in-vivo production systems in insect larvae have been made in the past, these could not help to overcome the problems described above (Sireesha 2006; Sridhar Kumar 2008). Hence, it has been proposed that the adoption of an alternative technology based on the viral propagation in insect cell cultures could help to overcome the problems associated with in vivo technology and enable the development of well standardized, controlled, and scalable production processes for insecticidal baculoviruses (Szewczyk et al. 2006). In the early part of the twentieth century, entomologists had a dream of utilizing insect cells grown in vitro as a tool for producing entomopathogens. These early experiments used a simple saline solution or hemolymph as the culture medium, and cultures could rarely be kept for more than a few days. A breakthrough occurred four decades ago when Grace (1962) successfully established long-term cultures of insect cells. Since then, over 500 continuous cell lines have been established from over 100 insect species (Lynn 1999). Insect pathologists have cells capable of replicating dozens of insect-specific viruses (Granados and McKenna 1995), while plant pathologists and vertebrate pathologists have cells capable of replicating viruses transmitted by insects (Mitsuhashi 1989). Mass production of the virus at reasonable costs is an important factor in the development of NPVs into a marketable product.

2.5.4   In Vitro Production

Baculoviruses can be produced in vitro in infected insect cells cultivated in bioreactors. In order to develop an economically feasible process to produce baculoviruses in insect cells, low-cost culture media that satisfy the nutritional demands of both uninfected and infected cells are needed to achieve high virus yields. Fetal calf serum is the most widely used additive in insect cell culture media. However, its high cost and batch-to-batch variability are the drawbacks in its utilization to sustain baculovirus production in large-scale processes. Therefore, a replacement for the fetal calf serum in insect cell culture media is the key step to develop a technically and economically feasible process to produce baculoviruses in vitro. Recent studies demonstrated that insect's cells are able to both proliferate permanently and replicate baculovirus in a lipid-free environment. However, in order to obtain a useful medium for technological applications, it will be necessary to optimize the composition of the multiple supplements and evaluate its performance in a lipid-supplemented environment. Insect cell cultures have been extensively utilized by Linda and Lua Steven Reid (2003) for means of production for heterologous proteins and bio-pesticides. *Spodoptera frugiperda* (*Sf*9) and *T. ni* (High Five™) cell lines have been widely used for the production of recombinant proteins; thus, metabolism of these cell lines has been investigated thoroughly over the recent years. NPV isolated from the alfalfa looper, *A. californica*, was replicated successfully and rapidly in a suspended ovarian cell line of the cabbage looper, *T. ni* (Vail and Jay 1973). Polyhedra were observed in the nucleus of cells within 20 h after inoculation. The cyto-pathological changes typical of nuclear polyhedrosis infections were observed, and an average of 64 polyhedra/cell was produced. These polyhedra were quantitatively as infectious to cabbage looper larvae as those produced in vivo. In addition, they were infective to *H. virescens*, *Pectinophora gossypiella*, *S. exigua*, *A. californica*, and *A. falcifera*. Bioassays have indicated that both *H. zea* and *H. armigera* viruses produced in vitro maintain biological activity (Suzanne 2009 and Szewczyk et al. 2006). Chakraborty et al. (1999) studied the in vitro production of virus from *H. armigera* (HaSNPV) and its possible use as a specific *Helicoverpa/Heliothis* larvicide. Growth kinetics of *H. zea* cells and virus OB yields were compared in three SF900II-based media, namely, SF900II (serum-free), SF900II +1 % serum, or SF900II +10 % serum. Viable cell densities were usually higher in the media supplemented with serum than in the serum-free medium; however, in the serum-free medium, cell diameters were 1.7 times greater (i.e., individual cell volumes were five times larger). Four new cell lines, designated as NTU-LY-1 to 4, respectively, were established from the pupal tissues of *Lymantria xylina* Swinhoe (Lepidoptera: Lymantriidae) (Wu and Wang 2006). These cell lines have been cultured approximately 80 passages during 2 years in TNM-FH medium supplemented with 8 % fetal bovine serum, at a constant temperature of 28 °C. Each line consists of three major morphological types: round cells, spindle-shaped cells, and giant cells.

Sundeep et al. (2005) developed two cell lines from the larval hemocyte and embryonic tissue of *H. armigera* and designated them as NIV-HA-1195 and NIV-HA-197, respectively. The NIV-HA-197 cell lines were found highly susceptible to HaSNPV, yielding a very high titer (2.88×10^7 NPV/ml) on the 10th postinfection day (PID). The HaSNPV OBs produced in vitro were highly virulent to the second and third instars *H. armigera* larvae causing cessation of feeding on the 2nd day and mortality in 6 days. This cell line is also found to be growing well in goat serum (GS)-supplemented medium producing a comparable yield of OBs. Goat serum, being cheap and locally available, will help in the large-scale production of *Ha*SNPV for use as a biopesticide in the future. The cell line NIV-HA-197 was found to be susceptible to the baculoviruses AcMNPV, SlMNPV, and HaSNPV (Sundeep et al. 2002). More than 90 % of the cells were infected by HaSNPV on the seventh PID, and 28.8×10^6 NPV/ml was yielded on the 10th PID. The in vitro-grown HaSNPV caused 100 % mortality, when fed to the second instar *H. armigera* larvae, in 6 days. Isoenzyme profile and results of 16S rRNA heteroduplex analysis clearly indicated the species specificity of the new cell line NIV-HA-1195 (Sundeep et al. 2002) and was also found susceptible to the baculoviruses, *Ac*NPV, *Sl*NPV, and the homologous *Ha*NPV. Pant et al. (2002) reported that the *H. armigera* cell line from the embryonic tissue was highly susceptible to *Ha*NPV (6.3×10^6 NPV/ml). These in vitro-grown *Ha*NPVs caused 100 % mortality in the second instar larvae. The cultures could grow as suspension culture on shakers and may find application for the in vitro production of wild-type/recombinant baculoviruses as bio-insecticides.

Nakat (2004, In vitro production of nuclear polyhedrosis virus of *Helicoverpa armigera* and *Spodoptera litura* and its field efficacy in Western Maharashtra. Department of Entomology, Mahatma Phule Krishi Vidyapeeth, Rahuri, Maharashtra, unpublished) standardized the procedure for monolayer and spinner culture of *Sf*-9, *Sf*-21, and *Ha*-197 cells. The growth curve of different cells in spinner culture was plotted on the basis of daily viable cell count. The cell line *Sf*-9 was susceptible for both the baculoviruses *Ac*MNPV and *Sl*NPV in monolayer. The cell line *Ha*-197 was found susceptible for *Ha*NPV in monolayer. For production of *Ha*NPV, *Ha*-197 cell line with *Ha*NPV wild-type MPKV strain was found to be efficient, more virulent, and infectious in both the cell lines. The procedure for the cell lysis was standardized to extract the POBs from infected cells with the addition of 0.1 % SDS and deep freezing at -20 °C followed by 15 min sonication resulting in good separations of POBs from 80 to 90 % infected cells. The field demonstrations of in vitro- and in vivo-produced *Sl*NPV and *Ha*NPV were conducted on capsicum, gerbera, rose, soybean, and chickpea. The effectiveness of in vitro-produced NPVs was found to be superior, causing larval mortality in the range of 78–100 % as compared to in vivo-produced virus which was 70–88 %.

The insect cell line, the culture medium, the bioreactor, the virus, the infection parameters, and the culture strategy are elements of the insect cell culture technology that must be optimized in order to develop in vitro production processes for insecticidal baculoviruses (Claus Juan et al. 2012). The cell line *Hz*-AM1 has been used widely to examine possible factors affecting the yields and the potency of *Ha*NPV. These factors include the medium, supplemented serum, cell density at infection, multiplicity of infection, viral strain, and passage effect (Chakraborty et al. 1996; Lua and Reid 2000; Ogemo et al. 2007). Increasing OB yields per cell in culture is the main challenge to enable commercialization of in vitro production of NPVs. Isolating clones from a heterogeneous cell population may allow development of a high virus-producing cell clone. An automated robotic clone-picking system to establish over 250 insect clones of an *H. zea* cell population to be screened for virus production has been carried out by Nguyen et al. 2011. The type and degree of passage effect are dependent on the cell lines and the virus species (Krell 1996). Homologous cell lines are desirable for the production of an NPV, whereas heterologous NPV infection of cell lines decreases the productivity and yields less virulent progeny viruses (Tompkins et al. 1988).

Although the production of insecticidal baculoviruses in insect cell cultures has been proposed as an alternative to overcome the limitations of the in vivo processes, so far no in vitro process could be even implemented on an industrial scale and baculovirus occlusion bodies are still produced in infected insect larvae (Claus Juan et al. 2012). Some factors that 25 years ago have hindered the development of large-scale production processes for baculoviruses in insect cell cultures, such as the sensitivity of insect cells to the stresses linked to the mechanical agitation in stirred tank reactors and to the bubble rupture in sparged bioreactors, have been resolved, and several cell lines can be cultivated today in industrial bioreactors of large volume to produce occlusion bodies or recombinant proteins.

In vitro propagation in susceptible insect cell lines is the best option for the commercial production of this virus. Recently, NIV (Pune) developed indigenous cell lines from four insect species, and their susceptibility to different NPVs was tested, and are commercially selling them to interested scientists (Pant et al. 2002). The cultures can grow as suspension culture on shakers and are found to be successful for in vitro production of wild-type/recombinant baculoviruses as bio-insecticides. However, most cell lines have not been sufficiently characterized with respect to certain issues such as (1) simplification of the composition of the culture medium, (2) possibility to obtain high volumetric yields of viral OBs, and (3) optimization of seed virus or budded virus or nonoccluded virus. Production related to economic feasibilities for entrepreneurs needs to be addressed.

2.6 Diagnostic and Quality Control Tests

The reliability of the product is crucial in ensuring acceptance and sustained use by the farmers. The issue of erratic performance of viral biocontrol agents has been recognized as a significant factor in limited successful commercialization (Lisansky 1997). It has been widely perceived that viral agents have not achieved a level of efficacy comparable with that of chemicals or other bio-pesticides such as *Bacillus thuringiensis* (Berliner).

Mass production of NPV insecticide is simple and widely produced even at farmer level. Although commercial production, quality, and storage were still contentious issues, NPV is multiplied on field-collected larvae and being applied on crops. Multiplying NPV on field-collected larvae was found to be easier and cost-effective compared to laboratory-reared larvae, but efficacy and quality of which may be affected due to contaminants such as bacteria and fungi.

The effectiveness of the viral insecticide is critically dependent on the concentration of POB, which is expressed as LE. Generally, a standard stock preparation consists of 1LE, i.e., 6×10^9 POBs/ml. NPV production methods have been well established in many developing countries. Appropriate, sensitive, and reliable serological tools (Kohler and Milstein 1975; Kelly et al. 1978; Towbin et al. 1979; Crook and Payne 1980; Smith and Summers 1981; Zhang and Kaupp 1988; Lu et al. 1995) are not available at this stage and will go a long way in the quality control of insect viruses in developing countries. Once developed, the tools would be of immense value to public and private entrepreneurs, such as state bio-pesticide production laboratories and regulatory agencies. In addition to this, the highly standardized, accurate, and sensitive diagnostic tools for NPV detection in field-collected larvae would be beneficial to pest management personnel, because early detection of NPV disease could make it possible to predict the occurrence of an imminent epizootic and thus alter the pest control tactics to be employed (Volkman and Falcon 1982).

As part of quality control during mass production of NPVs used for commercial viral insecticide preparations at ICRISAT, polyclonal antibodies were raised against purified polyhedrin (POB) protein preparations and used these antibodies to develop ELISA and Western blotting assays to detect NPVs. The sensitivity of the ELISA was 15 ng/ml of semi-purified viral protein or 30 ng/ml POBs from the NPV infected larval extracts. These antibodies are useful to diagnose the early stages of larval infection by NPV and also for the quantification of the NPVs during production of viral insecticides for *Ha*NPV, *Sl*NPV, and AmalNPV (Sridhar Kumar et al. 2007). For qualitative detection of NPVs in larval homogenates, Western immunoblotting and indirect immunofluorescence assay, and for quantitative detection direct antigen coating (DAC) and indirect competitive (IC)-ELISA tools were developed and evaluated (Sridhar Kumar 2008). Since, polyhedrin is the major component of NPV polyhedra, is coded by the virus, and its presence in larvae indicates the presence of NPV or an NPV infection. Similarly, the relationship between three NPVs isolated from the larvae of *H. armigera*, *S. exigua*, and *S. litura* in Taiwan was determined by assaying the polyhedrin in DAC-ELISA with polyhedrin polyclonal antisera specific to each polyhedrin (Tuan et al. 1999), and similarly, a monoclonal antibody-based DAC-ELISA was developed for the identification and differentiation of OpMNPV and OpSNPV and also for detection of their homologous polyhedrin in larval extracts with sensitivity of 100 ng/ml in the presence of host tissue extract, rather than 5 ng/ml in its absence (Quant et al. 1984). However, competition between insect and viral proteins for binding to ELISA plate surface has reduced the detection sensitivity of the DAC-ELISA, particularly when crude insect extracts were used. Since it was found that host tissue extract interfered with the assay, still we were able to determine its absolute sensitivity in the presence of unknown quantities of host tissue extract. To eliminate the competition between insect and viral proteins for binding sites in the ELISA plate surface in DAC-ELISA, we attempted to avoid the interference phenomenon by standardizing an IC-ELISA to estimate the polyhedrin content in insect extracts. Similarly, an IC-ELISA was standardized to evaluate the bio-safety of recombinant and wild-type NPV of *A. californica* (Ashour et al. 2007). Similarly, Crook and Payne (1980) examined the direct, indirect, and double antibody sandwich methods of ELISA for their ability to detect and discriminate between granulosis viruses from *Pieris brassicae*, *Agrotis segetum*, and *C. pomonella* and for their specificity in the presence of host material, and they concluded that the indirect method was the most sensitive and capable of detecting down to about 1 ng of dissolved capsules/ml compared with 10 ng/ml for the double antibody sandwich method and 25 ng/ml for the direct method and the double antibody sandwich method was more specific and showed greatest discrimination between different granulosis viruses.

Also the DAC and IC-ELISA tools were evaluated for their performance in quantification of POBs in commercial NPV preparations. The number of POBs present in the sample bottle was determined by extracting the total polyhedrin and compared with the standard regression graph of polyhedrin extracted from known number (estimated by microscopic counting) of POB standards such as 6×10^9 to 2.34×10^7 POBs/ml (1 LE to 0.0078 LE). These ELISA methods are sensitive to a minimum of approximately 4.6875×10^7 POBs/ml (0.015 LE/ml), which is little bit higher to the range (100–2,000) of previous reports (Crook and Payne 1980; Kelly et al. 1978; Longworth and Carey 1980; Shamim et al. 1994). The ELISA methods can be used to quantify infection, unlike other methods, and this feature may be applied to predict the potential inoculum required for future populations. Previously, it has been shown that the ELISA method can be used to quantify baculoviruses (Clark and Barbara 1987). Tuan et al. (1998) compared the visual counting of POBs by microscope, bioassay, SDS-PAGE, and ELISA for quality control of SlNPV products, and ELISA has proved to be better than SDS-PAGE. The ELISA results were comparable to light microscope counting of POBs (Sridhar Kumar 2008). The absorbance values suggest that the ELISA method can be used to accurately quantify virus POBs and virus infections from tissue homogenates. Recently, Thorne et al. (2007) reported that the alkali-treated POB lysates were assayed in IC-ELISA for estimation of total POBs in semi-purified commercial NPV samples or in insect tissue extracts with a sensitivity of minimum of approximately 850 POBs. Similarly, Parola et al. (2003) reported the estimation of granulin in commercial GV suspensions of *Epinotia aporema* GV by DAS-ELISA with a sensitivity of 0.53 ng/ml of purified OB suspensions; this represented 2.0×10^4 OBs/ml.

These diagnostic and quality control tests are convenient for routine detection and quantification of NPVs, and this technology will also be transferred to the bio-products agribusiness units for commercialization of NPV production. Also the ELISA and Western immunoblot assays can be successfully applied in bioassay experiments during optimization of conditions for the productivity and quality of NPVs to get the maximum virus yield as well as to reduce the development of bacterial contamination. In addition to this application of ELISA tools at field level, evaluation of the efficacy of NPVs is useful for ecological and epidemiological studies of NPVs during IPM programs and also during the surveys of their persistence and outbreaks of natural epizootics in the environment.

2.7   Way Forward

- Over the past 25 years, the research approach on NPVs has evolved toward being more ecologically holistic with industry's concerns.
- Although viral pesticides still represent a very small portion of plant protection at present, their role was considered significant.
- Though NPVs gained prominence as environmentally friendly alternatives to chemical insecticides, they still face a number of hurdles in their production, marketing, and utilization.
- Importance of effective multidisciplinary research, public, private, people partnerships.
- Need for in-depth knowledge among farmers, extension, and policy makers about bio-pesticides.

- Lack of effective regulations can lead to poor product quality, performance, and loss of user confidence.
- NPVs that can perform effectively in wider environments and larger host range have immense potential.
- Prioritize research for better integration of bio-agents into production systems, such as in rotating these with chemical pesticides and developing these into effective bio-models.

References

Abul-Nasr S, Elnagar S (1980) The susceptibility of the cotton leafworm, *Spodoptera littoralis* (Boisd) to different isolates of nuclear polyhedrosis virus. J Appl Entomol 90:289–292

Aizawa K (1963) The nature of infections caused by nuclear polyhedrosis viruses. In: Steinhaus EA (ed) Insect pathology: an advanced treatise, vol 1. Academic, New York, pp 381–412

Anderson RF, Shieh TR, Huang HT, Rogoff MH (1972) Production of a viral pesticide. Proc Symp Fermentat Technol Today 2:493–505

Angelini A, Labonne V (1970) A technique for the rearing of *Heliothis armigera* (Hub.) and a possibility of producing a nuclear virus disease in the Ivory Coast. Cotton Fibres Tropicales 25:501–504

Anon (2009) Biopesticides market to reach $ 1 billion in 2010. http://www.ien.com/article/biopesticides-market-to/8648

Arguad O, Croizier L, Lopez-Ferber M, Croizier G (1998) Two key mutations in the host-range specificity domain of the p143 gene of *Autographa californica* nucleopolyhedrovirus are required to kill *Bombyx mori* larvae. J Gen Virol 79:931–935

Armes NJ, Jadhav DR, Bond GS, King ABS (1992) Insecticide resistance in *Helicoverpa armigera* in South India. Pestic Sci 34:355–364

Arora R, Battu GS, Bath DS (1997) Comparative evaluation of some native isolates of a nuclear polyhedrosis virus against *Heliothis armigera* (Hübner). J Entomol Res 21:183–186

Ashour MB, Ragheb DA, EI-Sheikh EI-Sayed A, Gomaa EI-Adarosy A, Kamita Shizuo G, Hammock Bruce D (2007) Biosafety of recombinant and wild type nucleopolyhedrovirus as bioinsecticides. Environ Res Public Health 4:111–125

Attathom T, Chaeychomsri S, Mahattana C, Siriwattanagul W (1990) Technological development for the local production of nuclear virus of *Helicoverpa armigera*. In: Proceedings on V international colloquium on invertebrate pathology and microbial control, Adelaide, 20–24 August 1990

Backwad DG (1979) Studies on NPV infection in *Heliothis armigera* (Hubner) and *Anomis sabulifera*. M.Sc. (Ag.) Thesis, Marathwada Agriculture University, Parbhani, p 141

Banowetz GM, Fryer JL, Iwai PJ, Martignoni ME (1976) Effects of the douglas-fir tussock moth nucleopolyhedrosis virus (baculovirus) on three species of salmonid fish. Forest Service Research Paper PNW-214. U.S. Department of Agriculture, Washington, DC

Bansal OB, Bansal A, Behera AK, Das RH, Kumar M (1997) Characterization of the polyhedron gene of *Spodoptera litura* nuclear polyhedrosis virus. Virus Genes 14:175–180

Battu GS (1990) Yield levels of the occlusion bodies obtained from the Baculovirus infected *Hellula undalis* (Fabricius). Ind J Entomol 51:317–325

Battu GS (1992) Potential use of baculoviruses on Lepidopteran pests in Punjab. In: Ananthakrishnan TN (ed) Emerging trends in biological control of phytophagous insects. Oxford and IBH Company Private Ltd., New Delhi, pp 174–181

Battu GS, Ramakrishna N (1987) Investigations on the nuclear polyhedrosis virus of *Diacrisia oblique* (Walker): bioassay of viral activity. Entomon 12:81–84

Bell MR (1991) Effectiveness of microbial control of *Heliothis* spp. on early season wild geraniums. Field and field cage tests. J Econ Entomol 84:851–854

Bellotti AC (1999) Recent advances in cassava pest management. Annu Rev Entomol 44:345–370

Benz GA (1986) Introduction: historical perspective. In: Granados RR, Federici BA (eds) The biology of baculoviruses. CRC, Boca Raton, pp 1–35

Bergold GH (1947) Die Isolierung des Polyeder Virus und die Natur der Polyeder. Z Naturforsch 2B:122–143

Bergold GH (1948) Uber die Kapsel virus-krankheit. Z Naturforsch 3B:338–342

Bergold GH (1982) The nature of the nuclear polyhedrosis viruses. Biol Baculoviruses 413–453

Biji CP, Sudheendrakumar VV, Sajeev TV (2006) Influence of virus inoculation method and host larval age on productivity of the nucleopolyhedrovirus of the teak defoliator, *Hyblaea puera* (Cramer). J Virol Methods 133:100–104

Blissard GW, Rohrmann GF (1990) Baculovirus diversity and molecular biology. Annu Rev Entomol 35:127–155

Blissard GW, Black B, Crook N, Keddie BA et al (2000) Family Baculoviridae. In: Van Regenmortel MHV et al (eds) Virus taxonomy: seventh report of the International Committee on Taxonomy of Viruses. Academic, San Diego, pp 195–202

Braunagel SC, Russell WK, Rosas Acosta G, Russell DH, Summers MD (2003) Determination of protein composition of the occlusion-derived virus of *Autographa californica* nucleopolyhedrovirus. Proc Natl Acad Sci USA 100:9797–9802

Brown SE, Maruniak JE, Knudson DL (1984) Physical map of SeMNPV baculovirus DNA: an AcMNPV genomic variant. Virology 136:235–240

Bulach DM, Kumar CA, Zaia A, Liang BF, Tribe DE (1999) Group-II nucleopolyhedrovirus subgroups revealed by phylogenetic analysis of polyhedrin and DNA polymerase gene sequences. J Invertebr Pathol 73:59–73

Burges HD (1981) Microbial control of pests and plant diseases, 1970–1980. Academic, New York, p 949

Burges HD, Thompson EM (1971) Standardisation and assay of microbial insecticides. In: Burges HD, Hussey NW (eds) Microbial control of insects and mites. Academic, London, pp 591–622

Caballero P, Aldebis HK, Vargas-Osuna E, Santiago-Alvarez C (1992) Epizootics caused by a nuclear polyhedrosis virus (NPV) in populations of *Spodoptera exigua* in Southern Spain. Bio Sci Technol 2:45–149

Carter JB (1984) Viruses as pest control agents. Biotechnol Gen Eng Rev 1:375–405

Chakraborthy S, Monsour C, Teakle R, Reid S (1999) Yield, biological activity, and field performance of a wild-type *Helicoverpa* Nucleopolyhedrovirus produced in *H. zea* cell cultures. J Invertebr Pathol 73:199–205

Chakraborty S, Green Field P, Reid S (1996) *In-vitro* production studies with a wild type *Helicoverpa* baculovirus. Cytotechnology 22:217–224

Christian PD, Gibb N, Kasprzak AB, Richards A (2001) A rapid method for the identification and differentiation of *Helicoverpa* nucleopolyhedroviruses (NPV Baculoviridae) isolated from the environment. J Virol Methods 96:51–65

Clark M, Barbara D (1987) A method for quantitative analysis of ELISA data. J Virol Methods 15:213–222

Claus Juan D, Gioria Verónica V, Micheloud Gabriela A et al (2012) Production of insecticidal baculoviruses in insect cell cultures: potential and limitations. In: Soloneski S (ed) Insecticides – basic and other applications. In Tech. Available from: http://www.intechopen.com/books/insecticides

Croizier G, Ribeiro HCP (1992) Recombination as a possible major cause of genetic heterogeneity in *Anticarsia gemmatalis* nuclear polyhedrosis virus populations. Virus Res 26:183–196

Crook NE, Payne CC (1980) Comparison of three methods of ELISA for baculoviruses. J GenVirol 46:29–37

David WAL, Gardiner BOC (1966) Persistence of a granulosis virus of *Pieris brassicae* on cabbage leaves. J Invertebr Pathol 8:180–183

David WAI, Gardiner BOC (1967a) The effect of heat, cold, and prolonged storage on a granulosis virus of *Pieris brassicae*. J Invertebr Pathol 9:555–562

David WAL, Gardiner BOC (1967b) The persistence of a granulosis virus of *Pieris brassicae* in soil and in sand. J Invertebr Pathol 9:342–347

David WAL, Ellaby SV, Taylor G (1971) The stability of a purified granulosis virus of the European cabbageworm, *Pieris brassicae*, in dry deposits of intact capsules. J Invertebr Pathol 17:228–233

De Moraes RR, Maruniak JE (1997) Detection and identification of multiple baculoviruses using the polymerase chain reaction (PCR) and restriction endonuclease analysis. J Virol Methods 63:209–217

Doyle CJ, Hirst ML, Cory JS, Entwistle PF (1990) Risk assessment studies: detailed host range testing of wild-type cabbage moth, *Mamestra brassicae* (Lepidoptera: Noctuidae) nuclear polyhedrosis virus. Appl Environ Microbiol 56:2704–2710

Ebora RV, Shepard BM, Cadapan EP (1990) Mass propagation and factors affecting virulence of a nuclear polyhedrosis virus of *Spodoptera litura* (Fab.). Philippine J Biotechnol 2:138–148

Falcon LA (1971) Microbial control as a tool in integrated control programs. In: Huffaker CB (ed) Biological control. Plenum, New York, pp 346–364

Fauquet CM, Mayo MA, Maniloff J, Desselberger U, Ball LA (eds) (2004) Virus taxonomy. VII report of the ICTV. Elsevier/Academic, London, p 1258

Federici BA (1978) Baculovirus epizootic in a larval population of the clover cutworm, *Scotogramma trifolii*, in Southern California. Environ Entomol 7:423–427

Federici BA (1997) Baculovirus Pathogenesis. In: Miller LK (ed) The baculoviruses. Plenum, New York, pp 33–59

Federici BA, Hice RH (1997) Organization and molecular characterization of genes in the polyhedrin region of the *Anagrapha falcifera* multinucleocapsid NPV. Arch Virol 142:333–348

Figueiredo E, Munoz D, Escribano A, Mexia A, Vlak J, Caballero P (1999) Biochemical identification and comparative insecticidal activity of nucleopolyhedrovirus isolates pathogenic for *Heliothis armigera* larvae. J Appl Entomol 123:1–00

Funk CJ, Braunagel SC, Rohrmann GF (1997) Baculovirus structure. In: Miller LK (ed) The baculoviruses. Plenum, New York, pp 7–32

Geetha N, Rabindra RJ (1999) Genetic variability and comparative virulence of some geographic isolates of nuclear polyhedrosis virus of *Helicoverpa armigera* Hub. In: Ignacimuthu S, Sen A, Janarthana S (eds) Biotechnological application for integrated pest management. Oxford and IBH Publ Co. Pvt. Ltd., New Delhi, pp 65–78

Gelernter WD, Federici BA (1990) Virus epizootics in California populations of *Spodoptera exigua*: dominance of a single genotype. Biochem Syst Ecol 18:461–466

Gopali JB, Lingappa S (2001a) Refinement of technique for mass production of virus *from Helicoverpa armigera* (Hubner). Karnataka J Agric Sci 14:947–954

Gopali JB, Lingappa S (2001b) Evaluation of safety period for field use of virus (HaNPV) under different set of storage conditions. Karnataka J Agric Sci 14:1072–1074

Grace TD (1962) Establishment of four strains of cells from insect tissues grown in vitro. Nature 195:788–789

Granados RR, McKenna KA (1995) Insect cell culture methods and their use in virus research. In: Wood HA, Granados RR, Schuler ML et al (eds) Baculovirus expression systems and biopesticides. Wiley-Liss, New York, pp 13–39

Groner A (1986) Specificity and safety of baculoviruses. In: Granados RR, Federici BA (eds) The biology of baculoviruses, biological properties and molecular biology, vol I. CRC, Boca Raton, pp 177–202

Grzywacz D, McKinley D, Jones KA et al (1997) Microbial contamination in *Spodoptera littoralis* nuclear polyhedrosis virus produced in insects in Egypt. J Invertebr Pathol 69:151–156

Gudauskas RT, Cannerday D (1968) The effect of heat, buffer salt and H-ion concentration, and ultraviolet light on the infectivity of *Heliothis* and *Trichoplusia* nuclear polyhedrosis viruses. J Invertebr Pathol 12:405–411

Guo ZJ, Ge JQ, Wang D, Shao YM, Tang QY, Zang CX (2006) Biological comparison of two genotypes of *Helicoverpa armigera* single-nucleocapsid nucleopolyhedrovirus. BioControl. doi:10.1007/s1056-006-9008-6

Gupta RK, Raina JC, Monobrullah MD (2007) Optimization of In vivo production of nucleopolyhedrovirus in homologous host larvae of *Helicoverpa armigera*. J Entomol 4:279–288

Gupta RK, Raina JC, Arora RK, Bal K (2010) Selection and field effectiveness of nucleopolyhedrovirus isolates against *Helicoverpa armigera* (Hubner). Int J Virol 6:164–178

Harrap KA, Payne CC, Robertson JS (1977) The properties of three baculoviruses from closely related hosts. Virology 79:14–31

Harris JG (1997) Microbial insecticides, an industry perspective in microbial insecticides: novelty or necessity? Br Crop Protect Council Proc Monograph Ser 68:41–50

Harrison RL, Bonning BC (1999) The nucleopolyhedroviruses of *Rachiplusia ou* and *Anagrapha falcifera* are isolates of the same virus. J Gen Virol 80:2793–2798

Hedlund RC, Yendol WG (1974) Gypsy moth nuclear polyhedrosis virus production as related to inoculating time, dosage and larval weight. J Econ Entomol 67:61–63

Herniou EA, Olszewski JA, Cory JS, O' Reilly DR (2003) The genome sequence and evolution of baculoviruses. Annu Rev Entomol 48:211–234

Hink WF (1982) Production of *Autographa californica* nuclear polyhedrosis virus in cells from large scale suspension cultures. In: Kurstak E (ed) Microbial and viral pesticides. Marcel Dekker, New York, pp 493–506

Hooft van Iddekinge BJK, Smith GE, Summers MD (1983) Nucleotide sequence of polyhedrin gene of *Autographa californica* nuclear polyhedrosis virus. Virology 131:5561–5565

Hostetter DL, Puttler B (1991) A new broad host spectrum nuclear polyhedrosis virus isolated from a celery looper, *Anagrapha falcifera* (Kirby), (Lepidoptera: Noctuidae). Environ Entomol 20:1480–1488

Huber J (1985) Progress in the production of codling moth granulosis virus: comparison of several virus preparations. In: Production and application of biopesticides on a viral basis in orchards and vegetables. Expert meeting, Manfauet, 5–7 July 1985. Commission of European communities, General Directorate Agriculture

Huber J, Krieg A (1978) Studies with baculoviruses on mammals. Z Angew Zool 65:69–80

Hughes PR, Getting RR, McCarthy WJ (1983) Comparison of the time mortality response of *Heliothis zea* to 14 isolates of *Heliothis* nuclear polyhedrosis virus. J Invertebr Pathol 41:256–261

Hunter DK, Collier SJ, Hoffman DF (1975) Compatibility of malathion and the granulosis virus of the Indian meal moth. J Invertebr Pathol 25:389–390

Hunter-Fujita RF, Entwistle PF, Evans FH, Crook NE (1998) Insect virus and pest management. Wiley, Chichester, p 620

Ignoffo CM (1965) The nuclear polyhedrosis virus of *Heliothis zea* (Boddie) and *Heliothis virescens* (Fabricius) I-virus propagation and its virulence. J Invertebr Pathol 7:209–216

Ignoffo CM (1966a) Insect viruses. In: Smith CN (ed) Insect colonization and mass production. Academic, New York, pp 501–530

Ignoffo CM (1966b) Standardization of products containing insect viruses. J Invertebr Pathol 48:547–548

Ignoffo CM (1973) Effects of entomopathogens on vertebrates. Ann NY Acad Sci 217:141–172

Ignoffo CM, Boeing OP (1970) Compartmental disposable trays for rearing insects. J Econ Entomol 63:1696–1697

Ignoffo CM, Couch TL (1981) The nuclear polyhedrosis virus of *Heliothis species* as a microbial insecticide. In: Burges HD (ed) Microbial control of pests and plant diseases 1970–1980. Academic, London pp, pp 329–362

Ignoffo CM, Rafajko RR (1972) In vitro attempts to infect primate cells with the nucleopoly-hedrosis virus of *Heliothis*. J Invertebr Pathol 20:321–325

Ignoffo CM, Shapiro M (1978) Characteristics of baculovirus preparations processed from living and dead larvae. J Econ Entomol 71:186–188

Injac M (1973) The role of granulosis in regulating the density of population of fall webworm (*Hyphantria cunea* Drury) in Vojvodina, Yugoslavia. Zast Bilja 24:103–110

Jacques RP, Long DR (1978) Efficacy of mixture of *Bacillus thuringiensis*, viruses and chlordimeform against insects on cabbage. Can Entomol 110:443–448

Jayaraj S, Sathiah N (1993) *Heliothis* culture techniques. In: Proceedings of national training on mass multiplication of biocontrol agents, Tamil Nadu Agricultural University, Coimbatore, pp 1–4

Jehle JA, Fritsch E, Nickel A, Huber J, Backhaus H (1995) TC14.7: a novel Lepidopteran transposon found in *Cydia pomonella* granulosis virus. Virology 207:369–379

Jun T, Shohei O, Takayoshi I, Madoka N, Yasuhisa K (2007) Productivity and quality of polyhedral occlusion bodies of a nucleopolyhedrovirus harvested from *Spodoptera litura* (Lepidoptera: Noctuidae) larvae. Appl Entomol Zool 42:21–26

Kambrekar DN, Kulkarni KA, Giraddi RS (2007) Assessment of quality of *Ha*NPV samples produced by private firms. Karnataka J Agric Sci 20:417–419

Kamita SG, Maeda S (1997) Sequencing of the putative DNA helicase-encoding gene of the *Bombyx mori* nuclear polyhedrosis virus. Gene 190(1):173–9

Katagiri K (1981) Pest control by cytoplasmic polyhedrosis viruses. In: Burges HD (ed) Microbial control of pests and plant diseases 1970–1980. Academic, New York, pp 433–440

Kelly DC, Edwards ML, Evans HF, Robertson JS (1978) The use of the enzyme linked immuno-sorbent assay to detect a nuclear polyhedrosis virus in *Heliothis armigera larvae*. J Gen Virol 40:465–469

Kelly DC, Brown DA, Robertson JS, Harrap KA (1980) Biochemical, biophysical and serological properties of two singly enveloped nuclear polyhedrosis viruses from *H. armigera* and *H. zea*. Microbiologica 3:319–331

Kern M, Vaagt G (1996) Pesticide quality in developing countries. Pesticide Out Look 7–10

Kohler G, Milstein C (1975) Continuous cultures of fused cells secreting antibodies of predefined specificity. Nature 256:495–497

Kool M, Vonken JW, Van Lier FLJ, Tramper J, Vlak JM (1991) Detection and analysis of *Autographa californica* nuclear polyhedrosis virus mutants with defective interfering proper-ties. Virology 183:739–746

Kranthi KR, Jadhav DR, Kranthi S, Wanjari RR, Ali SS, Russel DA (2002) Insecticide resistance in five major insect pests of cotton in India. Crop Prot 21:449–460

Krell PJ (1996) Passage effect of virus interaction in insect cells. In: Vlak JM, De Gooigee CD, Tramper J et al (eds) Insect cell culture, fundamentals and applied aspects. Kluwer Academic, A.H. Dordrecht, pp 125–137

Lange M, Wang H, Hu Z, Jehle JA (2004) Towards a molecular identification and classification system of lepidopteran-specific baculoviruses. Virology 325:36–47

Lautenschlager RA, Kircher CH, Podgwaite JD (1977) Effect of nuclear polyhedrosis virus on selected mammalian predators of the gypsy moth. Forest Service Research Paper NE-377. U.S. Department of Agriculture, Washington, pp 9–29

Lee HH, Miller LK (1978) Isolation of genotypic variants of *Autographa californica* nuclear polyhedrosis virus. J Virol 27:754–767

Linda HL, Lua Steven Reid LS (2003) Growth, viral production and metabolism of a *Helicoverpa zea* cell line in serum-free culture. Cytotechnology 42:109–120

Lisansky S (1997) Microbial biopesticides, in microbial insecticides Novelty or Necessity? Br Crop Protect Council Proc Monograph Ser 68:3–10

Longworth JF, Carey GP (1980) The use of an indirect enzyme-linked immunosorbent assay to detect baculovirus in larvae and adults of *Oryctes rhinoceros* from Tonga. J Gen Virol 47:431–438

Lu ZQ, Hu GG, Chen LF, Yang F, Hang SB, Le YX, Su DM (1995) Use of monoclonal antibodies and ELISA to detect nuclear polyhedrosis virus (NPV) in the larvae of *Helicoverpa armigera*. Chinese J Biol Control 11:36–39

Lua HL, Reid S (2000) Serial passage of a *Helicoverpa armigera* nucleopolyhedrosis virus in *Helicoverpa zea* serum free suspension culture. J Gen Virol 81:2531–2543

Lynn DE (1999) Development of insect cell lines: virus susceptibility and applicability to prawn cell culture. Methods Cell Sci 21:173–181

Ma XC, Xu HJ, Tang MJ, Xiao Q, Hong Q, Zhang CX (2006) Morphological, phylogenetic and biological characteristics of *Ectropis oblique* single-nucleocapsid nucleopolyhedrovirus. J Microbiol 44:77–82

Maeda S, Mukohara Y, Kondo A (1990) Characteristically distinct isolates of the nuclear polyhedrosis virus from *S. litura*. J Gen Virol 71:2631–2639

Maskos CB, Miltenburger HG (1981) SDS-PAGE comparative studies on the polyhedral and viral polypeptides of the nuclear polyhedrosis viruses of *Mamestra brassicae, Autographa californica*, and *Lymantria-dispar*. J Invertebr Pathol 37:174–180

Matthiessen JN, Christian RL, Grace TDC et al (1978) Large scale field propagation and purification of the granulosis virus of potato moth, *Phthorimaea operculella* (Lep: Gel). Bull Entomol Res 68:385–391

Miller LK, Dawes KP (1978) Restriction endonuclease analysis to distinguish two closely related nuclear polyhedrosis viruses: *Autographa californica* MNPV and *Trichoplusia ni* MNPV. Appl Environ Microbiol 35:1206–1210

Mitsuhashi J (1989) Nutritional requirements of insect cells in vitro. In: Mitsuhashi J (ed) Invertebrate cell system applications. CRC, Boca Raton, pp 3–21, ISBN 0849343739

Monobrullah M (2003) Optical brighteners – pathogenicity enhancers of entomopathogenic viruses. Curr Sci 84:640–645

Monobrullah M, Nagata M (1999) Immunity of lepidopteran insects against baculoviruses. J Entomol Res 23:185–194

Monroe JE, McCarthy WJ (1984) Polypeptide analysis of genotypic variants of occluded *Heliothis* spp. Baculoviruses. J Invertebr Pathol 43:32–40

Moscardi F (1999) Assessment of the application of baculoviruses for control of Lepidoptera. Annu Rev Entomol 44:257–290

Moscardi F, Allen GE, Greene GL (1981) Control of the velvetbean caterpillar by nuclear polyhedrosis virus and insecticides and impact of treatments on the natural incidence of the entomopathogenic fungus *Nomuraea rileyi*. J Econ Entomol 74:48–485

Munoz D, Castillejo JI, Caballero P (1998) Two naturally occurring deletion mutants are parasitic genotypes in a *Spodoptera exigua* nucleopolyhedrovirus strain. Appl Environ Microbiol 64:4372–4377

Murillo R, Munoz D, Lipa JJ, Caballero P (2001) Biochemical characterization of three nucleopolyhedrovirus isolates of *Spodoptera exigua* and *Mamestra brassicae*. J Appl Entomol 125:267–270

Nakai M, Goto C, Kang W, Shikata M, Luque T, Kunimi Y (2003) Genome sequence and organization of a nucleopolyhedrovirus isolated from the smaller tea tortrix, *Adoxophyes honmai*. Virology 316:171–183

Narayanan K (1979) Studies on the nuclear polyhedrosis virus of gram pod borer, *Heliothis armigera* (Hubner) (Noctuidae: Lepidoptera). Ph.D. Thesis, Tamil Nadu Agriculture University, Coimbatore, p 204

Narayanan K (1985) Isolation, purification and inoculation of viral pathogens. In: Jayaraj S (ed) Microbial control and pest management. Tamil Nadu Agriculture University, Arul Jothi Printers, Coimbatore, pp 55–59

Nguyen Q, Qi YM, Wu YC et al (2011) In vitro production of *Helicoverpa* baculovirus biopesticides—automated selection of insect cell clones for manufacturing and systems biology studies. J Virol Methods 175:197–205

Odak SK, Srivastava DK, Misra VK et al (1984) Preliminary studies on the pathogenicity of *Bacillus thuringiensis* and nuclear polyhedrosis virus on *Heliothis armigera* host in the laboratory and in pot experiments. Legume Res 5:13–17

Ogemo JG, Chaeychomsri S, Kamiya K et al (2007) Cloning and comparative characterization of nucleopolyhedrosis viruses isolated from African bollworm, *Helicoverpa armigera* (Lepidoptera: Noctuidae) in different geographic regions. J Insect Biotechnol Sericol 76:39–49

Pant U, Sudeep AB, Athawale SS, Vipat VC (2002) Baculovirus studies in new, indigenous lepidopteran cell lines. Indian J Exp Biol 40:63–68

Parola AD, Cap AS, Glikmann G, Romanowski V (2003) An immunochemical method for quantification of *Epinotia aporema* granulovirus (EpapGV). J Virol Methods 112:13–21

Pawar VM, Thombre UT (1992) Prospects of baculovirus in integrated pest management of pulses. In: Ananthkrishnan TN (ed) Emerging trends in biocontrol of phytophagous insects. Oxford and IBH Publishing Company Private Limited, New Delhi, pp 253–258

Podgwaite JD, Bruen RB, Shapiro M (1983) Microorganisms associated with production lots of the nuclear polyhedrosis virus of the gypsy moth, *Lymantria dispar*. Entomophaga 28:9–16

Possee RD, Rohrmann GF (1997) Baculovirus genome organization and evolution. In: Miller LK (ed) The baculoviruses. Plenum, New York, pp 109–133

Powell JE, Robertson JL (1993) Status of rearing technology for cotton insects. In: King EG, Brown JM (eds) cotton insects and mites: characterization and management. Cotton Foundation, Memphis

Quant RL, Pearson MN, Rohrman GF et al (1984) Production of polyhedrin monoclonal antibodies for distinguishing two *Orgyia pseudotsugata* baculoviruses. Appl Environ Microbiol 48:732–736

Rabindra RJ (1992) Genetic improvement of baculoviruses. In: Ananthkrishnan TN (ed) Emerging Trends in Biological control of phytophagous insects. Oxford and F B H Publishing Cooperative Private Ltd., New Delhi, pp 183–186

Rabindra RJ, Subramanian TR (1974) Studies on the nuclear polyhedrosis of *Heliothis armigera* (Hubner). Madras Agric J 60:217–220

Rao GVR (2006) SATrends 64 ICRISAT Publication

Rao GVR, Meher KS (2004) Optimization of In vivo production of *Helicoverpa armigera* NPV and regulation of malodor associated with the process. Indian J Plant Protect 32:15–18

Rao GVR, Rupela OP, Rameshwar Rao V, Reddy YVR (2007) Role of biopesticides in crop protection: present status and future prospects. Indian J Plant Protect 35:1–9

Reardon R, Podgwaite JP, Zerillo RT (1996) GYPCHECK – the gypsy moth nucleopolyhedrosis virus product. USDA Forest Service Publication FHTET-96-16

Rivkin H, Mor M, Chejanovsky N (1998) Isolation, replication and polyhedrin gene sequence of an Israeli *Helicoverpa armigera* single nucleopolyhedrovirus. Virus Genes 17:11–19

Roder A, Punter J (1977) Interactions between nuclear polyhedrosis viruses and vertebrate cells. Zentralbl Bakteriol Parasitenkd Infektionskr Hyg Abt 1 Orig Reihe A 239:459 461

Rohrmann GF (1986) Evolution of occluded baculoviruses. In: Granados R, Federici B (eds) The biology of baculoviruses, vol 1. CRC, Boca Raton, pp 203–215

Rohrmann GF (1992) Baculovirus structural proteins. J Gen Virol 73:749–761

Rohrmann GF (1999) Nuclear polyhedrosis viruses. In: Webster RG, Granoff A (eds) Encyclopedia of virology, 2nd edn. Academic, London, pp 146–152

Rollinson WD, Hubbard HB, Lewis FB (1970) Mass rearing of the European pine sawfly for production of the nuclear polyhedrosis virus. J Econ Entomol 63:343–344

Rovesti L, Crook NE, Winstanely D (2000) Biological and biochemical relationships between the nucleopolyhedroviruses of *Mamestra brassicae* and *Heliothis armigera*. J Invertebr Pathol 75:2–8

Russell RLQ, Rohrmann GF (1990) A baculovirus polyhedron envelope protein: immuno gold localization in infected cells and mature polyhedra. Virology 174:177–184

Saxene H, Ahmad R (2005) NPV production, formulation and quality control. In: Saxena H, Rai AB, Ahmed R et al (eds) Recent advances in *Helicoverpa* management. Proceedings of the national symposium on *Helicoverpa* management: a national challenge, IIPR, pp 1–10

Senthil Kumar CM, Sathiah N, Rabindra RJ (2005) Optimizing the time of harvest of nucleopolyhedrovirus infected *Spodoptera litura* (Fabricius) larvae under In vivo production systems. Curr Sci 88:1682–1684

Shamim M, Baig M, Datta RK, Gupta SK (1994) Development of monoclonal antibody-based sandwich ELISA for the detection of nuclear polyhedra of nuclear polyhedrosis virus infection in *Bombyx mori L*. J Invertebr Pathol 63:151–156

Shapiro M, Bell RA (1981) Biological activity of *Lymantria dispar* nucleopolyhedrosis virus from living and virus killed larvae. Ann Entomol Soc Am 74:27–28

Shapiro M, Bell RA (1982) Production of gypsy moth, *Lymantria dispar* (L.) nucleopolyhedrosis virus, using carrageenans as dietary gelling agents. Ann Entomol Soc Am 75:43–45

Shapiro M, Ignoffo CM (1969) Nuclear polyhedrosis of *Heliothis*: stability and relative infectivity of virions. J Invertebr Pathol 14:130–134

Shapiro M, Ignoffo CM (1970) Nucleopolyhedrosis of *Heliothis*: activity of isolates from *Heliothis Zea*. J Invertebr Pathol 16:107–111

Shapiro M, Robertson JL, Injac MG, Katagiri K, Bell RA (1984) Comparative infectivities of gypsy moth (Lepidoptera: Lymantriidae) nucleopolyhedrosis virus isolates from North America, Europe and Asia. J Econ Entomol 77:153–156

Shapiro M, Robertson JL, Bell RA (1986) Quantitative and qualitative differences in gypsy moth (Lep: Lyman) NPV produced in different – aged larvae. J Econ Entomol 79:1174–1177

Sherman KE (1985) Considerations in the large scale and commercial production of viral insecticides. In: Karl M, Sherman KE (eds) Viral insecticides for biological control. Academic, London, pp 757–797

Shieh TR (1978) Characteristics of a viral pesticide, Elcar. In: Proceedings of international colloquium and invertebrate pathology, pp 91–194

Shieh TR (1989) Industrial production of viral pesticides. Adv Virus Res 36:315–343

Shieh TR, Bohmfalk GT (1980) Production and efficiency of baculoviruses. Biotechnol Bioeng 22:1357–1375

Shim HJ, Roh JY, Choi JY, Li MS, Woo SD, Oh HW, Boo KS, Je Y (2003) Isolation and characterization of a *Lymantria dispar* multinucleocapsid nucleopolyhedrovirus isolate in Korea. J Microbiol 41:306–311

Sireesha K (2006) Determination of efficacy of different *Helicoverpa armigera nucleopolyhedrovirus* (*Ha*NPV) strains and standardization of production procedures. Ph.D. Thesis, Department of Entomology, Agricultural College, Bapatla, Acharya N.G. Ranga Agricultural University, Rajendranagar, Hyderabad

Smith IRL, Crook NE (1988) In vivo isolation of baculovirus genotypes. Virology 166:240–244

Smith GE, Summers MD (1978) Analysis of baculovirus genomes with restriction endonucleases. Virology 89:517–527

Smith GE, Summers MD (1979) Restriction maps of five *Autographa californica* MNPV variants, *Trichoplusia ni* MNPV, and *Galleria mellonella* MNPV DNAs with endonucleases SmaI, KpnI, BamHI, SacI and EcoRI. J Virol 30:393–406

Smith GE, Summers MD (1981) Application of a novel radioimmunoassay to identify baculovirus structural proteins that share interspecies antigenic determinants. J Virol 39:125–137

Smith GE, Summers MD (1982) DNA homology among subgroup A, B and C baculoviruses. Virology 123:393–406

Smits PH, Vlak JM (1987) Biological activity of *Spodoptera exigua* nuclear polyhedrosis virus against *S. Exigua* larvae. J Invertebr Pathol 51:107–114

Somasekhar S, Jayapragasam M, Rabindra RJ (1993) Characterization of five Indian isolates of the nuclear polyhedrosis virus of *H. armigera*. Phytoparasitica 21:333–337

Sridhar Kumar Ch (2008) Development and evaluation of diagnostic tools for nucleopolyhe-droviruses infecting major lepidopteran pests of legume crops in the semi-arid tropics. Ph.D. Thesis, Department of Biochemistry, Acharya Nagarjuna University, Andhra Pradesh

Sridhar Kumar C, Sireesha K, Ranga Rao GV, Reddy AS, Waliyar F, Rambabu C, Kumar L (2007) Production of polyclonal antibodies for detection of *Nucleopolyhedrovirus* infecting *Helicoverpa armigera*. Indian J Plant Protect 35:210–212

Sridhar Kumar C, Ranga Rao GV, Sireesha K, Kumar PL (2011) Isolation and characterization of baculoviruses from three major lepidopteran pests in the semi-arid tropics of India. Indian J Virol 22:29–36

Stairs GR, Frasor T, Frasor M (1981) Changes in growth and virulence of a nuclear polyhedrosis virus from *Choristoneura fumiferana* after passage in *Trichoplusia ni* and *Galleria mellonella*. J Invertebr Pathol 38:230–235

Sudhakar S, Mathavan S (1999) Viral pesticides for environmental safety (*Helicoverpa armigera*) Baculovirus Genome analysis) In: Ignacimuthu A (eds) Biopesticides in insect pest management, pp 199–207

Summers MD, Egawa K (1973) Physical and chemical properties of *Trichoplusia ni* granulosis virus granulin. J Virol 12:1092–1103

Summers MD, Smith GE (1978) Baculovirus structural polypeptides. Virology 84:90–402

Summers MD, Smith GE, Knell JD, Burand JP (1980) Physical maps of *Autographa californica* and *Rachiplusia ou* nuclear polyhedrosis virus recombinants. J Virol 34:693–703

Sun XL, Peng H (2007) Recent advances in biological pest insects by using viruses in China. Virol Sin 22:158–162

Sundeep AB, Shouche YS, Mourya DT et al (2002) New *Helicoverpa armigera* Hbn cell line from larval hemocyte for baculovirus studies. Indian J Exp Biol 40:69–73

Sundeep AB, Mourya DT, Mishra AC (2005) Insect cell culture in research: Indian scenario. Indian J Med Res 121:725–738

Suzanne MT (2009) Baculovirus genes affecting host function. In Vitro Cell Dev Biol Anim 45:111–126

Szewczyk B, Hoyos-Carvajal L, Paluszek M et al (2006) Baculoviruses: re-emerging biopesticides. Biotechnol Adv 24:143–160

Tanada Y, Kaya H (eds) (1993) Insect pathology. Academic, San Diego

Taun SJ, Tang LC, Hou RF (1989) Factors effecting Pathogenicity of NPV preparations to the corn earworm, *Heliothis armigera* (Hubner). Entamophaga 34:541–549

Teakle RE, Byrne VS (1989) Nuclear polyhedrosis virus production in *H. armigera* infected at different larval ages. J Invertebr Pathol 53:21–24

Teakle RE, Jensen JM, Giles JE (1985) Susceptibility of *H. armigera* to a commercial nuclear polyhedrosis virus. J Invertebr Pathol 46:166–173

Thomas ED (1975) Normal virus levels and virus levels added for control. In: Summers M, Engler R, Falcon LA, Vail P (eds) Baculoviruses for insect pest control: safety considerations. Am Soc Microbiol, Washington, pp 87–89

Thorne CM, Otvos IS, Conder N, Levin DB (2007) Development and evaluation of methods to detect the *Choristoneura fumiferana* nuclear polyhedrosis virus. Chinese J Virol 4:61–64

Tompkins GJ, Daugherty EM, Adams JR, Diggs D (1988) Changes in the virulence of nuclear polyhedrosis virus when propagated in alternate Noctuid (Lepidoptera: Noctuidae) cell lines and hosts. J Econ Entomol 81:1027–1032

Towbin H, Staehlin T, Gordon J (1979) Electrophoretic transfer of proteins from polyacrylamide gels to nitrocellulose sheets: procedure and some applications. Proc Natl Acad Sci U S A 76:4350–4354

Tuan SJ, Chen WL, Kao SS (1998) In vivo mass production and control efficacy of S*podoptera litura* (Lepidoptera: Noctuidae) nucleopolyhedrovirus. Chinese J Entomol 18:101–116

Tuan SJ, Kao SS, Cheng DJ, Hou RF, Chao YC (1999) Comparison of the characterization and pathogenesis of three lepidopteran nucleopolyhedroviruses (HearNPV, SpeiNPV and SpltNPV) isolated from Taiwan. Chinese J Entmol 19:167–186

Vail PV, Jay DL (1973) Replication and infectivity of the nuclear polyhedrosis virus of the alfalfa looper, *Autographa californica*, produced in cells grown in vitro. J Invertebr Pathol 22:231–237

Van Lent JWM, Groenen JTM, Klinge-Roode EC et al (1990) Localization of the 34 kDa polyhedron envelope protein in *Spodoptera frugiperda* cells infected with *Autographa californica* nuclear polyhedrosis virus. Arch Virol 111:103–114

Vanderzant ES, Richardson CD, Fort SWJ (1962) Rearing of the bollworm on artificial diet. J Econ Entomol 55:140

Veenakumari K, Rabindra RJ, Srinivasa Naik CD et al (2006) Standardization of laboratory mass production of *Amsacta albistriga* nucleopolyhedrovirus. J Biol Control 20:183–190

Veenakumari K, Rabindra RJ, Srinivasa Naik CD, Shubha MR (2007) *In situ* field level mass production of *Amsacta albistriga* (Lepidoptera: Arctiidae) nucleopolyhedrovirus in a groundnut ecosystem in South India. Int J Trop Insect Sci 27:48–52

Vincent C, Andermatt M, Valéro J (2007) Madex® and VirosoftCP4®, viral biopesticides for coddling moth control. In: Vincent C, Goethel MS, Lazarovits G (eds) Biological control: a global perspective, Oxfordshire, UK, and Cambridge, USA: CAB International Virus taxonomy: seventh report of the international committee on taxonomy of viruses, pp 336–343

Vlak JM, Smith GE (1982) Orientation of the genome of *Autographa californica* nuclear polyhedrosis virus: a proposal. J Virol 41:1118–1121

Volkman LE, Falcon LA (1982) Use of monoclonal antibody in an enzyme-linked immunosorbent assay to detect the presence of *Trichoplusia ni* (Lepidoptera: Noctuidae) single nuclear polyhedrosis virus polyhedrin in *T. ni* larvae. J Econ Entomol 75:868–871

Volkman LE, Blissard GW, Friensen PD, Posse RD, Theilmann DA (1995) Baculovirudae. In: Murphy F, Fauquet CM, Bishop DHL, Jarvis AW, Martelly G, Mayo E, Summers M D (eds) Virus taxonomy: classification and nomenclature of viruses. Sixth report of the international committee on taxonomy of viruses. Arch Virol 10:104–113

Weiser J (1987) Patterns over place and time. In: Fuxa JR, Tanada Y (eds) Epizootiology of insect disease. Wiley, New York, pp 215–242

Wiegers F, Vlak JM (1984) Physical map of the DNA of a *Mamestra brassicae* nucleopolypoly-hedrovirus variant isolates from *Spodoptera exigua*. J Gen Virol 65:2011–2019

Wigley PJ (1976) The epizootiology of a nuclear polyhedrosis virus disease of the winter moth, *Operophtera burmata* L. at Wistman's Wood, Dartmoor. D.Phil. Thesis, University of Oxford

William CF, Payne CC (1984) The susceptibility of *Heliothis armigera* (Hubner) larvae to three nuclear polyhedrosis viruses. Ann Appl Biol 104:405–412

Williams T (1993) Covert iridovirus infection in blackfly larvae. Proc Roy Soc Lond B 251:225–230

Winstanley D, O'Reilly D (1999) Granuloviruses. In: Webster RG, Granoff A (eds) Encyclopedia of virology, 2nd edn. Academic, London, pp 69–84

Wolff JLC, Moracs RHP, Kitajima E, Leal EDS, Zanotto PMA (2002) Identification and characterization of a baculovirus from *Lonomia obliqua* (Lepidoptera: Saturniidae). J Invertebr Pathol 79:137–145

Wu C-Y, Wang C-H (2006) New cell lines from *Lymantria xylina* (Lepidoptera: Lymantriidae): characterization and susceptibility to baculoviruses. J Invertebr Pathol 93:186–191

Yendol WG, Hamlen RA (1973) Ecology of entomogenous viruses and fungi. Ann N Y Acad Sci 217:18–30

Zanotto PMD, Kessing BD, Maruniak JE (1993) Phylogenetic interrelationships among baculoviruses: evolutionary rates and host associations. J Invertebr Pathol 62:147–164

Zhang G, Kaupp WJ (1988) Use of monoclonal antibodies in an enzyme-linked immunosorbent assay to detect the *Choristoneura fumiferana* nuclear polyhedrosis virus. Chinese J Virol 4:61–64

Chapter 3
Bt Insecticidal Crystal Proteins: Role in Insect Management and Crop Improvement

K.Y. Srinivasa Rao, Debasis Pattanayak, and Rohini Sreevathsa

3.1 Introduction

Bacillus thuringiensis (*Bt*) is a gram-positive spore-forming bacterium naturally occurring in all types of soils and all types of terrain including beaches, desert and tundra habitats. Till date, 82 different serovars have been reported. *Bt* is a member of the Bacillaceae family and belongs to the *Bacillus cereus* group.

B. *thuringiensis* is the most commercially successful biological control agent of insect pests. According to Rowe and Margaritis (1987), toxin proteins from *Bt* strains are classified into nine different types based on their site of toxicity. They are α-exotoxin (phospholipase C), β-exotoxin (thermostable exotoxin), γ-exotoxin (toxic to sawflies), δ-endotoxin (protein parasporal crystal), louse factor exotoxin (active only against lice), mouse factor exotoxin (toxic to mice and Lepidoptera), water-soluble toxin, *Vip3A* (*Bt* vegetative insecticidal protein) and enterotoxin (produced by vegetative cells). Amongst several toxins produced by *Bt* strains, δ-endotoxins have been more efficiently utilised for protection of a variety of crops from various insect pests. In comparison to other commonly used insecticides, these biological agents are safe for both the pesticide users and consumers of treated crops. Among many bio-control agents, B. *thuringiensis*-based products have a major share, i.e. up to 97 % of the bio-insecticides used worldwide.

B. *thuringiensis* produces vegetative insecticidal proteins (VIP) during vegetative phase and produces cry and cyt toxins as parasporal crystals during sporulation phase; these crystals are commonly called as δ-endotoxins or pore-forming toxins (PFT) or inclusion lesions (ILs). These protein families show toxic activity against insects of different orders, viz., Lepidoptera, Coleoptera, Diptera and against nematodes. The vip, cry and cyt toxins are activated by host proteases.

K.Y. Srinivasa Rao • D. Pattanayak • R. Sreevathsa (✉)

National Research Centre on Plant Biotechnology, Indian Agricultural Research Institute, New Delhi 110012, India

e-mail: rohinisreevathsa@rediffmail.com

© Springer International Publishing Switzerland 2015   53

K.S. Sree, A. Varma (eds.), *Biocontrol of Lepidopteran Pests*, Soil Biology 43,
DOI 10.1007/978-3-319-14499-3_3

These activated proteins lyse midgut epithelial cells by inserting into the target membrane and form pores so that larvae die due to leakage of cell fluids. Among this group of proteins, members of the three-domain cry family are used worldwide for insect control, and their mode of action has been characterised. The other PFT, i.e. colicins, exotoxin A, diphtheria toxins can affect mammals, but Cry and Vip toxins have no effect on mammals due to lack of specific receptors. Cry toxins interact with specific receptors, viz., aminopeptidase N (APN) receptors, cadherin-like receptors and alkaline phosphatases (ALPs), located on the host cell surface and are activated by host proteases following receptor binding resulting in the formation of a pre-pore oligomeric structure that is insertion competent. In contrast, Cyt toxins directly interact with membrane lipids and insert into the membrane. Recent evidence suggests that Cyt synergise or overcome resistance to mosquitocidal Cry proteins by functioning as a Cry membrane-bound receptor. Several strains of B. thuringiensis can infect and kill insects due to the production of insecticidal toxins during vegetative and sporulation phases. In the present scenario, Bt is the only microbial insecticide in widespread use. The present review explains the relevance and utility of Bt cry genes in plant biotechnology.

3.2 Historical Perspective

In 1901, a Japanese biologist Shigetane Ishiwatari isolated a bacterium while working on the cause of sotto disease (killing large populations of silkworms), and he named it as B. sotto. Later in 1911, Ernst Berliner isolated a bacterium that killed a Mediterranean flour moth and named this bacterium as B. thuringiensis (Siegel 2000) because the moth was first identified in the German state Thuringia. He also reported that the existence of crystal within B. thuringiensis that spores and crystals were coated onto leaves, the caterpillars stopped feeding and died. This led to the discovery of insecticidal properties of the crystals and demonstrated that the dead flour moth caterpillars were found to be loaded with spores and crystals.

In 1927, Mattes isolated Bt strain discovered by Ernst Berliner and conducted subsequent field trials against the European corn borer (Ostrinia nubilalis). This work eventually led to the development of Sporeine in 1938 (Luthy et al. 1982) which was used as an insecticidal spray in France. Sporeine was the first commercial Bt insecticide derived from kurstaki strain HD1 which was active against lepidopteran pests (Baum et al. 1999). It was used primarily to kill flour moths and was registered as an insecticide in the United States in 1961. During this period, other Bt products such as Biobit, Dipel and Thuricide were derived from kurstaki HD1, although other strains were also used to tackle lepidopteran (kurstaki SA-11, kurstaki SA-12), dipteran (israelensis) and coleopteran (tenebrionis) pests (Kaur 2000). Until 1977, only 13 Bt strains had been described. All 13 subspecies were toxic only to certain species of lepidopteran larvae. In 1977, the first subspecies toxic to dipteran (flies) species was found, and in 1983, strains toxic to species of coleopterans (beetles) were first discovered. In the 1980s, the use of Bt increased

Table 3.1 Different *Bt* strains and its target insects

S. no.	Insecticide obtained from	Insecticide trade name	Target insect
1	*Bt kurstaki HD1*	Thuricide	Lepidoptera
2	*Bt kurstaki SA-12*	Costar	Lepidoptera
3	*Bt aizawai*	Florbac	Lepidoptera
4	*Bt israelensis*	Tekar	Diptera
5	*Bacillus sphaericus*	Vectolex	Diptera
6	*Bt tenebrionis*	Novodor	Coleoptera
7	*Bt tenebrionis*	Trident	Coleoptera

due to harmful effects on the environment by chemical pesticides. Now there are thousands of strains of *Bt*, and many of them have genes that encode unique toxic crystals in their DNA (Table 3.1). With the advancement in molecular biology, it soon became feasible to move the gene that encodes the toxic crystals into a plant. The first genetically engineered plants: potato, corn and cotton were registered in the Environmental Protection Agency (EPA), USA in 1995. Today, GM (genetically modified) crops including potato and cotton are planted throughout the world.

3.3 *Bt* Toxin Genes

The genes that produce toxic crystal proteins (Cry and Vip) during vegetative and sporulation phases of *B. thuringiensis* are called as *Bt* genes. There are mainly two different types of *Bt* toxins based on the phase of life cycle. The toxins produced during sporulation are Cry toxin proteins (Table 3.2), and the ones produced during vegetative phase are Vip toxin proteins (Fig. 3.1). Over time, many *Bt* toxins are identified and are being used in many industries and research organisations around the world (Schnepf et al. 1995; Federici 1999). In 1962, De Barjac first attempted to classify *Bt* isolates based on the flagellar (H) agglutination (De Barjac and Bonnefoi 1962). More than 67 H-serotypes and 8 nonflagellated biotypes are now available, and in many of these, the array of *Bt* toxin genes present is the same (Zeigler 1999). In 1989, Hofte and Whitely classified *Bt* toxins based on homology of toxin gene sequences and on the insecticidal activity into 42 *Bt* genes and 14 distinct types and grouped them into four major classes. These classes are cry I (Lepidoptera specific), cry II (Lepidoptera and Diptera specific), cry III (Coleoptera specific) and cry IV (Diptera specific).

In 1998, Crickmore et al. (Table 3.2) introduced a systematic nomenclature for classifying the *cry* and *vip* genes based on amino acid sequence of full-length gene products rather than their biological properties. The *cry* genes designated by Hofte and Whiteley (1989) have been retained, although the Roman numerals have been replaced by Arabic numbers, i.e. *cryII* is now *cry2*, followed by uppercase and lower case letters, i.e. *cry2Aa*, next followed by number, i.e. *cry2Aa1*. The classification is based on amino acid sequence homology and four hierarchical

Table 3.2 Recent classification of cry genes identified so far from *B. thuringiensis* (http://www.lifesci.sussex.ac.uk/home/neil_crickmore/Bt/toxins2.html)

Class	Subclass	Class	Subclass	Class	Subclass	Class	Subclass	Cry	Subclass
cry1	241	cry16	1	cry31	10	cry46	8	cry61	3
cry2	68	cry17	1	cry32	7	cry47	1	cry62	1
cry3	19	cry18	3	cry33	1	cry48	5	cry63	1
cry4	14	cry19	2	cry34	11	cry49	5	cry64	1
cry5	12	cry20	3	cry35	11	cry50	8	cry65	2
cry6	4	cry21	3	cry36	1	cry51	2	cry66	2
cry7	21	cry22	6	cry37	1	cry52	2	cry67	2
cry8	38	cry23	1	cry38	1	cry53	2	cry68	1
cry9	30	cry24	3	cry39	1	cry54	3	cry69	3
cry10	4	cry25	1	cry40	4	cry55	2	cyt1	12
cry11	7	cry26	1	cry41	4	cry56	2	cyt2	24
cry12	1	cry27	1	cry42	1	cry57	1		
cry13	1	cry28	2	cry43	4	cry58	1		
cry14	1	cry29	1	cry44	1	cry59	1		
cry15	1	cry30	11	cry45	1	cry60	6		

Fig. 3.1 Classification of *VIP* genes formed during vegetative phase of *B. thuringiensis* (http://www.lifesci.sussex.ac.uk/home/Neil_Crickmore/Bt/vip.html/)

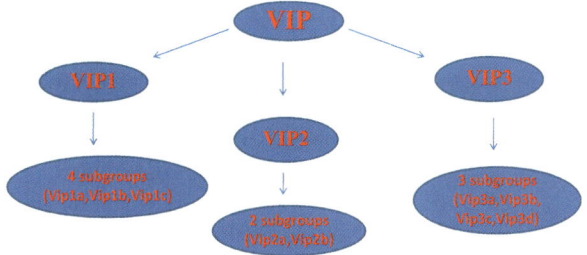

ranks consisting of numbers, capital letters, lower case letters and numbers (e.g. *cry1Aa1/vip1Aa1*), depending on its position in a phylogenetic tree.

Currently, there are many collections composed of thousands of *B. thuringiensis* isolates. Many of them may harbour novel *cry* genes, which may code for more potent toxins or may show new biological activities (Porcar and Caballero 2000). To date, more than 350 *cry* genes have been sequenced, which include major groups, *cry*1 to *cry*51, and the list keeps growing (Crickmore et al. 1998, http://www.lifesci.sussex.ac.uk/Home/Neil_Crickmore/*Bt*/ access). The search for new *cry* genes is of great importance, as this allows to find new toxins that could be more potent or specific to the target larvae (Fernández-Larrea 2002).

Vegetative insecticide proteins are second generation of insecticides. Their mode of action is similar to that of the other endotoxins but has no sequence homology with any of the other known endotoxins. Till now, three different types of *vip* genes are found, i.e. *vip1*, *vip2*, *vip3*. Among these genes, *vip3* type of genes are most abundant (67.4 %), followed by *vip2* (14.6 %) and *vip1* (8.1 %). *vip3* is specifically toxic to Lepidopteran pests, and *vip1* and *vip2* are specific to coleopteran pests.

vip genes can be classified based on amino acid sequence similarity into 3 groups, 9 subgroups, 25 classes and 82 subclasses (http://www.lifesci.sussex.ac.uk/home/Neil_Crickmore/*Bt*/vip.html/).

3.4 Structure of *Bt* δ-Endotoxin

Bt δ-endotoxins are globular protein molecules, which accumulate as protoxins in crystalline form during late stage of sporulation. Protoxins are liberated in the midgut after solubilisation and are cleaved off at C-terminal part to release 66 kDa (approx.) active N-terminal toxic molecules. The protoxin contains well-conserved cysteine residues (as many as 16 in *cry1Ac*), which helps in bridging the protoxin molecules through intermolecular disulphide bonds and thereby crystal formation.

Currently, three-dimensional protein structures have been determined through X-ray crystallography. The tertiary structure of δ-endotoxins is comprised of three distinct functional domains connected by a short conserved sequence. Each domain of δ-endotoxin has independent and inter-related functions in the larval midgut, which brings out colloid osmotic lysis (Knowles 1994). Phylogenetic analysis on the domains of δ-endotoxins revealed that domain I is the most conserved, and domain II is hyper variable among all δ-endotoxins (Bravo 1997). The analysis of domain III sequences revealed a different topology due to the domain III swapping among different types of toxins (Bravo 1997). The independent evolution of the three structural domains and domain III swapping among different toxins generated proteins with similar mode of action but with very different specificities (Bravo 1997).

The N-terminal domain (domain I) is a bundle of seven α-helices in which the central helix-α5 is hydrophobic and is encircled by six other amphipathic-α helices, and this helical domain is responsible for membrane insertion and pore formation. Hence, domain I is called as pore-forming toxin (PFT). Domain II consists of three antiparallel β-sheets with exposed loop regions, and domain III is a β-sandwich (Li et al. 1991; Grochulski et al. 1995; Morse et al. 2001; Galitsky et al. 2001; Boonserm et al. 2005, 2006). Exposed regions in domain II are involved in receptor binding (Bravo et al. 2007), and domain III are involved in maintaining stability of the total protein.

3.5   Mode of Action

Considerable amount of research on δ-endotoxin of *B. thuringiensis* has been devoted to understand the mode of action on susceptible insects. The main insecticidal activity of *Bt* is due to insecticidal crystalline inclusions formed during sporulation (Beard et al. 2008). In general, following ingestion, the crystalline inclusions are dissolved and then converted to active toxins by insect proteases. The active toxins bind to specific receptor sites and produce pores in the midgut epithelial membranes which results in loss of homeostasis and septicaemia, which are lethal to the insect (Broderick et al. 2006).

The crystalline proteins formed during sporulation phase of *B. thuringiensis* are protoxins, which do not have the toxic effect towards larvae. But when these nontoxic crystalline inclusions (protoxins) are ingested by larvae, they dissolve in high-pH (9.5) environment of larval midgut with the help of host proteases. However, many coleopterans have a neutral pH midgut, yet solubilisation of the coleopteran-specific toxins occurs (Koller et al. 1992). Host proteases enzymatically convert protoxins into smaller active toxins which are resistant to further protease digestion. These active toxins bind to unique receptors/binding sites on the epithelium cells in the midgut of susceptible insects. So far, four different receptor/ binding proteins have been described in different lepidopteran insects: a cadherin-like protein (CADR), a glycosylphosphatidylinositol (GPI)-anchored aminopeptidase N (APN), a GPI-anchored ALP and a 270 kDa glycoconjugate (Knight et al. 1994; Vadlamudi et al. 1995; Valaitis et al. 2001; Jurat-Fuentes and Adang 2004).

CADR is a transmembrane protein with cytoplasmic domain and extra cellular ectodomain. The ectodomain contains calcium-binding proteins, integrin interaction sequences and cadherin protein-binding sequences. Surface plasmon resonance experiments showed that the binding affinity of monomeric Cry1A toxins to the *Manduca sexta* CADR is in the range of 1 nM (Vadlamudi et al. 1995), while that of APN is in the range of 100 nM (Jenkins and Dean 2000).

Aminopeptidase N (APN) and ALP are external receptors of BBMV of larvae which were anchored by a GPI. Surface plasmon resonance binding studies of Cry1Ab mutants with pure *M. sexta* APN showed that domain II loop2 and loop3 are involved in APN recognition (Jenkins and Dean 2000). Cry1Ac domain III first interacts with APN GalNAc sugar moieties facilitating the subsequent interaction of domain II loop regions with another region in this receptor (Jenkins et al. 2000). In *M. sexta*, proteomic analysis of BBMV Cry1Ac-binding proteins revealed that ALP is a putative receptor molecule (McNall and Adang 2003).

After binding to the receptor, the toxin inserts irreversibly into the plasma membrane of the cell leading to lesion formation. There is a positive correlation between toxin activity and ability to bind brush border membrane vesicles epithelial cells lining the gut of larvae (Gill et al. 1992). The toxicity is correlated with receptor number rather than receptor affinity (Van Rie et al. 1989). The toxicity of *Bt* lies in the organisation of α-helices derived from domain I. After binding to the midgut epithelial cells, the α-helices can penetrate the apical membrane to form an ion channel (Knowles and Dow 1993). The formation of toxin-induced pores in the columnar cell apical membrane allows rapid fluxes of ions. The pores are K^+ selective (Sacchi et al. 1986), permeable to cations (Wolfersberger 1989), anions (Hendrickx and Estrada-Navarrete 1989) or to solutes such as sucrose irrespective of the charge. Carroll and Ellar (1993) observed that midgut permeability in the presence of Cry1Ac was altered for cations, anions, neutral solutes and water. Knowles and Dow (1993) suggested that *Bt* toxins lead to cessation of K^+ pump leading to swelling of columnar cells and osmotic lysis. The disruption of gut integrity leads to death of the insect through starvation or septicaemia. These pores possess both selective (only K^+ passes through) and nonselective (Na^+ and anions pass through) properties depending on the pH (Schwartz et al. 1993). The lepidopteran insect midgut is alkaline and the pores probably permit K^+ leakage. Formation of this cation selective channel destroys the membrane potentials (English and Slatin 1992) resulting in midgut necrosis, degeneration of peritrophic membrane and epithelium and ultimately bacterial septicaemia. These pores allow ions and water to pass freely into the cells, resulting in swelling, lysis and eventual death of the host.

3.6   Importance of *Bt* Toxins

Till now there are mainly three different applications that have been achieved by using *Bt* toxins. The first application is to control defoliator pests in forestry and crops; second application is to control mosquitoes because these are the vectors of many dangerous human diseases and third application is development of transgenic insect-resistant crops.

3.6.1   Bt *Toxin as Bioinsecticide Spray*

One of the most successful applications of *Bt* has been the control of lepidopteran defoliators by *Bt* sprays. The lepidopteran defoliators are pests of coniferous forests mainly in Canada and the USA. In both countries, the control of forests defoliators relies mostly on the use of *Bt* strain, *kurstaki HD1*, producing Cry1Aa, Cry1Ab, Cry1Ac and Cry2Aa toxins (van Frankenhuyzen 2000; Bauce et al. 2004). Successful application of *Bt* is highly dependent on proper timing, weather conditions and high dosage of spray applications. These factors combine to determine the probability of larvae ingesting lethal dose (van Frankenhuyzen 2000; Bauce et al. 2004). The use of *Bt* in the control of defoliators has resulted in a significant reduction in the use of chemical insecticides for pest control in the forests. Most *Bt* products which are used as insecticides are derived from *kurstaki strain HD1* (e.g. Biobit, Dipel and Thuricide), and other strains are used to tackle lepidopteran (*kurstaki SA-11*, *kurstaki SA-12*), dipteran (*israelensis*) and coleopteran (*tenebrionis*) pests (Kaur 2000) (Table 3.1).

In 1960, strain improvement led to the replacement of many of the early products with new *Bt* strains that were up to tenfold more potent than their predecessors. It can be achieved by creating new bacterial strains carrying unknown combinations of existing toxins by conjugation or direct transformation (Gonzalez et al. 1982; Kronstad et al. 1983; Carlton and Gonzalez 1985).

Below are two examples of developing combinational pesticides by conjugation process. First is Foil, a product which was achieved by conjugation process between strain *EG2424*, which produces Cry1Ac toxin from *Bt kurstaki*, which is active against the European corn borer, and Cry3A toxin from *Bt tenebrionis*, which is active against the Colorado potato beetle, *Leptinotarsa decemlineata* (Carlton and Gawron-Burke 1993). Second example is Tobaggi, a product obtained from *Bt* strain *NT0423* developed by Dongbu Hannong Chemicals, Korea. This strain had five known crystal protein genes *cry1Aa*, *cry1Ab*, *cry1C*, *cry1D* and *cry2A* and one new gene *cry 1Af1* (GenBank Accession No. U82003). It has a dual toxicity against lepidopteran larvae-like *Plutella xylostella*, *Spodoptera exigua* and *Hyphantria cunea* and dipteran larvae-like *Culex pipiens* and *Musca domestica*. Today, the *Bt* biopesticide market is dominated by Abbott Laboratories (Chicago, IL) (since the acquisition in 1995 of Novo Nordisk's biopesticide business) and Novartis (created through the merger in 1996 of Ciba and Sandoz), together accounting for >70 % of global production. The other 30 % is divided among approximately 30 companies with over 100 different *Bt* product formulations, most containing a single *Bt* toxin but some combining up to five.

There are some advantages and some disadvantages of using biopesticides as sprays. Advantages:

- The biopesticide has potent insecticidal activity.
- Has host specificity.
- Less costly and takes less time to develop bioinsecticide than chemical insecticide.
- Harmless to humans and other mammals.
- Biodegradable.

Disadvantages:

- Biopesticides are rapidly inactivated by UV, heat and extreme pH.
- It is easily removed from plant surface by wind and rain.
- These biopesticides are also susceptible to proteases in leaf exudates of plants.

3.6.2   Use of **Bt** *Toxin to Control Mosquitoes*

Many diseases of humans are caused through vectors such as mosquitoes. *Bt* toxins are highly active against mosquitoes like *A. aegypti*, which is a vector of dengue fever, *Simulium damnosum* is a vector of onchocerciasis and certain *Anopheles* species are vectors of malaria. *Bt* toxin is used as an alternative control method of mosquito and black fly populations (Becker 2000) due to its high insecticidal activity, lack of toxicity to nontarget organisms and lack of showing insect-resistant populations to chemical insecticides like DDT. In 1983, a control programme for the eradication of onchocerciasis was launched in 11 countries of Western Africa using *Bt* toxin. Presently, more than 80 % of this region is protected by *Bt* toxin applications and 20 % with the chemical larvicide, temephos. Furthermore, control of onchocerciasis has protected over 15 million children without the appearance of black fly resistance to *Bt* toxin (Guillet et al. 1990). This success of vector control using *Bt* toxin will certainly increase its use around the world.

3.7   **Bt** Transgenic Crops

The development of transgenic crops that produce *Bt* Cry toxins has been a major breakthrough in the substitution of chemical insecticides by environmental friendly alternatives. The main advantage of transgenic crops is that the *Bt* toxin is always produced in plant without any degradation. In 1987, first *Bt* transgenic plants were developed in tomato and tobacco (Barton et al. 1987; Fischhoff et al. 1987; Vaeck et al. 1987). The full-length or truncated *Bt* toxin genes (*cry1A*) were used in tomato and tobacco, and the gene was transformed from *Bt var. kurstaki HD1* which is toxic to lepidopteran pests. But expression of the toxin protein was very poor in the tobacco plants, and the mortality of *M. sexta* larvae was only 20 %.

In 1990, researchers at Monsanto made a significant advancement in the expression of *Bt* genes in plants (Perlak et al. 1990). They noticed that *Bt* genes were excessively AT rich in comparison with normal plant genes. This bias in nucleotide composition of the DNA could have a number of deleterious consequences to gene expression because AT-rich regions in plants are often found in introns or have a regulatory role in determining polyadenylation. In addition, plants have a tendency to use G or C in the third base of redundant codons, −A or T being rarer. By considering all these codon modifications, the low expression could be overcome in plants. The plants have been improved by engineering *cry* genes with a plant-biased codon usage by removal of putative splicing signal sequences and deletion of the carboxy-terminal region of the protoxin (Schuler et al. 1998).

Perlak et al. (1990) made a gene construct in which the first 1,359 nucleotides were derived from fully modified *cry1Ab* gene and the remaining sequence from partially modified −*cry1Ac* gene. The variant gene was placed under the control of *CaMV 35S* promoter containing a duplicated enhancer region. The toxin protein levels in transgenic tobacco and tomato harbouring these modified genes increased up to 100-fold over levels seen with the wild-type *Bt* gene in plants. Cotton variety Coker 312 was transformed, and the transgenic plants were shown to have total protection from *Trichoplusia ni* (cabbage looper), *S. exigua* and *Helicoverpa zea* (cotton bollworm).

In the past few years, more than 30 plant species have been transformed by using a range of modified *Bt* genes (Schuler et al. 1998; de Maagd et al. 1999). The use of insect-resistant crops has diminished considerably the use of chemical pesticides in areas where these transgenic crops are planted (Qaim and Zilberman 2003). The first commercial crop was potato; the transgenic potato harbouring *cry3A* gene from *Bt var. tenebrionis* was shown to protect the crop against Colorado potato beetle in the field much more efficiently than Cry3A topical sprays. Many commercially useful crops like rice, maize and cotton have since been developed (Perlak et al. 1993) (Table 3.3).

In 1995, the US EPA approved the first registration of *Bt* potato, corn and cotton crops. The first to reach the market was Monsanto's NewLeaf potato variety expressing *cry3A* gene, followed by two transgenic corn hybrids expressing *cry1Ab* gene to protect against the European corn borer, i.e. KnockOut by Syngenta (Basel, Switzerland) and NatureGard by Mycogen (both containing event 176). Monsanto released the cotton varieties Bollgard and Ingard (events 531, 757 and 1076) expressing a modified Cry1Ac toxin. NewLeaf potato and its successors (NewLeaf Y and NewLeaf Plus) were withdrawn from the market in 2002, and corn varieties containing event 176 were later withdrawn and replaced with more profitable products. Commercially important crops harbouring *Bt* genes and their resistance have been shown in Table 3.4.

Table 3.3 Some important Cry toxins used initially to generate GM crops

S. no.	*Bt* toxin	Crop	Target pest
1	Cry1A	Tomato	Pinworm
2	Cry1A	Tobacco	*Helicoverpa zea*
3	Cry1A	Potato	Tuber moth
4	Cry3A	Potato	Colorado potato beetle
5	Cry1A	Cotton	Pink bollworm
6	Cry1A	Maize	European corn borer
7	Cry1A	Rice	Stem borer

Table 3.4 Use of transgenic *Bt* gene in commercially important crops against insect pests

S. no.	*Bt* gene	Active against insect pest	Crop
1	*cry1Ab, cry1Ac*	Bollworms	Cotton
2	*cry1Ab, cry9C*	European corn borer	Maize
3	*cry3Aa*	Colorado potato beetle	Potato
4	*cry1Ab*	Tuber moth	Potato
5	*cry1Ac, cry2Aa*	Yellow stem borer	Rice
6	*cry1Ab*	Eight lepidopterans	Rice
7	*cry1Ac*	Yellow stem borer	Basmati rice
8	*cry1Ac*	Fruit borer	Tomato
9	*cry1Ab*	Fruit borer	Brinjal
10	*cry1Ac*	Diamondback moth	Cabbage
11	*cry1C*	Diamondback moth	Broccoli
12	*cry1Ac*	Diamondback moth	Canola
13	*cry1C*	Beet armyworm	Alfalfa
14	*cry1Ac*	Leaf miner	Coffee
15	*cry1Ac*	Cornstalk borer	Peanut

Since 1996, plants have been modified with short sequences of genes from *Bt* to express the crystal protein. With this method, plants themselves can produce the proteins and protect themselves from insects without any external *Bt* sprays or synthetic pesticide sprays. In 1999, 29 million acres of *Bt* corn, potato and cotton were grown globally. It has been estimated that by using *Bt* protected cotton, the United States was able to save approximately $92 million.

The genetic engineering companies with interests in *Bt* plants include the following:

Agracetus (USA), Agricultural Genetics Co Ltd (UK), Agrigenetics Advanced Sciences Co (USA subsidiary of Lubrizol), Ciba-Geigy (Switzerland, now Novartis), DeKalb (USA), Monsanto (USA), Plant Genetic Systems (Belgium) and Sandoz (Switzerland, now Novartis). By 1997, these companies were field-testing at least 18 different *Bt* crops.

From 1998 onwards, the *Bt* crop production has increased because the EPA had approved an insect-resistant tomato line expressing *cry1Ac*. In 2002, the Herculex corn variety was developed jointly by Pioneer Hi-Bred and Dow Agrosciences

Table 3.5 Events approval for field trials of GM crops happened in 2013

S. no.	Gene/event	Trait	Crop	Company name
1	cp4epsps/MON 88913	Herbicide tolerance	RRF Cotton	Maharashtra Hybrid Seeds Company Ltd.
2	Events Bt11, GA21 and stack of Bt11 × GA21	Insect resistance and herbicide tolerance	Corn	Syngenta Biosciences Pvt. Ltd.
	Bt11, GA21 and stack event of Bt11 × GA21	Insect resistance and herbicide tolerance		Syngenta Biosciences Pvt. Ltd.
	Bt11 and GA21	Insect resistance and herbicide tolerance		Syngenta Biosciences Pvt. Ltd.
	cry2Ab2 and cry1A.105 genes (Event MON 89034)	Insect resistance		Monsanto India Ltd.
3	cp4epsps (Event NK603)	Herbicide tolerance	Herbicide-tolerant maize	Monsanto India Ltd.
4	Stacked events, namely, GHB119 (cry2Ae/PAT) & T304-40 (cry1Ab/PAT) containing cry1Ab, cry2Ac and bar	Insect resistance	TwinLink® Cotton	Bayer Bioscience Pvt Ltd
5	2mepsps (Event GHB 614)	Herbicide tolerance	Herbicide-tolerant Glytol cotton	Bayer Bioscience Pvt Ltd

expressing *cry1F*, which protects plants from black cutworm (*Agrotis ipsilon*) and European corn borer. After that EPA approved Monsanto's Bollgard II cotton variety expressing two *Bt* toxins *cry1Ac* and *cry2Ab*. The first stacked variety developed by Monsanto YieldGard Plus expressing *cry1Ab1* and *cry3Bb1* was released in 2003. Some of events with *Bt* gene, which were given permission for field trials in India during 2013, are shown in Table 3.5. At present, the USA has the largest area for *Bt* crops followed by Argentina, Brazil, India, Canada, China and other countries (Table 3.6). Further, *Bt* crops are grown in 28 countries, and the total area globally under *Bt* crop cultivation in 2013 was 175 million hectares. Some examples of commercially important *Bt* crops are *Bt* corn, *Bt* cotton and *Bt* rice.

Bt cotton: The cotton crop in the USA is damaged by lepidopteran larvae, particularly by cotton bollworm (*H. zea*), pink bollworm (*Pectinophora gossypiella*) and tobacco budworm (*Heliothis virescens*). To reduce the damage by Helicoverpa larvae, the US government approved cultivation of *Bt* cotton. Firstly a single transformation line Monsanto 531 was developed by Monsanto

Table 3.6 Current scenario of GM crops in the world (total 28 countries)

S. no.	Country	Area grown by GM crops (in million ha) 2012	2013	Major crops
1	USA	69.5	70.1	Maize, soya bean, cotton, sugar beet, papaya, squash, canola
2	Brazil	36.6	40.3	Soya bean, maize, cotton
3	Argentina	23.9	24.4	Soya bean, maize, cotton
4	India	10.8	11.0	Cotton
5	Canada	11.6	10.8	Canola, maize, soya bean, sugar beet
6	China	4	4.2	Rice, cotton, papaya, tomato, sweet pepper
7	Other countries	13.9	14.4	
Total	28 countries	170.3	175.2	

Source: Clive James, February 17, 2014

and marketed as Bollgard cotton. These Bollgard cotton lines contain a synthetic *cry1Ac* gene derived by an enhanced *CaMV 35S* promoter. By using these lines in the USA, a remarkable difference in the production has been observed, because of this the acreage of *Bt* cotton has risen from 12 % in 1996 to 35 % in 2000. Since then, there has been a constant raise in the adoption of the technology in various countries worldwide. Major thrust is from developing countries with the resource-poor farmers being highly benefited from the technology in general and *Bt* cotton specifically (James 2012). The cultivation of *Bt* cotton has reduced the number of pesticide sprays on the cotton crop, thereby improving farmers' health. Benefits have also been visualised economically in various countries (Brookes and Barfoot 2013). The extent of adoption of the technology demonstrates that technology is here to stay.

Bt corn was also developed to control lepidopteran insects. Corn is mostly damaged by European corn borer complex (*Ostrinia nubilalis*), southwestern corn borer [*Diatraea grandiosella* (Dyar)] and sugar cane borer [*Diatraea saccharalis* (Fabricus)] (Shelton et al. 2002). As for the European corn borer (*O. nubilalis*), as the name suggests, the larvae damage maize crops by tunnelling into the central pith of stalks. The European corn borer-resistant maize has been developed to reduce the damage of corn by corn borer larvae. Two different companies, i.e. Novartis and Monsanto, developed three different lines showing resistance against corn borer larvae. The transgene *cry1Ab* was used in all three different lines with different promoters. *Bt*11 line commercially called as YieldGard was developed by Novartis, Mon810 event was developed by Monsanto and another line *Bt*176 event, commercially called as KnockOut, was developed by Novartis. Among these three lines, the first two had *cry1Ab* construct controlled by constitutive promoter, and the knock-out line was developed by bombarding with two separate *cry1Ab* constructs controlled by maize PEP carboxylase promoter. By using these lines, the production of corn increased drastically and the growing area of corn also increased from 5 % in

1996 to 25 % in 2000. By 2012, *Bt* maize occupied 7.5 million hectares, up by 1.5 million hectares from 2011 with a growth of 25 % and equivalent to 4 % of global biotech area (James 2012). Of late, there have been a large number of single and stacked events of corn: *cry1A.105* (MON89034), *cry34Ab1* (59122), *cry35Ab1* (59122), *cry1F* (1507), *cry2Ab* (MON89034), *cry3Bb1* (MON863 and MON88017), *mcry3A* (MIR604) and *vip3A* (MIN162). The corn which was genetically modified to produce VIP was first approved in the USA in 2010.

Bt rice was first developed by China and field trials of *Bt* rice were conducted in China in 1998. A number of *Bt* lines were developed containing modified *cry1A*, *cry1Ab* and *cry1Ac* genes (Huang et al. 2007). Rice stem borer is a major group of pest of rice and causes annual losses of 11.5 billion yuan (US$1.69 billion). In 1989, scientists from Chinese Academy of Agricultural Sciences (CAAS) by using PEG generated the first insect-resistant genetically modified rice plant with *cry* genes under the control of *CaMV 35S* promoter. Later in 2009, China developed two commercially important *Bt* rice lines containing *cry1Ab/Ac* gene with the names Huahui No. 1 and Bt Shanyou 63. The Chinese government is currently undertaking trials on insect-resistant cultivars. The benefit of this is that the farmers do not need to spray their crops with pesticides to control fungal, viral or bacterial pathogens. In comparison, conventional rice is sprayed three or four times per growing season. Benefits of growing *Bt* rice include increased yield and thereby boost in the economy. It is expected that India, Indonesia and the Philippines would carry out the cultivation of genetically modified rice.

3.8 Importance of Resistance Management Strategy

Development of resistance by the target pest or pathogen is an important aspect of pest management (Bates et al. 2005; van der Salm et al. 1994). In the past, it has been observed that mosquito could develop resistance to DDT, a chemical insecticide. This can happen in the case of biopesticides as well. There are many factors that are involved in the development of resistance against insecticides, i.e. in a population of insects, if some insects carry genes for resistance, they produce progenies with the resistant gene.

To avoid problems with resistance of insects against insecticides in transgenic crops, farmers have to follow resistance management requirements set by EPA. These include the use of more than one transgene (pyramiding), domain swapping, crop rotation and refuge planting as a part of resistant management in transgenic farming. Pyramiding is a process of stacking two or more transgenes in a population by conventional crossing between two different transgenic lines. Domain swapping or domain engineering is a process to increase the toxicity of *cry* genes by producing chimeric proteins. For example, *cry1AcF* gene, the first and second domain of *cry1Ac* gene and third domain of *cry1F*, is constructed by domain engineering. It shows effective resistance against a wide range of insects.

Crop rotation is a method to combat resistance of insects against GM crops. Due to crop rotation, the pressure of one specific strain of *Bt* on a specific insect will be minimised due to less time availability for development of resistance.

Refuge planting is a method of planting in the field in which *Bt* crops are planted with alternating rows of non-*Bt* crops. The reason behind this type plantation is according to laws of genetics, insects that have developed resistance to *Bt* have more chances of mating with an insect that has not developed resistance to *Bt* and that the progenies of insects do not become resistant to *Bt*. The EPA has set some guidelines for refuge planting in refuge area, i.e.:

- Growers may plant up to 80 % area of their *Bt* crop and at least 20 % must be planted with non-*Bt* crop.
- Refuge area must be within, adjacent to or near the *Bt* crop. It must be within half mile from *Bt* crop.
- If refuge areas are strips within a field, the strips should be at least four rows.

There are different types of refuge planting, which include block design, linear block design, bracket design, border (perimeter) design and strips (split planter). In block design, one block in the *Bt* field must be planted with non-*Bt* plants. In linear block design, one side (20 %) of the *Bt* field must be planted with non-*Bt*. In bracket design, both sides of the *Bt* field must be planted with non-*Bt* plants. In border design, all sides of *Bt* crop must be accompanied with non-*Bt* plants. In strips design, the non-*Bt* plantation has to be done between *Bt* plantation as strips.

3.9 Conclusion

The *Bt* ICPs have infused tremendous potential in helping the plants combat various insect pests. The *Bt* cotton saga is proof enough for the success of these genes. There is a continued increase in the adoption of the technology worldwide with various food and nonfood crops being transformed. Further, fishing for effective cry genes is also still on the move. The studies and the effective demonstrations of the technology will have immense value based on the acceptance of the technologies and products. An initial boost will enable various transgenics in the pipeline to be introduced to mankind. Care should be however taken in practising the right insect management practices for the successful utilisation of the painstaking efforts.

References

Barton KA, Whiteley HR, Yang NS (1987) *Bacillus thuringiensis* δ-endotoxin expressed in transgenic Nicotiana tabacum provides resistance to lepidopteran insects. Plant Physiol 85: 1103–1109

Bates SL, Zhao JZ, Roush RT, Shelton AM (2005) Insect resistance management in GM crops: past, present and future. Nat Biotechnol 23:57–62

Bauce E, Carisey N, Dupont A, van Frankenhuyzen K (2004) *Bacillus thuringiensis* subsp kurstaki (Btk) aerial spray prescriptions for balsam fir stand protection against spruce budworm (Lepidoptera: Tortricidae). J Econ Entomol 97:1624–1634

Baum JA, Johnson TB, Carlton BC (1999) *Bacillus thuringiensis*: natural and recombinant bioinsecticide products. In: Hall FR, Menn JJ (eds) Biopesticides: use and delivery. Humana Press, Totowa, pp 189–210

Beard CE, Court L, Mourant RG, James B, Van Rie J, Masson L, Akhurst RJ (2008) Use of a cry1Ac-resistant line of *Helicoverpa armigera* (Lepidoptera: Noctuidae) to detect novel insecticidal toxin genes in *Bacillus thuringiensis*. Curr Microbiol 57:175–180

Becker N (2000) Bacterial control of vector-mosquitoes and black flies. In: Entomopathogenic bacteria: from laboratory to field application. Kluwer Academic, Dordrecht, pp 383–396

Boonserm P, Davis P, Ellar DJ, Li J (2005) Crystal structure of the mosquito-larvicidal toxin Cry4Ba and its biological implications. J Mol Biol 348:363–382

Boonserm P, Mo M, Angsuthanasombat C, Lescar J (2006) Structure of the functional form of the mosquito larvicidal Cry4Aa toxin from *Bacillus thuringiensis* at a 2.8-Å resolution. J Bacteriol 188:3391–3401

Bravo A (1997) Phylogenetic relationships of *Bacillus thuringiensis* delta-endotoxin family proteins and their functional domains. J Bacteriol 179:2793–2801

Bravo A, Gill SS, Soberon M (2007) Mode of action of *Bacillus thuringiensis* Cry and Cyt toxins and their potential for insect control. Toxicon 49:423–435

Broderick NA, Raffa KF, Handelsman J (2006) Midgut bacteria required for Bacillus thuringiensis insecticidal activity. Proc Natl Acad Sci U S A 103:15196–15199

Brookes G, Barfoot P (2013) GM crops: global socio-economic and environmental impacts 1996–2011. PG Economics Ltd, Dorchester

Carlton BC, Gawron-Burke C (1993) Genetic improvement of *Bacillus thuringiensis* for bioinsecticide development. In: Kim L (ed) Advanced engineered pesticides. Marcel Dekker, New York, pp 43–61

Carlton BC, Gonzalez JN (1985) Plasmids and delta-endotoxin production in different subspecies of *Bacillus thuringiensis*. In: Hoch JA, Setlow P (eds) Molecular biology of microbial differentiation. American Society for Microbiology, Washington, DC, pp 246–252

Carroll J, Ellar DJ (1993) An analysis of *Bacillus thuringiensis* δ-endotoxin action on insect-midgut-membrane permeability using a light scattering assay. Eur J Biochem 214:771–778

Crickmore N, Zeigler DR, Feitelson J, Schnepf E, Van Rie J, Lereclus D, Baum J, Dean DH (1998) Revision of the nomenclature for the *Bacillus thuringiensis* pesticidal crystal proteins. Microbiol Mol Biol Rev 62:807–813

De Barjac H, Bonnefoi A (1962) Essai de classification biochimique et serologique de 24 souches de *Bacillus* du type *B. thuringiensis*. Entomophaga 7:5–31

De Maagd RA, Bakkar PL, Masson L, Adang MJ, Sangandala S, Stiekema W, Bosch D (1999) Domain III of the *Bacillus thuringiensis* delta-endotoxin Cry1Ac is involved in binding to *Manduca sexta* brush border membranes and to its purified amino peptidase N. Mol Microbiol 31:463–471

English L, Slatin SL (1992) Mini-review. Mode of action of Delta-endotoxins from *Bacillus thuringiensis*: a comparison with other bacterial toxins. Insect Biochem Mol Biol 22:1–7

Federici BA (1999) Bacillus thuringiensis in biological control. In: Bellows TS, Fisher TW (eds) Handbook of biological control: principles and applications, Chapter 21. Academic, San Diego, pp 575–593

Fernández-Larrea O (2002) Tecnologías de producción de *Bacillus thuringiensis*. Manejo Integrado de Plagas y Agroecología 64:110–115

Fischhoff DA, Bowdish KS, Perlak FJ, Marrone PG, McCoormick SM, Niedermeyer JG, Dean DA, Kusano KK, Mayer EJ, Rochester DE, Rogers SG, Fraley RT (1987) Insect tolerant transgenic tomato plants. Biotechnology 5:807–813

Galitsky N, Cody V, Wojtczak A, Ghosh D, Luft JR, Pangborn W, English L (2001) Structure of the insecticidal bacterial δ-endotoxin CryBb1 of *Bacillus thuringiensis*. Acta Crystallogr D Biol Crystallogr D 57: 1101–1109

Gill SS, Cowles EA, Pietrantonio PV (1992) The mode of action of *Bacillus thuringiensis* endotoxins. Annu Rev Entomol 37:615–636

Gonzalez JM, Brown BJ, Carlton BC (1982) Transfer of *Bacillus thuringiensis* plasmids coding for delta-endotoxin among strains of *B. thuringiensis* and *B. cereus*. Proc Natl Acad Sci USA 79: 6951–6955

Grochulski P, Masson L, Borisova S, Pusztai-Carey M, Schwartz JL, Brousseau R, Cygler MJ (1995) *Bacillus thuringiensis* CryIA(a) insecticidal toxin: crystal structure and channel formation. Mol Biol 254:447–464

Guillet P, Kurtak DC, Philippon B, Meyer R (1990) Use of Bacillus thuringiensis israelensis for Onchocerciasis control in West Africa. In: de Barjac H, Sutherland D (eds) Bacterial control of mosquitoes and black flies: biochemistry, genetics, and applications of Bacillus thuringiensis israelensis and Bacillus sphaericus. Rutgers Univ. Press, New Brunswick, pp 187–201

Hendrickx ME, Estrada-Navarrete FD (1989) A checklist of the species of pelagic shrimps (Penaeoidea and Caridea) from the eastern Pacific with notes on their zoogeography and depth distribution. CalCoFi Rep 30:104–111

Hofte H, Whiteley HR (1989) Insecticidal crystal proteins of *Bacillus thuringiensis*. Microbiol Rev 53:242–255

Huang F, Leonard BR, Andow DA (2007) Sugarcane borer (Lepidoptera: Crambidae) resistance to transgenic *Bacillus thuringiensis* maize. J Econ Entomol 100:164–171

James C (2012) Global status of commercialised biotech D GM Crops: 2012. ISAAA Brief 44. ISAAA, Ithaca

Jenkins JL, Dean DH (2000) Exploring the mechanism of action of insecticidal proteins by genetic engineering methods. In: Setlow JK (ed) Genetic engineering: principles and methods. Plenum, New York, p 33

Jenkins JL, Lee MK, Valaitis AP, Curtiss A, Dean DH (2000) Bivalent sequential binding model of a *Bacillus thuringiensis* toxin to gypsy moth aminopeptidase N receptor. J Biol Chem 275: 14423–14431

Jurat-Fuentes JL, Adang MJ (2004) Characterization of a Cry1Ac-receptor alkaline phosphatase susceptible and resistant *Heliothis virescens* larvae. Eur J Biochem 271:3127–3135

Kaur S (2000) Molecular approaches towards development of novel *Bacillus thuringiensis* biopesticides. World J Microbiol Biotechnol 16:781–793

Knight P, Crickmore N, Ellar DJ (1994) The receptor for *Bacillus thuringiensis* CryIA(c) delta-endotoxin in the brush border membrane of the lepidopteran *Manduca sexta* is aminopeptidase N. Mol Microbiol 11:429–436

Knowles BH (1994) Mechanism of action of *Bacillus thuringiensis* Insecticidal d-endotoxins. Adv Insect Physiol 4:275–308

Knowles BH, Dow JAT (1993) The crystal d-endotoxins of *Bacillus thuringiensis*: models for their mechanism of action on the insect gut. Bioessays 15:469–476

Koller CN, Bauer LS, Hollingworth RM (1992) Characterization of the pH-mediated solubility of *Bacillus thuringiensis* var. *san diego* native delta endotoxin crystals. Biochem Biophys Res Commun 184:692–699

Kronstad J, Schnepf WHE, Whiteley HR (1983) Diversity of locations for *Bacillus thuringiensis* crystal protein genes. J Bacteriol 154:419–428

Li J, Carroll J, Ellar DJ (1991) Crystal structure of insecticidal d-endotoxin from *Bacillus thuringiensis* at 2.5 Å resolutions. Nature 353:815–821

Luthy P, Cordier JL, Fischer HM (1982) *Bacillus thuringiensis* as a bacterial insecticide: basic considerations and application. In: Kurstak E (ed) Microbial and viral pesticides. Marcel Dekker, New York/Basel, pp 35–74

McNall RJ, Adang MJ (2003) Identification of novel *Bacillus thuringiensis* Cry1Ac binding proteins in *Manduca sexta* midgut through proteomic analysis. Insect Biochem Mol Biol 33: 999–1010

Morse RJ, Yamamoto T, Stroud RM (2001) Structure of Cry2Aa suggests an unexpected receptor binding epitope. Structure 9:409–417

Perlak FJ, Deaton RW, Armstrong RL, Fuchs RL, Sims SR, Greenplate JT, Fischhoff DA (1990) Insect resistant cotton plants. Biotechnology 8:939–943

Perlak FJ, Stone TB, Muskopf YM, Petersen LJ, Parker GB, McPherson SA, Wyman J, Love S, Reed G (1993) Genetically improved potatoes: protection from damage by Colorado potato beetles. Plant Mol Biol 22:313–321

Porcar M, Caballero P (2000) Molecular and insecticidal characterization of a *Bacillus thuringiensis* strain isolated during a natural epizootic. J Appl Microbiol 89:309–316

Qaim M, Zilberman D (2003) Yield effects of genetically modified crops in developing countries. Science 299:900–902

Rowe GE, Margaritis A (1987) Bioprocess developments in the production of bioinsecticides by *Bacillus thuringiensis*. CRC Crit Rev Biotechnol 6:87–127

Sacchi VF, Parenti P, Hanozet GM, Giordana B, Luethy P, Wolfersberger MG (1986) *Bacillus thuringiensis* toxin inhibits K-gradient-dependent amino acid transport across the brush border membrane of *Pieris brassicae* midgut cells. FEBS Lett 204:213–218

Schnepf E, Crickmore N, Rie JV, Lereclus D, Baum J, Feitelson J, Zeigler DJ, Dean DH (1995) *Bacillus thuringiensis* and its pesticidal crystal proteins. Microbiol Mol Biol Rev 62:775–806

Schuler TH, Poppy GM, Kerry BR, Denholm I (1998) Insect-resistant transgenic plants. Trends Biotechnol 16:168–175

Schwartz JL, Garneau L, Savaria D, Masson L, Brousseau R, Rousseau E (1993) Lepidopteran-specific crystal toxins from *Bacillus thuringiensis* form cation- and anion-selective channels in planar lipid bilayers. J Membr Biol 132:53–62

Shelton AM, Zhao JZ, Roush RT (2002) Economic, ecological, food safety, and social consequences of the deployment of Bt transgenic plants. Annu Rev Entomol 47:845–881

Siegel JP (2000) Bacteria. In: Lacey LL, Kaya HK (eds) In field manual of techniques in invertebrate pathology. Kluwer, Dordrecht, pp 209–230

Vadlamudi RK, Weber E, Ji I, Ji TH, Bulla LA (1995) Cloning and expression of a receptor for an insecticidal toxin of *Bacillus thuringiensis*. J Biol Chem 270:5490–5494

Vaeck M, Reynaerts A, Hofte H, Jansens S, Beukeleer MD, Dean C, Zabeau M, Montagu MV, Leemans J (1987) Transgenic plants protected from insect attack. Nature 328:33–37

Valaitis AP, Jenkins JL, Lee MK, Dean DH, Garner KJ (2001) Isolation and partial characterization of Gypsy moth BTR-270 an anionic brush border membrane glycoconjugate that binds *Bacillus thuringiensis* Cry1A toxins with high affinity. Arch Insect Biochem Physiol 46: 186–200

van der Salm T, Bosch D, Honée G, Feng L, Munsterman E, Bakker P, Stiekema WJ, Visser B (1994) Insect resistance of transgenic plants that express modified *Bacillus thuringiensis* cryIA (b) and cryIC genes: a resistance management strategy. Plant Mol Biol 26:51–59

Van Frankenhuyzen K (2000) Application of *Bacillus thuringiensis* in forestry. In: Charles JF, Delécluse A, Nielsen-LeRoux C (eds) Entomopathogenic bacteria: from laboratory to field application. Kluwer Scientific publishers, Dordrecht, Netherlands, pp. 371

Van Rie J, Jansens S, Hofte H, Degheele D, Van Mellaert H (1989) Specificity of *Bacillus thuringiensis* δ-endotoxins. Eur J Biochem 186:239–247

Wolfersberger MG (1989) Neither barium nor calcium prevents the inhibition by *Bacillus thuringiensis* δ-endotoxin of sodium or potassium gradient-dependent amino acid accumulation by tobacco hornworm midgut brush border membrane vesicles. Arch Insect Biochem Physiol 12:267–277

Zeigler DR (1999) *Bacillus* Genetic Stock Center catalog of strains. Part 2: *Bacillus thuringiensis* and *Bacillus cereus*, 7th edn. *Bacillus* Genetic Stock Center, Columbus

Chapter 4
Identification and Characterization of Receptors for Insecticidal Toxins from *Bacillus thuringiensis*

Ricardo A. Grande-Cano and Isabel Gómez

4.1 Introduction

During the stationary phase of the life cycle of *Bacillus thuringiensis* (*Bt*) is produced a parasporal body (crystal) containing insecticidal proteins called Cry toxins, which are used for development of transgenic crops and spray formulations for control of pest insects (Van Rie 2000). Although *Bt* has been used for two decades as a commercial biopesticide, the mode of action has only been described recently. Cry proteins are included into the pore-forming toxins (PFTs), one of the largest classes of bacterial toxins (Bravo et al. 2011). Many sequences of *cry* genes have been identified, including 56 families (Cry1–Cry56) and 180 subtypes (Cry1Aa, Cry4Ba, etc.). Of all of these proteins, some structures have been resolved by X-ray crystallography. It is interesting to note that the sequences are different but the structural topology is conserved, indicating a similar mechanism of action. All contain three structural domains: Domain I contains seven α-helixes and is involved in the oligomerization of the toxin and its insertion into the membrane, and Domains II and III are made up of β-sheets and are involved in the attachment to the receptors (Soberon et al. 2010).

The mechanism of toxicity begins when the insect ingests food contaminated with spores; the crystal is then solubilized in the midgut, releasing protoxins which are activated by proteases, and the toxic fragment binds in sequential fashion to receptors located in the microvilli. The first receptor is a cadherin-like protein (CADP), whose interaction allows the oligomerization of the toxin. This oligomeric

R.A. Grande-Cano
Unidad Universitaria de Secuenciación Masiva de DNA-UNAM, Av. Universidad 2001, Col. Chamilpa, Cuernavaca, Morelos 62210, Mexico

I. Gómez (✉)
Departamento de Microbiología Molecular, Instituto de Biotecnología-UNAM, Av. Universidad 2001, Col. Chamilpa, Cuernavaca, Morelos 62210, Mexico
e-mail: isabelg@ibt.unam.mx

© Springer International Publishing Switzerland 2015
K.S. Sree, A. Varma (eds.), *Biocontrol of Lepidopteran Pests*, Soil Biology 43,
DOI 10.1007/978-3-319-14499-3_4

structure increases its affinity to the second receptor, either aminopeptidase-N (APN) or alkaline phosphatase (ALP), anchored by glycosylphosphatidyl-inositol (GPI) to the lipid raft membrane (Pacheco et al. 2009; Arenas et al. 2010). Finally, the oligomer is inserted into the lipid raft, forming holes in the membrane, allowing swelling and death of the intestinal cells (Zhuang et al. 2002; Bravo et al. 2004). When the intestinal epithelium is destroyed, the spores are able to access the hemolymph, a source rich in nutrients for its growth. Many species of insects are susceptible; one of the main characteristics of Cry toxins is its specificity which is determined by its recognition of the receptors, although the solubilization and activation steps due to the physical and chemical environment of the midgut of the insects are also important determining factors (Pigott and Ellar 2007; Soberon et al. 2009; Bravo et al. 2013).

As mentioned previously, Cry toxins are highly selective and kill only a limited number of insect species. This selectivity is mainly due to the interaction of Cry toxins with larval proteins located in the midgut epithelial cells. The crucial role of this receptor binding for toxicity is emphasized by the observation that insects selected for resistance to a Cry toxin often have no or reduced binding capacity for that toxin (Ferre and Van Rie 2002). A major research effort has taken place in the identification of insect proteins that bind Cry toxins and mediate toxicity. Among these, two major types of receptors have been identified: transmembrane proteins, such as cadherins, and proteins anchored to the membrane such as the GPI-anchored proteins that have been proposed to be involved in the action of Cry toxins (Pigott and Ellar 2007).

After it was demonstrated that specific high-affinity toxin-binding sites are present in the midgut of susceptible insects, efforts to identify and clone these molecules have been intensified. Several putative Cry toxin receptors have since been reported, of which the best characterized are the aminopeptidase N (APN) receptors (Knight et al. 1994; Sangadala et al. 1994; Gill et al. 1995) and the cadherin-like receptors (Vadlamudi et al. 1993; Nagamatsu et al. 1998; Gahan et al. 2001; Chen et al. 2014) identified in lepidopterans. In nematodes, glycolipids are believed to be an important class of Cry toxin receptors (Griffitts et al. 2005). Other putative receptors include ALPs (Jurat-Fuentes and Adang 2004; Fernandez et al. 2006; Arenas et al. 2010; Zuniga-Navarrete et al. 2013), a 270-kDa glycoconjugate (Valaitis et al. 2001), and a 252-kDa protein (Hossain et al. 2004).

4.2   Toxin Receptor by Binding Blot Overlay

Radioligand binding assay remains the most sensitive quantitative technique to measure binding parameters of affinity and receptor density and is widely used to characterize receptors and determine their anatomical distribution. In saturation experiments, tissue sections, cultured cells, or homogenates are incubated with an increasing concentration of a radiolabeled ligand. Analysis using iterative nonlinear curve-fitting programs, such as Scatchard, measures the affinity of the labeled ligand for a receptor (equilibrium dissociation constant, K_D), receptor density (B_{max}), and hill slope (nH). The affinity and selectivity of an unlabeled ligand to compete for the binding of a fixed concentration of a radiolabeled ligand to a receptor are determined using a competition-binding assay. Kinetic assays measure the rate of association to or dissociation from a receptor from which a kinetic K_D may be derived (Maguire et al. 2012).

In principle, a blot overlay is similar to a Western blot. For both procedures, samples are run on SDS–PAGE gels, transferred to nitrocellulose or PVDF, and then overlaid with a Cry soluble protein that may bind to one or more immobilized proteins on the blot. In the case of a Western blot, the overlaid protein is antibody. In the case of a blot overlay, the overlaid protein is Cry toxin labeled with biotin or radiolabeled usually with I^{125}. The overlaid probe can be detected either via incubation with an antibody (this method is often referred to as a "Far Western blot"), via incubation with streptavidin (if the probe is biotinylated), or via autoradiography if the overlaid probe is radiolabeled.

4.2.1   Using One-Dimensional Electrophoresis

Toxin-binding proteins from several insects have been identified by ligand blot analysis of brush border membrane (BBM) vesicles (BBMVs) prepared from the midguts of the insects (Garczynski et al. 1991). The protein blot overlay assay is a powerful technique for identifying proteins transferred to an immobilizing matrix. It is based on the specific binding of labeled ligands with proteins of interest blotted onto the membrane, followed by detection of the complexes formed.

Specific protein receptors involved in Cry toxin killing of target insects has been known since the mid-1980s. Studies of the binding of radiolabeled Cry toxins in suspensions of insect midgut proteins isolated by various procedures have generated a rather extensive list of putative receptor molecules (Hofmann et al. 1988; Van Rie et al. 1989; Lee et al. 1992; Denolf et al. 1993; Escriche et al. 1997; Simpson et al. 1997) without simultaneously identifying the binding protein(s) in question by the use of midgut protein sodium dodecyl sulfate (SDS)–polyacrylamide gel electrophoresis (PAGE) blots. SDS–PAGE blots, when incubated with radiolabeled toxins, at least permit a visual estimation of both the number of proteins involved and their molecular masses. Ligand blots of *Manduca sexta* midgut proteins have been used successfully to identify and partially characterize CADP and aminopeptidase as insect binding proteins for the Cry1A lepidopteran-specific toxins (Garczynski et al. 1991; Vadlamudi et al. 1993, 1995; Martinez-Ramirez et al. 1994; de Maagd et al. 1996; Francis and Bulla 1997; Keeton and Bulla 1997) and the polyphagous pest *Heliothis virescens* (Gill et al. 1995; Luo et al. 1997; Oltean et al. 1999).

4.2.2   Using Two-Dimensional Electrophoresis

Two-dimensional gel electrophoresis is a powerful and widely used method for the analysis of complex protein mixtures extracted from cells, tissues, or other biological samples. This technique separates proteins in two steps, according to two independent properties: the first dimension is isoelectric focusing (IEF), which separates proteins according to their isoelectric points (pI); the second dimension is SDS–PAGE, which separates proteins according to their molecular weights. In this way, complex mixtures consisting of thousands of different proteins can be resolved and the relative amount of each protein can be determined (Rabilloud and Lelong 2011).

The insect midgut brush border has distinct structural elements; for this reason, the identification of the proteome of BBMVs is a necessary step in defining potential Cry toxin receptors. Because the insect midgut is the primary target site for Bt toxins, several groups are focusing on constitutive expression changes in the midgut proteins of resistant versus susceptible larvae. For that purpose, a comparative analysis of BBM proteins using 2D DIGE (two-dimensional differential in-gel electrophoresis) with mass spectrometry-based proteomic identification can result in unique protein profiles and insight into functional processes of toxin–receptor interactions. The goal of two-dimensional electrophoresis is to separate and display all gene products present. As the only method currently available, which is capable of simultaneously separating thousands of proteins, it has been very useful in the identification of new putative receptors for Cry toxins.

Proteomic approaches have been previously used to identify novel Cry toxin-binding proteins (McNall and Adang 2003) or compare proteomes from Cry-susceptible and -resistant insects (Candas et al. 2003) or cell lines (Liu et al. 2004). In these studies, the increased resolving power of 2D electrophoresis allowed for identification of proteins that were not successfully resolved using traditional SDS–PAGE electrophoresis. One of the most interesting proteins confirmed with this kind of methodology is the ALP, involved in Cry1Ac binding with *H. virescens* midgut (Jurat-Fuentes and Adang 2007). Many other proteins have been detected by 2D electrophoresis but they need to be confirmed in vivo by other strategies (McNall and Adang 2003; Krishnamoorthy et al. 2007; Bayyareddy et al. 2009; Gai et al. 2013).

4.3   Toxin-Affinity Chromatography

This technique involves the elution of a single protein from an affinity column after prior elution of nonspecifically adsorbed proteins. Cry toxins are coupled to Sepharose (an insoluble, large, pore-sized chromatographic matrix). The proteins from the midgut are incubated and bind to the Cry toxins covalently bound to the matrix. To elute the bound molecules from the affinity matrix, the toxin–receptor interaction is destabilized by incubation with SDS solubilization buffer according to the procedure of Laemmli (Laemmli 1970). After electrophoresis, eluted proteins can be recovered by electro-elution to sequence. Several receptors for different Cry toxins have been identified with this methodology. Cadherin from *M. sexta* was isolated using immobilized Cry1Ab toxin (Vadlamudi et al. 1993); also amino-peptidase was identified in combination with overlay assay with Cry1Ac toxin (Knight et al. 1994) and with Cry1Ab toxin (Denolf et al. 1993). More recently, a similar strategy was used to identify an ALP as receptor for Cry11A toxin in *Aedes aegypti* mosquito (Fernandez et al. 2006).

4.4 Immunohistochemical Detection of Toxin-Binding Receptors

In situ screening of histological sections could reveal the cell-specific distribution of Cry toxin receptors in target tissues using histochemical or immunohisto-chemical staining. Because of the superior morphology provided by formalin-fixed paraffin-embedded tissues, this has become the medium of choice for most research studies with the Cry toxins. The peroxidase-labeled antibody method, introduced in 1968, was the first practical application of antibodies to paraffin-embedded tissues and overcame some of the limitations of earlier fluorescence antibody methods (Nakane 1968). These pioneering studies using enzyme labels instead of fluorescent dyes opened the door to the development of modern methods of immunohistochemistry. In 1981 a new generation of immunohistochemical methods emerged with the advent of the avidin–biotin methods, which remains widely used today (Hsu et al. 1981). All avidin–biotin methods rely on the strong affinity of avidin or streptavidin for the vitamin biotin.

The cytology and ultrastructure of the midgut cells of *M. sexta* and other lepidopteran larvae was reported in the 1980s (Endo and Nishiitsutjiuwo 1980). This made it easier to study the toxic effects of Bt toxins on the insect midgut. After intoxication with spore preparations from Bt, it was observed that the epithelial cells of the midgut swell shortly after ingestion of the Bt toxins. Eventually, the cells burst and release their cytoplasmatic content into the midgut lumen (Lane et al. 1989).

Immunocytochemical localization of toxin-fed larvae has demonstrated that the insecticidal proteins accumulate in the microvilli of the gut epithelial cells (Rausell et al. 2000). Histopathological studies have shown that one of the most rapid effects of Bt Cry intoxication is a striking change in the microvilli targeted by the Bt Cry toxins. A reduction of binding sites is a hallmark of development of resistance to Bt, because toxicity depends on the availability of specific high-affinity receptors. Receptor expression levels have been shown to correlate with Cry toxin activity. For example, in *M. sexta*, the three identified protein receptors are expressed in the anterior, middle, and posterior regions of the midgut (Chen et al. 2005).

4.5   Expression of Receptors in Cell Lines

Insect cell lines have been used to study Cry toxin membrane insertion and channel formation (Vachon et al. 1995). Cry toxins cause a cytotoxic response in some insect cell lines, and frequently cell swelling served as a marker of cytotoxicity (Pigott and Ellar 2007). Various types of insect cell lines have also been studied for their toxicity response to different Cry toxins (Johnson 1994), and it was found that *Sf*-9 cells showed very low toxicity to Cry1Ab and showed maximum toxicity with Cry1C (Kwa et al. 1998). Cell lines not susceptible to Cry1 toxins can be used to test the receptor function in vivo. The effectiveness of using live insect cells expressing putative receptor proteins and demonstrating the ability of the protein to function as a receptor to Bt toxin has been shown by Nagamatsu et al. (1999). They have been able to demonstrate swelling and lysis of *Sf*-9 cells expressing cadherin protein of *Bombyx mori* after overlay with Cry1Ab toxin. Purified membranes from COS cells expressing Bt-R_1 (cadherin from *M. sexta*) bound all three Cry1A toxins in binding assays and ligand blots (Keeton and Bulla 1997). Furthermore, expression of Bt-R_1 on the surface of COS7 cells led to toxin-induced cell toxicity as monitored by immunofluorescence microscopy with fixed cells (Dorsch et al. 2002). *Drosophila S*-2 cell line constitutively expressing *M. sexta* cadherin gene showed heavy damage and lysis when overlaid with Cry1Ab (Hua et al. 2004).

4.6   RNAi

RNA interference has become an effective and important tool to study the functional relevance of various proteins and genes in an organism. As described in detail in various reviews, RNAi by dsRNA results in sequence-specific posttranscriptional degradation of the target mRNA (Agrawal et al. 2003). The principle behind the use of RNAi technology is that entry of exogenous dsRNA into a cell can activate an innate defense process that results into the breakdown of the dsRNA and the degradation of any endogenous RNA molecule having the same nucleotide sequence. Therefore, RNAi provides a means of protection for eukaryotic cells against aberrant RNAs. The dsRNA thus induces a sequence-specific silencing of gene expression at the posttranscriptional level (Hakim et al. 2010). This property has led to the generation and analysis of the phenotypic effect(s) caused by targeted knockdown of specific mRNAs. In the past few years, several studies have explored the potential of RNAi as an innovative, promising strategy for controlling a number of agriculturally important insect pests (Baum et al. 2007; Gordon and Waterhouse 2007).

The success of oral RNAi in vivo is determined primarily by the presence of an uptake mechanism for dsRNA in the target cells, but little is known about the mechanism of dsRNA uptake by the insect midgut and its further spread throughout the insect body (Hakim et al. 2010).

Successful RNAi experiments have been carried out in a number of lepidopteran species to date (Terenius et al. 2011). The first lepidopteran RNAi publications appeared in 2002; one reported the knockdown of a pigment gene following dsRNA injection into *B. mori* embryos (Quan et al. 2002); another targeted a pattern recognition protein, hemolin, in *Hyalophora cecropia* embryos by heritable RNAi (Bettencourt et al. 2002); and a third targeted a putative *B. thuringiensis* toxin receptor in *Spodoptera litura* larvae (Rajagopal et al. 2002).

Feeding of dsRNA is an even more attractive approach than hemocoel injection because it is noninvasive. Interest for this approach received a great boost after the high-profile publications of its feasibility in several pest insect species, including the lepidopteran *H. armigera* (Baum et al. 2007; Mao et al. 2007). Feeding of dsRNA has been applied with greatest effect in *Plutella xylostella*, *S. exigua*, *M. sexta*, and *Ostrinia nubilalis* (Bautista et al. 2009; Tian et al. 2009; Whyard et al. 2009; Yang et al. 2010; Flores-Escobar et al. 2013). Some targets successfully blocked by oral delivery of dsRNA are involved in the mode of action of Cry toxins, such as aminopeptidase in *H. armigera*, *S. litura*, *S. exigua*, and *M. sexta*, cadherin in *S. exigua* and *M. sexta*, and ALP in *M. sexta* (Rajagopal et al. 2002; Sivakumar et al. 2007; Flores-Escobar et al. 2013; Park and Kim 2013; Ren et al. 2013, 2014).

4.7 Transcriptome as Tool to Identify New Proteins Involved in the Mechanism of Action of Cry Toxins

RNA-Seq is a set of methodologies that allows us to analyze the transcriptome of a tissue or set of cells in a given time and under specific growth conditions, using the technologies of next-generation massive sequencing (Mutz et al. 2013). The term transcriptome was coined nearly two decades ago by Charles Auffray to define the set of transcripts (mRNA, small RNA, lncRNA, rRNA, etc.) into a set of cells or tissues (McGettigan 2013). Historically, large-scale transcriptome studies were based mainly on microarray analysis and supported in many cases by RT-PCR experiments. The first report for monitoring the expression of genes in parallel using a microarray as we know today was described by Schena et al. (1995). Technical limitations of this technology relative to RNA-Seq studies have enabled the latter to develop at an incredible speed. For example, to perform microarray studies, it is necessary to know the sequence of the organism to design the microarray. In comparison, studies of RNA-Seq do not require prior knowledge of the genome of the organism to perform the experiment. In addition, due to the large number of related sequences in organisms, microarray studies have been a problem because of the cross hybridization between these sequences. A further advantage is the high resolution of up to one base pair in RNA-Seq experiments which allows to know in great detail and with high confidence the 5′ and 3′ ends of the transcripts, which are unable with studies with microarrays. This last point is important because it allows us to know with high accuracy the regulatory regions governing expression of a gene. A fundamental disadvantage of microarray studies upon RNA-Seq studies is that due to the nature of the analogic signal, it is difficult to quantify the high and low species of the RNA expressed. The wide dynamic range of RNA-Seq experiments also allow us to determine better the differential gene expression (Nagalakshmi et al. 2008) compared with microarray experiments, since the latter have a more limited dynamic range. Experiments of RNA-Seq have also allowed the identification of new forms of alternative splicing, results that would be impossible to obtain via microarray experiments since the latter can tell us what is being expressed in a given time but not the structure of that transcript. Finally, there is a high reproducibility of results between studies of RNA-Seq in biological and technical replicates compared to microarray-based studies.

RNA-Seq analysis begins with the preparation of a suitable sample of total RNA. To this end, currently there are a large number of commercially available kits that help us to obtain high-quality RNA from limited quantities of biological samples and even to obtain different RNA species from the same preparation. Once the RNA has been purified, an evaluation of the integrity of the material is necessary to proceed with the construction libraries. The sample analysis is performed through the use of devices like 2100 Bioanalyzer or 2200 TapeStation from Agilent that allow us to analyze quickly and accurately the concentration and integrity of the RNA samples, the latter via *RNA integrity number* (RIN) calculation (Schroeder et al. 2006). After obtaining an adequate sample of RNA, the next step is the construction of the library. To this end, several methods have been developed that allow us to build general or strand-specific libraries from a specific amount of genetic material (Levin et al. 2010).

Some of the studies of differential gene expression in insects using massive sequencing technologies have been made using the 454 platform of Roche (Park et al. 2009; Pauchet et al. 2010; Oppert et al. 2012); however, even when the reads are large in average size which facilitates the assembly and subsequent analysis, they are relatively few compared to those obtained from massive sequencing experiments using other platforms such as Illumina platforms Genome Analyzer GAIIx and HiSeq (He et al. 2012; Ma et al. 2012; Lei et al. 2014). The sequence data obtained should be analyzed using a limited number of programs currently available on the Internet. Most of these programs are specific to running in Linux environments, which require adequate computer infrastructure. When a reference genome is available, the data can be aligned with the help of programs like Bowtie (Langmead et al. 2009), SOAP (Li et al. 2008) or BWA (Li and Durbin 2009). Meanwhile, when a reference genome is not available, the data can be assembled de novo using SOAPdenovo (Li et al. 2009), SMART, or Trinity (Grabherr et al. 2011), among several others (Garber et al. 2011).

With the development of the new technologies for massive sequencing (Margulies et al. 2005; Bentley et al. 2008; Rothberg et al. 2011), RNA-Seq studies have become extremely important, especially for those non-model organisms where no genomic information is available for the transcriptome analysis. Currently, many pests of economically important crops are managed using chemical insecticides with high efficiency. However, the indiscriminate use of these products brings many consequences at different levels, from potential hazards to the environment and food to the development of resistance by organisms against many of the chemicals used in the manufacture of insecticides. By this reason, the pest management using biological control is being increasingly used.

Even when the mechanism of action of Cry toxins and their interaction with their receptors in some insect models is known in detail, the absence of genomic information in most of them has delayed further study, i.e., to understand in a better way the mechanism and thereby applying in a more suitable way the biological mechanism to control pests attacking economically important crops. For this reason, studies focused on the generation of genomic information of organisms affecting crop plants are required.

In an analysis of RNA-Seq complemented with microarray experiments (using the 454 platform of Roche) performed on *Tenebrio molitor*, a stored grain pest, intoxicated with toxin Cry3Aa of Bt, Oppert et al. (2012) found that the expression of the three best characterized receptors in insects to Bt toxin, cadherin, aminopeptidase N, and ALP, occurs in a very particular way. No differences were observed in the expression pattern of the cadherin and aminopeptidase N receptors amongst the control larvae and larvae intoxicated with the Cry3Aa after 24 h of intoxication. However, transcripts associated with the enzyme ALP, also proposed to be acting as a receptor to the Cry1Ac toxin in Lepidoptera, were observed only in the larvae intoxicated with the Cry3Aa toxin. Interestingly, increased levels of ALP were observed in Cry1Ab-resistant European corn borer, *O. nubilalis* (Khajuria et al. 2009). In a subsequent analysis using a custom script for function assignment, enrichment for biological processes was obtained, specifically an enrichment of genes associated with transport, suggesting that once the larva is intoxicated it tries to maintain ionic homeostasis (Oppert et al. 2012).

The diamondback moth (*P. xylostella*) is a widely distributed insect that belongs to the order Lepidoptera. Currently, it has great economic importance because this insect is a plague of cruciferous vegetables like cabbage, broccoli, and mainly cauliflower, important crops for human consumption. With the indiscriminate use of synthetic insecticides, the diamondback moth became the first pest to develop resistance to DDT; in addition, this was also the first insect to develop resistance in the field to *B. thuringiensis* bacteria (Talekar and Shelton 1993). In a transcriptome analysis of the midgut of two resistant strains of *P. xylostella* and one sensitive to Cry1Ac toxin, Lei et al. found that when the insects are challenged with Cry1Ac toxin, expression of 34 sequences related to ABC transporters, two unigenes related to aminopeptidase, and four related to cadherin, all of them possibly associated to the resistance to Cry1Ac was observed. Fifteen sequences out of 34 related to ABC transporters were common to both strains resistant to Cry1Ac. In addition, six differentially expressed unigenes (DEU) coding to aminopeptidase and cadherin in the resistant strains of *P. xylostella* had differential expression levels of 2–10-fold compared to susceptible strain (Lei et al. 2014). In a previous transcriptome analysis using the Genome Analyzer IIx (GAIIx) of Illumina and Sanger sequencing, He et al. found that genes coding for CADP, aminopeptidase N, and intestinal mucin were the three most abundant transcripts related to resistance to Bt in diamondback moth in four different stages of development and two insecticide-resistant strains (He et al. 2012).

Leptinotarsa decemlineata also known as the Colorado potato beetle is an insect native to Mexico and the Southeastern USA. This insect is an important plague of potato because both larvae and adult organisms feed on potato leaves and several other related species in the Solanaceae family as *Solanum rostratum* and *S. angustifolium*, causing drastic damage to potato yields. Even though *L. decemlineata* was exposed to *S. tuberosum* since 1820, the change to potato feed was documented until 1859 in the USA (Casagrande 1987). The pest was accidentally established in France in 1920, from where it had spread through Europe, becoming a problem in Asia Minor, Iran, Central Asia, and western China at the end of 20th century (Grapputo et al. 2005). Previous studies have shown that the ADAM metalloprotease serves as a receptor for the Cry3Aa toxin in *L. decemlineata* (Ochoa-Campuzano et al. 2007). Additionally, the Cry3Aa toxin also binds to calmodulin in a calcium-independent manner (Ochoa-Campuzano et al. 2012) and to cadherin fragments (Park et al. 2009). By using a pyrosequencing strategy on the 454-FLX platform of Roche, Kumar et al. identified 621 contigs involved in insecticide resistance; of these three contigs belonged to metallo-protease ADAM, 10 to cadherin, and 98 to calmodulin, among others (Kumar et al. 2014).

Bt var. israelensis (Bti) produces a set of proteins during sporulation, which are toxic against the mosquito *A. aegypti*. In an analysis of BBM proteins using 2D DIGE and microarray transcriptome analysis supplemented with RT-PCR reactions against Bti toxin receptors, ALP, cadherin, and aminopeptidase, of an *A. aegypti* strain resistant to Bti Cry toxins which was obtained in the laboratory after 30 generations of selection with leaf litter containing Bti and a susceptible strain, Tetreau et al. found that even though they detected an altered expression of pro-teases and Cry toxin receptor proteins by the two methods used, very little overlapping results were found. The reason for this may be due to technical limitations inherent in each method. It is expected that the repertoire of proteins anchored to the membrane, those obtained for the assay of 2D DIGE, might be very different to that coded in transcriptome assay (Tetreau et al. 2012).

The rice borer worm (*Chilo suppressalis*) is an insect belonging to the order Lepidopteran and is an important pest for this cereal. Although rice is a staple crop for many countries in Asia, very little genomic information is available on this pest. In an analysis of midgut transcriptome using the Illumina Genome Analyzer II platform complemented by an analysis of proteome, Ma et al. generated a total of 37,040 contigs from 39 million Illumina reads. From the total assembled contigs obtained, they were able to identify 16 for cadherin-like transcripts, 27 for APNs, and 11 for ALPs. This correlates with proteome analysis, specifically by Western blot analysis against the proteins that bind to Cry1Ac. Using this technique, they were able to detect two isoforms of APN but were unable to detect cadherin probably due to limitations of the technique to resolve proteins with high molecular weight (Ma et al. 2012).

M. sexta is a moth that belongs to the order Lepidoptera; defoliators of tobacco, tomato, and other plants of the Solanaceae family characterize this group of insects and are present over much of the American continent. Because of its large size, *M. sexta* has served as a model organism to study the mechanism of action of Cry toxins from Bt (Shields and Hildebrand 2001; Kanost et al. 2004; Pigott and Ellar 2007). Even for this important model organism, there was very little genetic information available before 2008, when Zou et al. through a transcriptome analysis using the first generation of pyrosequencing technology of 454 identified a large number of ESTs associated with hemolymph and immune system which considerably increased the genomic information available for this organism (Zou et al. 2008). Further analysis using the second-generation technology of 454 massive sequencing device and conventional Sanger sequencing, Pauchet et al. identified over 387,000 ESTs from the *M. sexta* midgut transcriptome. Functional analysis of these data allowed to group contigs in functions related to digestion, detoxification, Bt toxin binding, peritrophic membrane metabolism, and innate immunity. Among the candidates for Bt toxin binding, 19 contigs were identified as coding for APNs, of which 12 were previously known and 7 genes coding for completely new APNs (Pauchet et al. 2010).

4.8 Conclusions and Future Prospects

Cry toxins are highly selective against their target insect; they present a complex mechanism that involves interaction with several receptors. However, insects may become resistant to these proteins and alternatives to counteract this potential problem must be generated soon. The cellular distribution of receptors in the target tissues and its identification and characterization are very important in the study of the mode of action of Cry toxins produced by *B. thuringiensis*. Several in vitro and in vivo methodologies have been used with interesting results. The combination with the new era of genomics and proteomics promises great advances in the identification of functional proteins as receptors. This information will be useful to design new strategies for pest control.

Acknowledgments Our thanks to PAPIIT/UNAM IN213514 for the support in our research work.

References

Agrawal N, Dasaradhi PV, Mohmmed A, Malhotra P, Bhatnagar RK, Mukherjee SK (2003) RNA interference: biology, mechanism, and applications. Microbiol Mol Biol Rev 67:657–685
Arenas I, Bravo A, Soberon M, Gomez I (2010) Role of alkaline phosphatase from *Manduca sexta* in the mechanism of action of *Bacillus thuringiensis* Cry1Ab toxin. J Biol Chem 285: 12497–12503

Baum JA, Bogaert T, Clinton W, Heck GR, Feldmann P, Ilagan O, Johnson S, Plaetinck G, Munyikwa T, Pleau M, Vaughn T, Roberts J (2007) Control of coleopteran insect pests through RNA interference. Nat Biotechnol 25:1322–1326

Bautista MA, Miyata T, Miura K, Tanaka T (2009) RNA interference-mediated knockdown of a cytochrome P450, CYP6BG1, from the diamondback moth, *Plutella xylostella*, reduces larval resistance to permethrin. Insect Biochem Mol Biol 39:38–46

Bayyareddy K, Andacht TM, Abdullah MA, Adang MJ (2009) Proteomic identification of *Bacillus thuringiensis* subsp. israelensis toxin Cry4Ba binding proteins in midgut membranes from *Aedes* (*Stegomyia*) *aegypti* Linnaeus (Diptera, Culicidae) larvae. Insect Biochem Mol Biol 39: 279–286

Bentley DR, Balasubramanian S, Swerdlow HP, Smith GP, Milton J, Brown CG, Hall KP, Evers DJ, Barnes CL, Bignell HR, Boutell JM, Bryant J, Carter RJ, Cheetham RK, Cox AJ, Ellis DJ, Flatbush MR, Gormley NA, Humphray SJ, Irving LJ, Karbelashvili MS, Kirk SM, Li H, Liu X, Maisinger KS, Murray LJ, Obradovic B, Ost T, Parkinson ML, Pratt MR, Rasolonjatovo IM, Reed MT, Rigatti R, Rodighiero C, Ross MT, Sabot A, Sankar SV, Scally A, Schroth GP, Smith ME, Smith VP, Spiridou A, Torrance PE, Tzonev SS, Vermaas EH, Walter K, Wu X, Zhang L, Alam MD, Anastasi C, Aniebo IC, Bailey DM, Bancarz IR, Banerjee S, Barbour SG, Baybayan PA, Benoit VA, Benson KF, Bevis C, Black PJ, Boodhun A, Brennan JS, Bridgham JA, Brown RC, Brown AA, Buermann DH, Bundu AA, Burrows JC, Carter NP, Castillo N, Chiara ECM, Chang S, Neil Cooley R, Crake NR, Dada OO, Diakoumakos KD, Dominguez-Fernandez B, Earnshaw DJ, Egbujor UC, Elmore DW, Etchin SS, Ewan MR, Fedurco M, Fraser LJ, Fuentes Fajardo KV, Scott Furey W, George D, Gietzen KJ, Goddard CP, Golda GS, Granieri PA, Green DE, Gustafson DL, Hansen NF, Harnish K, Haudenschild CD, Heyer NI, Hims MM, Ho JT, Horgan AM, Hoschler K, Hurwitz S, Ivanov DV, Johnson MQ, James T, Huw Jones TA, Kang GD, Kerelska TH, Kersey AD, Khrebtukova I, Kindwall AP, Kingsbury Z, Kokko-Gonzales PI, Kumar A, Laurent MA, Lawley CT, Lee SE, Lee X, Liao AK, Loch JA, Lok M, Luo S, Mammen RM, Martin JW, McCauley PG, McNitt P, Mehta P, Moon KW, Mullens JW, Newington T, Ning Z, Ling Ng B, Novo SM, O'Neill MJ, Osborne MA, Osnowski A, Ostadan O, Paraschos LL, Pickering L, Pike AC, Chris Pinkard D, Pliskin DP, Podhasky J, Quijano VJ, Raczy C, Rae VH, Rawlings SR, Chiva Rodriguez A, Roe PM, Rogers J, Rogert Bacigalupo MC, Romanov N, Romieu A, Roth RK, Rourke NJ, Ruediger ST, Rusman E, Sanches-Kuiper RM, Schenker MR, Seoane JM, Shaw RJ, Shiver MK, Short SW, Sizto NL, Sluis JP, Smith MA, Ernest Sohna Sohna J, Spence EJ, Stevens K, Sutton N, Szajkowski L, Tregidgo CL, Turcatti G, Vandevondele S, Verhovsky Y, Virk SM, Wakelin S, Walcott GC, Wang J,Worsley GJ, Yan J, Yau L, Zuerlein M, Rogers J, Mullikin JC, Hurles ME, McCooke NJ, West JS, Oaks FL, Lundberg PL, Klenerman D, Durbin R, Smith AJ (2008) Accurate whole human genome sequencing using reversible terminator chemistry. Nature 456:53–59

Bettencourt RO, Terenius O, Faye I (2002) Hemolin gene silencing by ds-RNA injected into Cecropia pupae is lethal to next generation embryos. Insect Mol Biol 11:267–271

Bravo A, Gomez I, Conde J, Munoz-Garay C, Sanchez J, Miranda R, Zhuang M, Gill SS, Soberon M (2004) Oligomerization triggers binding of a *Bacillus thuringiensis* Cry1Ab pore-forming toxin to aminopeptidase N receptor leading to insertion into membrane microdomains. Biochim Biophys Acta 1667:38–46

Bravo A, Likitvivatanavong S, Gill SS, Soberon M (2011) *Bacillus thuringiensis*: a story of a successful bioinsecticide. Insect Biochem Mol Biol 41:423–431

Bravo A, Gomez I, Porta H, Garcia-Gomez BI, Rodriguez-Almazan C, Pardo L, Soberon M (2013) Evolution of *Bacillus thuringiensis* Cry toxins insecticidal activity. Microb Biotechnol 6: 17–26

Candas M, Loseva O, Oppert B, Kosaraju P, Bulla LA Jr (2003) Insect resistance to *Bacillus thuringiensis*: alterations in the indianmeal moth larval gut proteome. Mol Cell Proteomics 2: 19–28

Casagrande RA (1987) The Colorado potato beetle: 125 years of mismanagement. Bull ESA 33:142–150

Chen J, Brown MR, Hua G, Adang MJ (2005) Comparison of the localization of *Bacillus thuringiensis* Cry1A delta-endotoxins and their binding proteins in larval midgut of tobacco hornworm, *Manduca sexta*. Cell Tissue Res 321:123–129

Chen RR, Ren XL, Han ZJ, Mu LL, Li GQ, Ma Y, Cui JJ (2014) A cadherin-like protein from the beet armyworm *Spodoptera exigua* (Lepidoptera: Noctuidae) is a putative Cry1Ac receptor. Arch Insect Biochem Physiol 86:58–71

de Maagd RA, van der Klei H, Bakker PL, Stiekema WJ, Bosch D (1996) Different domains of *Bacillus thuringiensis* delta-endotoxins can bind to insect midgut membrane proteins on ligand blots. Appl Environ Microbiol 62:2753–2757

Denolf P, Jansens S, Peferoen M, Degheele D, Van Rie J (1993) Two different *Bacillus thuringiensis* delta-endotoxin receptors in the midgut brush border membrane of the European corn borer, *Ostrinia nubilalis* (Hubner) (Lepidoptera: Pyralidae). Appl Environ Microbiol 59:1828–1837

Dorsch JA, Candas M, Griko NB, Maaty WS, Midboe EG, Vadlamudi RK, Bulla LA Jr (2002) Cry1A toxins of *Bacillus thuringiensis* bind specifically to a region adjacent to the membrane-proximal extracellular domain of BT-R(1) in *Manduca sexta*: involvement of a cadherin in the entomopathogenicity of *Bacillus thuringiensis*. Insect Biochem Mol Biol 32:1025–1036

Endo Y, Nishiitsutjiuwo J (1980) Mode of action of *Bacillus thuringiensis* δ-endotoxin. Histopathological changes in the silkworm. J Invertebr Pathol 36:90–103

Escriche B, Ferre J, Silva FJ (1997) Occurrence of a common binding site in *Mamestra brassicae*, *Phthorimaea operculella*, and *Spodoptera exigua* for the insecticidal crystal proteins CryIA from *Bacillus thuringiensis*. Insect Biochem Mol Biol 27:651–656

Fernandez LE, Aimanova KG, Gill SS, Bravo A, Soberon M (2006) A GPI-anchored alkaline phosphatase is a functional midgut receptor of Cry11Aa toxin in *Aedes aegypti* larvae. Biochem J 394:77–84

Ferre J, Van Rie J (2002) Biochemistry and genetics of insect resistance to *Bacillus thuringiensis*. Annu Rev Entomol 47:501–533

Flores-Escobar B, Rodriguez-Magadan H, Bravo A, Soberon M, Gomez I (2013) Differential role of *Manduca sexta* aminopeptidase-N and alkaline phosphatase in the mode of action of Cry1Aa, Cry1Ab, and Cry1Ac toxins from *Bacillus thuringiensis*. Appl Environ Microbiol 79:4543–4550

Francis BR, Bulla LA Jr (1997) Further characterization of BT-R1, the cadherin-like receptor for Cry1Ab toxin in tobacco hornworm (*Manduca sexta*) midguts. Insect Biochem Mol Biol 27: 541–550

Gahan LJ, Gould F, Heckel DG (2001) Identification of a gene associated with Bt resistance in *Heliothis virescens*. Science 293:857–860

Gai Z, Zhang X, Wang X, Peng J, Li Y, Liu K, Hong H (2013) Differential proteomic analysis of *Trichoplusia ni* cells after continuous selection with activated Cry1Ac toxin. Cytotechnology 65:425–435

Garber M, Grabherr MG, Guttman M, Trapnell C (2011) Computational methods for transcriptome annotation and quantification using RNA-seq. Nat Methods 8:469–477

Garczynski SF, Crim JW, Adang MJ (1991) Identification of putative insect brush border membrane-binding molecules specific to *Bacillus thuringiensis* delta-endotoxin by protein blot analysis. Appl Environ Microbiol 57:2816–2820

Gill SS, Cowles EA, Francis V (1995) Identification, isolation, and cloning of a *Bacillus thuringiensis* CryIAc toxin-binding protein from the midgut of the lepidopteran insect *Heliothis virescens*. J Biol Chem 270:27277–27282

Gordon KH, Waterhouse PM (2007) RNAi for insect-proof plants. Nat Biotechnol 25:1231–1232

Grabherr MG, Haas BJ, Yassour M, Levin JZ, Thompson DA, Amit I, Adiconis X, Fan L, Raychowdhury R, Zeng Q, Chen Z, Mauceli E, Hacohen N, Gnirke A, Rhind N, di Palma F, Birren BW, Nusbaum C, Lindblad-Toh K, Friedman N, Regev A (2011) Full-length transcriptome assembly from RNA-Seq data without a reference genome. Nat Biotechnol 29: 644–652

Grapputo A, Boman S, Lindstrom L, Lyytinen A, Mappes J (2005) The voyage of an invasive species across continents: genetic diversity of North American and European Colorado potato beetle populations. Mol Ecol 14:4207–4219

Griffitts JS, Haslam SM, Yang T, Garczynski SF, Mulloy B, Morris H, Cremer PS, Dell A, Adang MJ, Aroian RV (2005) Glycolipids as receptors for *Bacillus thuringiensis* crystal toxin. Science 307:922–925

Hakim RS, Baldwin K, Smagghe G (2010) Regulation of midgut growth, development, and metamorphosis. Annu Rev Entomol 55:593–608

He W, You M, Vasseur L, Yang G, Xie M, Cui K, Bai J, Liu C, Li X, Xu X, Huang S (2012) Developmental and insecticide-resistant insights from the de novo assembled transcriptome of the diamondback moth, *Plutella xylostella*. Genomics 99:169–177

Hofmann C, Vanderbruggen H, Hofte H, Van Rie J, Jansens S, Van Mellaert H (1988) Specificity of *Bacillus thuringiensis* delta-endotoxins is correlated with the presence of high-affinity binding sites in the brush border membrane of target insect midguts. Proc Natl Acad Sci U S A 85: 7844–7848

Hossain DM, Shitomi Y, Moriyama K, Higuchi M, Hayakawa T, Mitsui T, Sato R, Hori H (2004) Characterization of a novel plasma membrane protein, expressed in the midgut epithelia of *Bombyx mori*, that binds to Cry1A toxins. Appl Environ Microbiol 70:4604–4612

Hsu SM, Raine L, Fanger H (1981) Use of avidin-biotin-peroxidase complex (ABC) in immunoperoxidase techniques: a comparison between ABC and unlabeled antibody (PAP) procedures. J Histochem Cytochem 29:577–580

Hua G, Jurat-Fuentes JL, Adang MJ (2004) Bt-R1a extracellular cadherin repeat 12 mediates *Bacillus thuringiensis* Cry1Ab binding and cytotoxicity. J Biol Chem 279:28051–28056

Johnson DE (1994) Cellular toxicities and membrane binding characteristics of insecticidal crystal proteins from *Bacillus thuringiensis* toward cultured insect cells. J Invertebr Pathol 63: 123–129

Jurat-Fuentes JL, Adang MJ (2004) Characterization of a Cry1Ac-receptor alkaline phosphatase in susceptible and resistant *Heliothis virescens* larvae. Eur J Biochem 271:3127–3135

Jurat-Fuentes JL, Adang MJ (2007) A proteomic approach to study Cry1Ac binding proteins and their alterations in resistant *Heliothis virescens* larvae. J Invertebr Pathol 95:187–191

Kanost MR, Jiang H, Yu XQ (2004) Innate immune responses of a lepidopteran insect, *Manduca sexta*. Immunol Rev 198:97–105

Keeton TP, Bulla LA Jr (1997) Ligand specificity and affinity of BT-R1, the *Bacillus thuringiensis* Cry1A toxin receptor from *Manduca sexta*, expressed in mammalian and insect cell cultures. Appl Environ Microbiol 63:3419–3425

Khajuria C, Zhu YC, Chen MS, Buschman LL, Higgins RA, Yao J, Crespo AL, Siegfried BD, Muthukrishnan S, Zhu KY (2009) Expressed sequence tags from larval gut of the European corn borer (*Ostrinia nubilalis*): exploring candidate genes potentially involved in *Bacillus thuringiensis* toxicity and resistance. BMC Genomics 10:286

Knight PJ, Crickmore N, Ellar DJ (1994) The receptor for *Bacillus thuringiensis* CrylA(c) delta-endotoxin in the brush border membrane of the lepidopteran *Manduca sexta* is aminopeptidase N. Mol Microbiol 11:429–436

Krishnamoorthy M, Jurat-Fuentes JL, McNall RJ, Andacht T, Adang MJ (2007) Identification of novel Cry1Ac binding proteins in midgut membranes from *Heliothis virescens* using proteomic analyses. Insect Biochem Mol Biol 37:189–201

Kumar A, Congiu L, Lindstrom L, Piiroinen S, Vidotto M, Grapputo A (2014) Sequencing, de novo assembly and annotation of the Colorado Potato Beetle, *Leptinotarsa decemlineata*, transcriptome. PLoS One 9:e86012

Kwa MS, de Maagd RA, Stiekema WJ, Vlak JM, Bosch D (1998) Toxicity and binding properties of the *Bacillus thuringiensis* delta-endotoxin Cry1C to cultured insect cells. J Invertebr Pathol 71:121–127

Laemmli UK (1970) Cleavage of structural proteins during the assembly of the head of bacteriophage T4. Nature 227:680–685

Lane NJ, Harrison JB, Lee WM (1989) Changes in microvilli and Golgi-associated membranes of lepidopteran cells induced by an insecticidally active bacterial δ-endotoxin. J Cell Sci 93: 337–347

Langmead B, Trapnell C, Pop M, Salzberg SL (2009) Ultrafast and memory-efficient alignment of short DNA sequences to the human genome. Genome Biol 10:R25

Lee MK, Milne RE, Ge AZ, Dean DH (1992) Location of a *Bombyx mori* receptor binding region on a *Bacillus thuringiensis* delta-endotoxin. J Biol Chem 267:3115–3121

Lei Y, Zhu X, Xie W, Wu Q, Wang S, Guo Z, Xu B, Li X, Zhou X, Zhang Y (2014) Midgut transcriptome response to a Cry toxin in the diamondback moth, *Plutella xylostella* (Lepidoptera: Plutellidae). Gene 533(1):180–187

Levin JZ, Yassour M, Adiconis X, Nusbaum C, Thompson DA, Friedman N, Gnirke A, Regev A (2010) Comprehensive comparative analysis of strand-specific RNA sequencing methods. Nat Methods 7:709–715

Li H, Durbin R (2009) Fast and accurate short read alignment with Burrows–Wheeler transform. Bioinformatics 25:1754–1760

Li R, Li Y, Kristiansen K, Wang J (2008) SOAP: short oligonucleotide alignment program. Bioinformatics 24:713–714

Li R, Yu C, Li Y, Lam TW, Yiu SM, Kristiansen K, Wang J (2009) SOAP2: an improved ultrafast tool for short read alignment. Bioinformatics 25:1966–1967

Liu K, Zheng B, Hong H, Jiang C, Peng R, Peng J, Yu Z, Zheng J, Yang H (2004) Characterization of cultured insect cells selected by *Bacillus thuringiensis* crystal toxin. In Vitro Cell Dev Biol Anim 40:312–317

Luo K, Sangadala S, Masson L, Mazza A, Brousseau R, Adang J (1997) The *Heliothis virescens* 170 kDa aminopeptidase functions as "receptor A" by mediating specific *Bacillus thuringiensis* Cry1A delta-endotoxin binding and pore formation. Insect Biochem Mol Biol 27:735–743

Ma W, Zhang Z, Peng C, Wang X, Li F, Lin Y (2012) Exploring the midgut transcriptome and brush border membrane vesicle proteome of the rice stem borer, *Chilo suppressalis* (Walker). PLoS One 7:e38151

Maguire JJ, Kuc RE, Davenport AP (2012) Radioligand binding assays and their analysis. Methods Mol Biol 897:31–77

Mao YB, Cai WJ, Wang JW, Hong GJ, Tao XY, Wang LJ, Huang YP, Chen XY (2007) Silencing a cotton bollworm P450 monooxygenase gene by plant-mediated RNAi impairs larval tolerance of gossypol. Nat Biotechnol 25:1307–1313

Margulies MM, Egholm M, Altman WE, Attiya S, Bader JS, Bemben LA, Berka J, Braverman MS, Chen YJ, Chen Z, Dewell SB, Du L, Fierro JM, Gomes XV, Godwin BC, He W, Helgesen S, Ho CH, Irzyk GP, Jando SC, Alenquer ML, Jarvie TP, Jirage KB, Kim JB, Knight JR, Lanza JR, Leamon JH, Lefkowitz SM, Lei M, Li J, Lohman KL, Lu H, Makhijani VB, McDade KE, McKenna MP, Myers EW, Nickerson E, Nobile JR, Plant R, Puc BP, Ronan MT, Roth GT, Sarkis GJ, Simons JF, Simpson JW, Srinivasan M, Tartaro KR, Tomasz A, Vogt KA, Volkmer GA, Wang SH, Wang Y, Weiner MP, Yu P, Begley RF, Rothberg JM (2005) Genome sequencing in microfabricated high-density picolitre reactors. Nature 437:376–380

Martinez-Ramirez AC, Gonzalez-Nebauer S, Escriche B, Real MD (1994) Ligand blot identification of a *Manduca sexta* midgut binding protein specific to three *Bacillus thuringiensis* Cry1A-type ICPs. Biochem Biophys Res Commun 201:782–787

McGettigan PA (2013) Transcriptomics in the RNA-seq era. Curr Opin Chem Biol 17:4–11

McNall RJ, Adang MJ (2003) Identification of novel *Bacillus thuringiensis* Cry1Ac binding proteins in *Manduca sexta* midgut through proteomic analysis. Insect Biochem Mol Biol 33: 999–1010

Mutz KO, Heilkenbrinker A, Lonne M, Walter JG, Stahl F (2013) Transcriptome analysis using next-generation sequencing. Curr Opin Biotechnol 24:22–30

Nagalakshmi U, Wang Z, Waern K, Shou C, Raha D, Gerstein M, Snyder M (2008) The transcriptional landscape of the yeast genome defined by RNA sequencing. Science 320: 1344–1349

Nagamatsu Y, Toda S, Yamaguchi F, Ogo M, Kogure M, Nakamura M, Shibata Y, Katsumoto T (1998) Identification of *Bombyx mori* midgut receptor for *Bacillus thuringiensis* insecticidal CryIA(a) toxin. Biosci Biotechnol Biochem 62:718–726

Nagamatsu Y, Koike T, Sasaki K, Yoshimoto A, Furukawa Y (1999) The cadherin-like protein is essential to specificity determination and cytotoxic action of the *Bacillus thuringiensis* insecticidal CryIAa toxin. FEBS Lett 460:385–390

Nakane PK (1968) Simultaneous localization of multiple tissue antigens using the peroxidase-labeled antibody method: a study on pituitary glands of the rat. J Histochem Cytochem 16: 557–560

Ochoa-Campuzano C, Real MD, Martinez-Ramirez AC, Bravo A, Rausell C (2007) An ADAM metalloprotease is a Cry3Aa *Bacillus thuringiensis* toxin receptor. Biochem Biophys Res Commun 362:437–442

Ochoa-Campuzano C, Sanchez J, Garcia-Robles I, Real MD, Rausell C, Sanchez J (2012) Identification of a calmodulin-binding site within the domain I of *Bacillus thuringiensis* Cry3Aa toxin. Arch Insect Biochem Physiol 81:53–62

Oltean DI, Pullikuth AK, Lee HK, Gill SS (1999) Partial purification and characterization of *Bacillus thuringiensis* Cry1A toxin receptor A from *Heliothis virescens* and cloning of the corresponding cDNA. Appl Environ Microbiol 65:4760–4766

Oppert B, Dowd SE, Bouffard P, Li L, Conesa A, Lorenzen MD, Toutges M, Marshall J, Huestis DL, Fabrick J, Oppert C, Jurat-Fuentes JL (2012) Transcriptome profiling of the intoxication response of *Tenebrio molitor* larvae to *Bacillus thuringiensis* Cry3Aa protoxin. PLoS One 7:e34624

Pacheco S, Gomez I, Arenas I, Saab-Rincon G, Rodriguez-Almazan C, Gill SS, Bravo A, Soberon M (2009) Domain II loop 3 of *Bacillus thuringiensis* Cry1Ab toxin is involved in a "ping pong" binding mechanism with *Manduca sexta* aminopeptidase-N and cadherin receptors. J Biol Chem 284:32750–32757

Park Y, Kim Y (2013) RNA interference of cadherin gene expression in *Spodoptera exigua* reveals its significance as a specific Bt target. J Invertebr Pathol 114:285–291

Park Y, Abdullah MA, Taylor MD, Rahman K, Adang MJ (2009) Enhancement of *Bacillus thuringiensis* Cry3Aa and Cry3Bb toxicities to coleopteran larvae by a toxin-binding fragment of an insect cadherin. Appl Environ Microbiol 75:3086–3092

Pauchet Y, Wilkinson P, Vogel H, Nelson DR, Reynolds SE, Heckel DG, ffrench-Constant RH (2010) Pyrosequencing the *Manduca sexta* larval midgut transcriptome: messages for digestion, detoxification and defence. Insect Mol Biol 19:61–75

Pigott CR, Ellar DJ (2007) Role of receptors in *Bacillus thuringiensis* crystal toxin activity. Microbiol Mol Biol Rev 71:255–281

Quan GX, Kanda T, Tamura T (2002) Induction of the white egg 3 mutant phenotype by injection of the double-stranded RNA of the silkworm white gene. Insect Mol Biol 11:217–222

Rabilloud T, Lelong C (2011) Two-dimensional gel electrophoresis in proteomics: a tutorial. J Proteomics 74:1829–1841

Rajagopal R, Sivakumar S, Agrawal N, Malhotra P, Bhatnagar RK (2002) Silencing of midgut aminopeptidase N of *Spodoptera litura* by double-stranded RNA establishes its role as *Bacillus thuringiensis* toxin receptor. J Biol Chem 277:46849–46851

Rausell C, De Decker N, Garcia-Robles I, Escriche B, Van Kerkhove E, Real MD, Martinez-Ramirez AC (2000) Effect of *Bacillus thuringiensis* toxins on the midgut of the nun moth *Lymantria monacha*. J Invertebr Pathol 75:288–291

Ren XL, Chen RR, Zhang Y, Ma Y, Cui JJ, Han ZJ, Mu LL, Li GQ (2013) A *Spodoptera exigua* cadherin serves as a putative receptor for *Bacillus thuringiensis* Cry1Ca toxin and shows differential enhancement of Cry1Ca and Cry1Ac toxicity. Appl Environ Microbiol 79: 5576–5583

Ren XL, Ma Y, Cui JJ, Li GQ (2014) RNA interference-mediated knockdown of three putative aminopeptidases N affects susceptibility of *Spodoptera exigua* larvae to *Bacillus thuringiensis* Cry1Ca. J Insect Physiol 67C:28–36

Rothberg JM, Hinz W, Rearick TM, Schultz J, Mileski W, Davey M, Leamon JH, Johnson K, Milgrew MJ, Edwards M, Hoon J, Simons JF, Marran D, Myers JW, Davidson JF, Branting A, Nobile JR, Puc BP, Light D, Clark TA, Huber M, Branciforte JT, Stoner IB, Cawley SE, Lyons M, Fu Y, Homer N, Sedova M, Miao X, Reed B, Sabina J, Feierstein E, Schorn M, Alanjary M, Dimalanta E, Dressman D, Kasinskas R, Sokolsky T, Fidanza JA, Namsaraev E, McKernan KJ, Williams A, Roth GT, Bustillo J (2011) An integrated semiconductor device enabling non-optical genome sequencing. Nature 475:348–352

Sangadala S, Walters FS, English LH, Adang MJ (1994) A mixture of *Manduca sexta* amino-peptidase and phosphatase enhances *Bacillus thuringiensis* insecticidal CryIA(c) toxin binding and 86Rb(+)-K+ efflux in vitro. J Biol Chem 269:10088–10092

Schena M, Shalon D, Davis RW, Brown PO (1995) Quantitative monitoring of gene expression patterns with a complementary DNA microarray. Science 270:467–470

Schroeder A, Mueller O, Stocker S, Salowsky R, Leiber M, Gassmann M, Lightfoot S, Menzel W, Granzow M, Ragg T (2006) The RIN: an RNA integrity number for assigning integrity values to RNA measurements. BMC Mol Biol 7:3

Shields VD, Hildebrand JG (2001) Recent advances in insect olfaction, specifically regarding the morphology and sensory physiology of antennal sensilla of the female sphinx moth *Manduca sexta*. Microsc Res Tech 55:307–329

Simpson R, Burgess EPJ, Markwick NP (1997) *Bacillus thuringiensis* delta-endotoxin binding sites in two Lepidoptera, *Wiseana* spp. and *Epiphyas postvittana*. J Invertebr Pathol 70: 136–142

Sivakumar S, Rajagopal R, Venkatesh GR, Srivastava A, Bhatnagar RK (2007) Knockdown of aminopeptidase-N from *Helicoverpa armigera* larvae and in transfected Sf21 cells by RNA interference reveals its functional interaction with *Bacillus thuringiensis* insecticidal protein Cry1Ac. J Biol Chem 282:7312–7319

Soberon M, Gill SS, Bravo A (2009) Signaling versus punching hole: how do *Bacillus thuringiensis* toxins kill insect midgut cells? Cell Mol Life Sci 66:1337–1349

Soberon M, Pardo L, Munoz-Garay C, Sanchez J, Gomez I, Porta H, Bravo A (2010) Pore formation by Cry toxins. Adv Exp Med Biol 677:127–142

Talekar NS, Shelton AM (1993) Biology, ecology, and management of the diamondback moth. Annu Rev Entomol 38:275–301

Terenius O, Papanicolaou A, Garbutt JS, Eleftherianos I, Huvenne H, Kanginakudru S, Albrechtsen M, An C, Aymeric JL, Barthel A, Bebas P, Bitra K, Bravo A, Chevalier F, Collinge DP, Crava CM, de Maagd RA, Duvic B, Erlandson M, Faye I, Felfoldi G, Fujiwara H, Futahashi R, Gandhe AS, Gatehouse HS, Gatehouse LN, Giebultowicz JM, Gomez I, Grimmelikhuijzen CJ, Groot AT, Hauser F, Heckel DG, Hegedus DD, Hrycaj S, Huang L, Hull JJ, Iatrou K, Iga M, Kanost MR, Kotwica J, Li C, Li J, Liu J, Lundmark M, Matsumoto S, Meyering-Vos M, Millichap PJ, Monteiro A, Mrinal N, Niimi T, Nowara D, Ohnishi A, Oostra V, Ozaki K, Papakonstantinou M, Popadic A, Rajam MV, Saenko S, Simpson RM, Soberon M, Strand MR, Tomita S, Toprak U, Wang P, Wee CW, Whyard S, Zhang W, Nagaraju J, Ffrench-Constant RH, Herrero S, Gordon K, Swevers L, Smagghe G (2011) RNA interference in Lepidoptera: an overview of successful and unsuccessful studies and implications for experimental design. J Insect Physiol 57:231–245

Tetreau G, Bayyareddy K, Jones CM, Stalinski R, Riaz MA, Paris M, David JP, Adang MJ, Despres L (2012) Larval midgut modifications associated with Bti resistance in the yellow fever mosquito using proteomic and transcriptomic approaches. BMC Genomics 13:248

Tian H, Peng H, Yao Q, Chen H, Xie Q, Tang B, Zhang W (2009) Developmental control of a lepidopteran pest *Spodoptera exigua* by ingestion of bacteria expressing dsRNA of a non-midgut gene. PLoS One 4:e6225

Vachon V, Paradis MJ, Marsolais M, Schwartz JL, Laprade R (1995) Ionic permeabilities induced by *Bacillus thuringiensis* in Sf9 cells. J Membr Biol 148:57–63

Vadlamudi RK, Ji TH, Bulla LA Jr (1993) A specific binding protein from *Manduca sexta* for the insecticidal toxin of *Bacillus thuringiensis* subsp. berliner. J Biol Chem 268:12334–12340

Vadlamudi RK, Weber E, Ji I, Ji TH, Bulla LA Jr (1995) Cloning and expression of a receptor for an insecticidal toxin of *Bacillus thuringiensis*. J Biol Chem 270:5490–5494

Valaitis AP, Jenkins JL, Lee MK, Dean DH, Garner KJ (2001) Isolation and partial character-ization of gypsy moth BTR-270, an anionic brush border membrane glycoconjugate that binds *Bacillus thuringiensis* Cry1A toxins with high affinity. Arch Insect Biochem Physiol 46: 186–200

Van Rie J (2000) *Bacillus thuringiensis* and its use in transgenic insect control technologies. Int J Med Microbiol 290:463–469

Van Rie J, Jansens S, Hofte H, Degheele D, Van Mellaert H (1989) Specificity of *Bacillus thuringiensis* delta-endotoxins. Importance of specific receptors on the brush border membrane of the mid-gut of target insects. Eur J Biochem 186:239–247

Whyard S, Singh AD, Wong S (2009) Ingested double-stranded RNAs can act as species-specific insecticides. Insect Biochem Mol Biol 39:824–832

Yang Y, Zhu YC, Ottea J, Husseneder C, Leonard BR, Abel C, Huang F (2010) Molecular characterization and RNA interference of three midgut aminopeptidase N isozymes from *Bacillus thuringiensis*-susceptible and -resistant strains of sugarcane borer, *Diatraea saccharalis*. Insect Biochem Mol Biol 40:592–603

Zhuang M, Oltean DI, Gomez I, Pullikuth AK, Soberon M, Bravo A, Gill SS (2002) *Heliothis virescens* and *Manduca sexta* lipid rafts are involved in Cry1A toxin binding to the midgut epithelium and subsequent pore formation. J Biol Chem 277:13863–13872

Zou Z, Najar F, Wang Y, Roe B, Jiang H (2008) Pyrosequence analysis of expressed sequence tags for *Manduca sexta* hemolymph proteins involved in immune responses. Insect Biochem Mol Biol 38:677–682

Zuniga-Navarrete F, Gomez I, Pena G, Bravo A, Soberon M (2013) A *Tenebrio molitor* GPI-anchored alkaline phosphatase is involved in binding of *Bacillus thuringiensis* Cry3Aa to brush border membrane vesicles. Peptides 41:81–86

Chapter 5
Non-Bt Soil Microbe-Derived Insecticidal Proteins

Leela Alamalakala, Srinivas Parimi, Sandip Dangat, and Bharat R. Char

5.1 Introduction

Although soil microbes have contributed immensely to world agriculture through their potential biochemical attributes in the areas of insect pest management, crop nutrient enhancement [increasing nutrient availability by plant growth-promoting rhizobacteria (PGPR), phosphate-solubilizing bacteria (PSB)], and soil fertility management [through metal detoxification, mycorrhizal-helping bacteria (MHB), and arbuscular mycorrhizal fungi (AMF)] (Khan 2005), innovations in biotechnology have opened new vistas for enhancing the contribution of the soil microbial diversity to agricultural productivity. For many decades, pest control programs in agriculture and public health have relied heavily on the use of broad-spectrum chemical insecticides. However, the use of chemical insecticides came under scrutiny since the early 1960s when the environmental classic *Silent Spring* was published (Carson 1962), leading to a paradigm shift in insect pest management strategies and emphasizing the need to identify non-chemical pest control strategies that are insect-specific and environmentally safe. Naturally occurring microbial entomopathogens, such as bacteria, fungi, viruses, and nematodes, are effective non-chemical alternatives for the suppression and management of insect pests causing economic losses in different crops (Lacey et al. 2001; Lacey and Kaya 2007; Shahid et al. 2012). Characterization of the genes and genomes of these entomopathogens has facilitated the identification and deployment of novel insecticidal genes for crop protection. Of all entomopathogens, bacteria have been the most extensively used organisms to date, and overwhelming commercial success was achieved with *Bacillus thuringiensis* (*Bt*) (Firmicutes: *Bacillaceae*) toxins.

Lepidoptera, the second largest insect order, comprised of moths and butterflies, represents a diverse and important group of insect pests that affect commercial

L. Alamalakala (✉) • S. Parimi • S. Dangat • B.R. Char
Maharashtra Hybrid Seeds Company Limited, Dawalwadi, PO Box 76, Jalna 431203, India
e-mail: Leela.Alamalakala@mahyco.com

© Springer International Publishing Switzerland 2015   89
K.S. Sree, A. Varma (eds.), *Biocontrol of Lepidopteran Pests*, Soil Biology 43,
DOI 10.1007/978-3-319-14499-3_5

agriculture, causing widespread economic damage on food and fiber crop plants, fruit trees, forests, and stored grains. The larval stage of the moths is detrimental to an array of economically valuable crops including cotton, tobacco, tomato, corn, sorghum, pulses, and wheat (Srinivasan et al. 2006). Some examples of lepidopteran pests include: the cotton bollworms, *Helicoverpa armigera* (Hübner) and *Helicoverpa zea* (Boddie) (Lepidoptera: Noctuidae); the gypsy moth, *Lymantria dispar* (Linnaeus) (Lepidoptera: Lymantriidae), a voracious defoliator of Palaearctic and Nearctic forests (Reineke et al. 1999); the diamondback moth (DBM), *Plutella xylostella* (Linnaeus) (Lepidoptera: Plutellidae), a pest of cole crops (Talekar and Shelton 1993); brinjal shoot and fruit borer, *Leucinodes orbonalis* (Guenée) (Lepidoptera: Crambidae); and okra shoot and fruit borer, *Earias vittella* (Fabricius) (Lepidoptera: Noctuidae) which causes 69 % loss in marketable yield in okra (Radake and Undirwade 1981). Until recently, the control of pests in agriculture has mostly relied on the intensive application of broad-spectrum synthetic insecticides, with about 40 % targeted to the control of lepidopteran insects (Brooke and Hines 1999). Over the years the application of insecticides has led to the development of insecticide-resistant insects, destruction of natural enemies as well as harmful effects on humans and the environment. Therefore an urgent need was felt for alternative control strategies that reduce dependence on conventional insecticides. The interest in biopesticides started growing significantly, as a result of the withdrawals of many synthetic pesticides and the high cost of developing new ones. In this scenario, it is important to note that the global pesticide market is growing at a compound annual growth rate (CAGR) of 3.6 % and the value is expected to reach $51 billion in 2014. The biopesticide segment which represents a strong growth arena in the global pesticide market is expected to grow at a 15.6 % CAGR from $1.6 billion to $3.3 billion in 2014 (BCC Research 2010; Ruiu et al. 2013). Thus, the direct application of entomopathogens as biological control agents or deploying GM crops developed using novel entomotoxic proteins provides a good market opportunity that can be captured by the industry. Agriculture was perceived to benefit from futuristic eco-friendly strategies such as the use of natural enemies, autocidal control methods such as sterile insect technique (SIT) and F1 sterility, and transgenic plants expressing entomotoxic proteins (Fitt 1994; Gatehouse et al. 1994; Haq et al. 2004; Saour 2014).

B. thuringiensis (Bt) is a gram-positive bacterium that is found in a variety of ecological niches such as soil, water, plant surfaces, stored cereals and dead insects (Federici and Siegel 2008). The bacteria form spores containing proteinaceous crystals known as Cry or Cyt proteins (also known as δ-endotoxins), as well as VIPs (vegetative insecticidal proteins) exhibiting potent insecticidal activity (Sanahuja et al. 2011). Different strains of *Bt* produce different types of insect-toxic virulence factors, and the activity of these virulence factors toward Lepidoptera, Diptera, Coleoptera, Hymenoptera, Homoptera, Orthoptera, and Mallophaga insect orders have been reported (Schnepf et al. 1998). Bt was released as a biopesticide (ICP and viable spores) since 1951 (Steinhaus 1951) and formulations based on Bt (67 registered products and more than 450 formulations) occupied the key position accounting for nearly 90 % of the total biopesticide sales worldwide (Neale 1997). Bt products used for managing lepidopteran pests were primarily derived from Bt *Kurstaki* HD-1 strain (e.g., Biobit, Dipel, and Thuricide) and to a lesser extent from *Kurstaki* SA-11 and *Kurstaki* SA-12 strains (Kaur 2000). However, Bt had limited use as a foliar insecticide due to the short window of effectiveness as a result of which multiple sprays had to be undertaken which led to increase in the amount of product for application and fuel needed for spraying. The sprays also had little impact on cryptic pests (Sanchis 2011). These inherent limitations of topical Bt pesticides were overcome by introducing Bt *cry* genes into target crops thereby enhancing plant health and conferring plant protection (Sanahuja et al. 2011).

Transgenic crops protected from insect pests have become an integral part of insect pest management (IPM) with over 58 Mha planted worldwide in 2010 (James 2010; Baum et al. 2012). A number of transgenic crops including corn, cotton, rice and soybean harboring *Bt* genes are cultivated commercially since 1996 (Huang et al. 2007; Sanahuja et al. 2011). *Bt* cotton, in particular, has provided effective control of several lepidopteran pest species including tobacco budworm, *Heliothis virescens* (Fabricius) (Lepidoptera: Noctuidae), pink bollworm, *Pectinophora gossypiella* (Saunders) (Lepidoptera: Gelechiidae), and the cotton bollworm, *H. armigera* (Hübner) (Lepidoptera: Noctuidae), resulting in increased yield, reduced frequency of insecticide applications, and area-wide suppression of the same primary insect pest in other crops (Perlak et al. 2001; Carrière et al. 2003; Jackson et al. 2003; Wu et al. 2008; James 2010). The widespread cultivation of Bt cotton varieties can contribute to a resurgence in beneficial arthropod populations necessary for successful IPM (Head et al. 2005; Naranjo 2005; Whitehouse et al. 2005). Although Bt-derived biopesticides used as foliar sprays or expressed in plants through genetic engineering are environmentally safe and effective, their use is still restricted due to problems of limited host range and the potential for the development of resistance and cross resistance in key pests due to continuous use, thus necessitating the discovery of novel insecticidal genes with improved activity and host range. Field evolved resistance to Bt crops has been reported for populations of several insect pests (Gassmann et al. 2011).

Insecticidal nematodes in the genera *Steinernema* and *Heterorhabditis*, each carrying a specific genus of bacteria, are the only insect-parasitic nematodes possessing an optimal balance of biological control attributes (Poinar 1979; Bedding et al. 1993), and they have been used for the biological control of soil-dwelling pests that include weevils and lepidopteran insects (Wang and Li 1987; Klein 1990). However, factors such as cost, shelf life, handling, mixing, compatibility, and profit margins to manufacturers and distributors have contributed to the failure of entomopathogenic nematodes (EPN) especially for large-scale agriculture applications, as they failed to penetrate many markets or gain significant market share in the current markets (Lacey and Georgis 2012). The widespread adoption and success of Bt crops and the associated risk of resistance development to Bt protein(s) have stimulated the research and development of more environmentally responsible alternatives. Thus, a significant research need was centered toward the characterization of insecticidal proteins, called the toxin complex (Tc) proteins and other virulence factors from bacteria that are symbionts of EPNs as they can be used for the development of a new generation of GM crops that are protected against a wider spectrum of insect pests (Bowen and Ensign 1998).

Sequencing, annotation, and screening of the genomes of entomopathogenic bacteria such as *Photorhabdus*, *Xenorhabdus*, and *Pseudomonas* spp. have begun to reveal previously unidentified insecticidal toxins. The presence of orthologues of a multitude of insecticidal genes (*tc*, *mcf1* and *mcf2*, *xaxAB*) in different species of insect-pathogenic bacteria indicates that a large amount of genetic transfer occurs between these species, presumably within a shared environmental niche (Hinchliffe et al. 2010), thus providing an excellent source of novel candidates, which can be used as potential alternatives to the insecticidal proteins derived from *B. thuringiensis*. The toxin complex proteins from *Photorhabdus luminescens* have also been transferred into plants and tested for their activity on different insect pests. The intent of this book chapter is to provide a comprehensive review of virulence factors produced by gram-negative entomopathogenic bacteria that demonstrate potential toxicity toward lepidopteran pests so that they could be successfully exploited for plant protection.

5.2 A Vast Arsenal of Insecticidal Toxins Derived from Gram-Negative Bacteria

Although deployment of crops expressing insecticidal proteins has led to effective control of insect pests and reduced the use of chemicals for insect control, very few candidate proteins have been commercially used for crop protection. These proteins typically control limited ranges of pest species and are predominantly from gram-positive bacteria and derived mostly from *B. thuringiensis*. Entomopathogenic gram-negative bacteria also produce toxins that are harmful to insects and can be used to augment the list of genes used in developing pest control products. Members in the Enterobacteriaceae such as *Photorhabdus*, *Xenorhabdus*, *Serratia*, *Pseudomonas*, and *Yersinia* spp. produce insecticidal toxins with toxicity similar to that of Bt toxins. Initial studies showed that these bacteria were highly effective when fed to a range of pest species belonging to at least three orders of insects including the Lepidoptera. Based on their targeted tissue, these toxins can be categorized into three types: (a) cytotoxins, (b) digestive toxins, and (c) neurotoxins (Castagnola and Stock 2014). This section describes the virulence factors associated with different gram-negative bacteria, their activity spectrum, and mode of action.

5.2.1 Insect Virulence Factors Produced by Photorhabdus Species

The *Photorhabdus* genus currently consists of three species: *P. luminescens*, *Photorhabdus temperata*, and *Photorhabdus asymbiotica*, found in a symbiotic association with an insect-pathogenic soil nematode of the genus *Heterorhabditis*. Several subspecies are recognized (Fischer-Le Saux et al. 1999). The genus *Photorhabdus* and the two species *P. luminescens* and *P. temperata* have been the subject of intensive study by entomologists and agricultural scientists in view of their insect pathogenicity and their potential for the development of novel biopesticides and insect-resistant transgenic plants. *P. asymbiotica* is a human pathogen that has been recovered from human clinical specimens from the USA and Australia and is currently considered an emerging human pathogen model system (Gerrard et al. 2004; Gerrard et al. 2006).

Photorhabdus are gram-negative, bioluminescent, motile bacteria of the family Enterobacteriaceae which live in an obligate mutualistic association with insect-parasitic *Heterorhabditis* nematodes, which invade and kill insects in the soil (Waterfield et al. 2009). The infective juvenile (IJ) nematode exists as a free-living, non-feeding individual in the soil and actively seeks out and colonizes the insect prey in the soil. The IJ nematode enters a potential victim either through respiratory spiracles, the mouth, or the anus. The symbiotic bacteria vectored by these IJs are then regurgitated from the nematode intestine into the open circulatory system of the insect prey. The bacteria colonize the anterior midgut of the insect initially, undergo rapid multiplication, and subsequently kill the insect within 1–2 days. As the bacterial population reaches a high level, the insect cadaver becomes red in color and visibly bioluminescent (Bowen and Ensign 1998). During the growth in the insect prey, the bacteria release a plethora of virulence factors to kill the insect and produce antibiotics with antifungal and antibacterial activities that probably prevent the invasion of the cadaver by other microorganisms resulting in ideal conditions for the growth and reproduction of the nematode (Paul et al. 1981; Akhurst 1982).

P. luminescens appears to encode numerous putative (ffrench-Constant et al. 2000) and proven (Bowen et al. 1998; Waterfield et al. 2001) insect virulence factors in its genome. The *Photorhabdus* genome is organized into genomic islands relating both to pathogenicity and to symbiosis. Genomic islands involved in pathogenicity are called "pathogenicity islands" (PAIs). PAIs are unstable regions that are present in the pathogen but absent from non-pathogens (Hacker and Kaper 2000) and are often inserted next to tRNA genes and have differing GC content from the rest of the genome (Waterfield et al. 2002; ffrench-Constant et al. 2003). Functional analysis of genomic islands facilitated the identification of a diversity of anti-invertebrate virulence factors from *P. luminescens* (Daborn et al. 2002; Waterfield et al. 2002). Multiple copies of "toxin complex" (*tc*) genes (Bowen et al. 1998; Waterfield et al. 2001) inserted at an AspV tRNA were detected in the first unique island. The *tc* genes encode high molecular weight, multi-subunit, orally active insecticidal toxins first characterized in insect pathogens *Photorhabdus* and *Xenorhabdus* spp. (Bowen et al. 1998; Waterfield et al. 2001), but now seen in a range of pathogens, including those of humans. Some Tc's have demonstrated oral toxicity to insects making them potential candidates for insect pest control. A second island inserted at a Phe tRNA was found to encode the novel toxin "makes caterpillars floppy" or Mcf, a large toxin with little similarity to known proteins (Daborn et al. 2002). A third island with a skewed GC content contains a gene encoding a cytotoxic necrotizing factor (CNF)-like toxin, designated *Pnf*. The specific role of *Pnf* in *Photorhabdus* is unknown (Buetow et al. 2001; Waterfield et al. 2002). Two copies of a macrophage-toxin-like encoding gene similar to that found in pathogenic strains of *Escherichia coli* are carried by a fourth island that also contains an *rhs* element and a CP4-like integrase gene. This island is linked to the *phlAB* hemolysin locus and, probably forms a part of larger region involved in pathogenicity. Lastly, a fifth island encodes a type III secretion system (TTSS), and the order of genes in the TTSS island is similar to that in *Yersinia pestis*, and these genes are probably important in the interaction of *Photorhabdus* with its invertebrate hosts (ffrench-Constant et al. 2000; Waterfield et al. 2002; Silva et al. 2002).

5.2.1.1 The Toxin Complexes (Tc's) of *Photorhabdus*

The Tc proteins produced by *P. luminescens* are an important class of secreted toxins, with an estimated molecular weight of 1,000,000 and with no detectable protease, phospholipase, or hemolytic activity but showing a trace lipase activity. The Tc is a large, multimeric complex comprising of several protein subunits ranging in size from 30 to 200 kDa, some of which are found to be lethal when fed to or injected into the hemolymph of *Manduca sexta* larvae and several other insect species (Bowen and Ensign 1998). Purification of the active protein complex revealed the presence of four distinct protein "toxin complexes" which were termed Tca, Tcb, Tcc, and Tcd, and the genes corresponding to these Tc proteins were cloned from strain W14 (Bowen et al. 1998). The different Tc's are encoded at discrete PAIs in the *Photorhabdus* genome where multiple *tc* gene copies are found (Wilkinson et al. 2009). Although all Tc proteins show injectable toxicity to *M. sexta*, majority of the oral toxicity of *tc* genes toward lepidopteran insect pests was found to be mediated by *tca* and *tcd* genes as shown via gene knockout studies (Bowen et al. 1998). Sequence analysis of the *tca*, *tcb*, *tcc* and *tcd* loci revealed a high degree of similarity between loci, and despite the apparent complexity of the loci, the individual genes within these loci could be grouped into three basic types of genetic elements: the *tcdA*-like or [A], the *tcaC*-like or [B], and the *tccC*-like or [C] (ffrench-Constant and Waterfield 2005). These groupings suggest similar roles of their encoded proteins within the assembled toxin complex, and a representative of [A], [B], and [C] was required for full toxicity (Waterfield et al. 2005a). The *tca* locus of *P. luminescens* W14 consists of three open reading frames (ORFs), *tcaA*, *tcaB*, and *tcaC* with the [A] subunit of the Tc encoded by *tcaA* and *tcaB*, the

[B] subunit encoded by *tcaC*, and the [C] subunit encoded by *tccC* gene from the *tcc* locus. A fourth ORF the *tcaZ* associated with the *tca* locus is encoded in the opposite orientation, and the function of this protein is not yet known. The *tcb* locus consists of a single [A] gene, *tcbA* and the *tcc* locus consists of an [A] encoded by *tccA* and *tccB*, and a [C] encoded by *tccC*. The *tcd* locus is the largest of the four *tc* loci and consists of four [A] genes, *tcdA1-A4*; two [B] genes, *tcdB1–tcdB2*; and four [C] genes, *tccC2–tccC5* (Hinchliffe et al. 2010). Orthologues of *tc* encoding genes are widespread in gram-negative bacteria (*Xenorhabdus*, *Serratia*, and *Yersinia*) and also present in some gram-positive bacteria (*Paenibacillus*). The association of these loci with transposase-like or bacteriophage-like genes, indicates that they are highly mobile and can be transferred between species (Hinchliffe et al. 2010).

The role of proteins encoded by the *tc loci* in *Photorhabdus* biology is ambiguous despite efforts undertaken to understand the structure, function, genetics, and mode of action of these proteins. The Tc's of *Photorhabdus* spp. appear to be very undiscriminating in their activity, with demonstrable toxicity toward a wide spectrum of insect species (Hinchliffe et al. 2010); therefore, their use as candidates for crop protection may be limited unless their effects on non-target organisms (NTO), especially the beneficial arthropods are investigated. The insecticidal *tc's* have been shown to be preferentially expressed at low temperatures (<15 °C). Analysis of the expression and insecticidal activity of the protein subunits of the *P. luminescens* W14 *tcd* locus revealed that the [A] subunit itself possessed a low level of toxicity which is potentiated by [BC], a complex formed by [B] and [C] when expressed together. The [BC] complex demonstrated mild oral toxicity toward *M. sexta*, whilst the [B] and [C] subunits individually were not orally toxic. The [C] subunit appeared to play an important role in the complex as its presence was observed to be necessary for oral toxicity of Tca and Tcd. A thorough understanding of specific protein interactions is therefore very crucial for these proteins to be successfully used for insect pest control.

The mechanism of action of the Tc's is not well understood. Although the toxicity of the Tc proteins has been demonstrated on specific model insects and cultured cells, comprehensive information on how these effects are mediated by the proteins is currently unavailable. The ingestion of purified *P. luminescens* Tca by *M. sexta* led to the complete destruction of the midgut epithelium leading to cessation of feeding and eventual starvation of the insect host. Tca also showed characteristic, midgut-specific histopathology in *M. sexta* that included the apical swelling of the columnar cells in the epithelium of the anterior midgut and blebbing of the vesicles into the gut lumen (Blackburn et al. 1998). No pathological effects were observed on any other tissues indicating the gut specificity of the toxin. Liu et al. (2006) reported that a *P. luminescens* Tca-like toxin (PL toxin) caused channel formation in the midguts and permeabilized unilamellar lipid vesicles of *M. sexta* in a pH-dependent manner. However, structural studies of XptA1 indicated that the protein binds to the brush border membrane vesicles but does not form pores in the membrane. It was therefore hypothesized that the [BC] probably aids in the insertion of the [A] tetramer into the membrane (Hinchliffe et al. 2010). Toxin A protein (283 kDa) was expressed in *Arabidopsis thaliana* using a synthetic plant-codon-optimized variant of *tcd*A, and the insecticidal efficacy of the protein was tested for control of feeding insects (Liu et al. 2003). Transgenic plants expressing more than 700 ng/mg of extractable protein were found to be highly toxic to *M. sexta*, and the toxin A purified from transgenic plants had a strong inhibitory effect on the growth of southern corn rootworm. In the best transgenic *Arabidopsis* line, high toxin A expression and insect resistance were found to be consistent for at least five generations in all progeny (Liu et al. 2003). These results indicate that the tc proteins from *Photorhabdus* may open a new route to transgenic pest control.

5.2.1.2 *Photorhabdus* Insect-Related Binary Toxins, PirA/B

Two genetic loci (*plu4093–plu4092 and plu4437–plu4436*) sharing significant sequence similarity with a putative juvenile hormone esterase (JHEs) of *Leptinotarsa decemlineata* (Vermunt et al. 1997; Duchaud et al. 2003) were identified from the genome sequence of Pl TT01. JHEs regulate metamorphosis by inactivating the juvenile hormones involved in maintaining the insect in a larval state. The inappropriate activation of the insect endocrine machinery by JHE-like proteins may therefore be an effective strategy for insect control (Bonning and Hammock 1996). The *plu4093–plu4092* and *plu4437–plu4436* genetic loci were renamed as "*Photorhabdus* insect-related" (Pir) proteins, with PirA referring to products of *plu4093/4437* homologs and PirB to products of *plu4092/4436* homologs (Waterfield et al. 2005b). Pir proteins are binary toxins having both injectable (Waterfield et al. 2005b) and oral toxicity (Blackburn et al. 2006) toward insects from the orders Diptera and Lepidoptera. Waterfield et al. (2005b) demonstrated that each of the genes in the *P. luminescens* loci was required for toxicity when injected into larvae of *Galleria mellonella* L. (Lepidoptera: Pyralidae), but the combination was not sufficient to cause mortality in *M. sexta* L. (Lepidoptera: Sphingidae) by either injection or oral administration. The oral activity of Pir A/B tested against the diamondback moth, *P. xylostella*, demonstrated that the midgut is the primary site of action. The pathology observed was similar to those seen with other gut-active toxins, but consistent effects were noticed in the posterior midgut. *P. xylostella* was found to be 300-fold more susceptible to Pir toxins than other insect species tested (Blackburn et al. 2006). However, these proteins had no effect on the growth or mortality of *H. virescens* F. (Lepidoptera: Noctuidae), *M. sexta* L. (Lepidoptera: Sphingidae), or *L. dispar* L. (Lepidoptera: Lymantriidae) larvae on oral delivery (Blackburn et al. 2006). Based on insect bioassays, it can be concluded that the Pir A/B proteins may not be broadly useful as insecticidal proteins. Although the PirB protein shares some sequence similarities with the δ-endotoxins from *B. thuringiensis*, no significant difference was observed in the responses of the susceptible *P. xylostella* larvae (lab colony) and the commercially available Cry 1A-resistant strain to the PirB protein. The PirA/B proteins have been shown to lack the esterase activity and evidence presented by Crosland et al. (2005) suggested that these proteins are related to leptinotarsin, a neurotoxic protein present in the hemolymph of several *Leptinotarsa* species (Hsiao and Fraenkel 1969). Consequently the mechanism for the potential insecticidal activity appears to be the destructive effects on the neural tissue upon injection (Castagnola and Stock 2014).

5.2.1.3 The Makes Caterpillars Floppy (Mcf1 and Mcf2) Toxins

The Mcf toxin is a high molecular weight protein (324 kDa) that was found to facilitate the persistence of *E. coli* expressing this gene within the insect host and kill the insects (Daborn et al. 2002). The predicted amino acid sequence of the 8.8-Kb *mcf1* gene fragment cloned from *P. luminescens* subsp. *akhurstii* strain W14 showed only partial homology to known proteins; however, it carried a BH3 domain, a domain found in pro-apoptotic proteins (Budd 2001). The Mcf toxins are potent toxins that are active upon injection and induce apoptosis via the mitochondrial pathway in insect phagocytes, helping the bacteria avoid phagocytosis. These proteins destroy the columnar and goblet cells of the insect midgut epithelium, causing the caterpillar to lose body turgor due to impaired osmoregulation, and become "floppy" (Daborn et al. 2002). This toxin also promotes apoptosis in mammalian tissue culture cells (Dowling et al. 2004). A second Mcf1-like ORF (*mcf2*) which also caused loss of body turgor when injected into *M. sexta* larvae was identified during end sequencing of Pl W14 cosmid library (Waterfield et al. 2003). The two mcf proteins are 77.5 % identical across the majority of their lengths only differing in their N-terminal regions (Hinchliffe et al. 2010). While Mcf1 contained a long 900 amino acid N-terminal region with no similarity to other proteins in the database, a shorter 300 amino acid N-terminal region (a HopA1-like region) containing a domain showing similarity to several type III secreted proteins was found in Mcf2. The region of similarity between Mcf1 and Mcf2 contains a BH3-like domain, two domains found in RTX-like toxins, and a large domain found in the *Clostridium difficile* binary toxins (Hinchliffe et al. 2010). Comparisons of available sequence data from different *Photorhabdus* strains revealed that copies of *mcf1* are always present, suggesting that it may be the dominant insect-killing toxin. However, it is very likely that *mcf2* is only present in a subset of strains. Toxins like Mcf which act on both the gut and insect immune system represent a promising, yet underexploited avenue for future insecticide development (Daborn et al. 2002).

5.2.1.4 Txp40 Toxin

A novel 42 kDa secreted protein encoded by the toxin gene *txp40* ($txp40_{V16}$, identified from *P. luminescens* strain V16) and initially identified in *Xenorhabdus nematophila* (A24tox, $txp40_{A24}$) was found to be part of a genomic island involved in pathogenicity and highly conserved and widespread among the *Photorhabdus* strains (Brown et al. 2006). Txp40 protein was found to have hemolymph toxicity and was effective against a range of lepidopteran species (*G. mellonella*, *H. armigera*, and *Plodia interpunctella*) and the dipteran species *Lucilia cuprina*. The protein exhibited significant cytotoxicity in vitro against two dipteran cell lines

(*Aedes aegypti* and *Drosophila melanogaster* cell line S2) and two lepidopteran cell lines (*Spodoptera* cell lines Sf9 and Sf21), but not against a mammalian cell line. The broad insecticidal activity of the Txp40 toxin suggests that the toxin has a target that is common to many different insects (Brown et al. 2006). Gut histology studies of *H. armigera* showed that the midgut and fat body are the targets and the toxin caused a significant decrease in midgut intercellular adhesion, degradation of the peritrophic matrix lining of the midgut cells, and degradation of the fat body nuclei. Although the selective toxicity of Txp40 against a broad spectrum of lepidopteran insect pests, and the lack of toxicity against mammalian cell lines makes it a good candidate for pest control, the protein has to retain its insect-toxic properties (it should not be degraded by the insect gut enzymatic machinery) upon oral delivery for it to be a potential option for developing transgenic crops.

5.2.1.5   An Array of Other Insecticidal Proteins from *Photorhabdus*

Photorhabdus Insecticidal Toxin

Photorhabdus insecticidal toxin (Pit), a probable toxin from *P. luminescens* showing 30 % amino acid sequence similarity to a fragment of a 13.6 kDa insecticidal crystal protein gene of *B. thuringiensis*, demonstrated injectable toxicity to the larvae of *G. mellonella* and *Spodoptera litura*. However, ingestion of purified Pit protein caused an inhibition of growth of *S. litura* and *H. armigera* larvae, but did not cause larval mortality. The hemocoel insecticidal activity of Pit was comparable with other hemocoel toxins such as Txp40 of *Photorhabdus* (Li et al. 2009).

Photorhabdus Virulence Cassettes

Photorhabdus virulence cassettes (PVCs) are phage-like loci found as repetitive cassettes in the genome of *Photorhabdus* and contain putative toxin effector genes. PVCs are functional homologues of the prophage-like locus on the pADAP plasmid of *Serratia entomophila* (Yang et al. 2006). Recombinant expression of various PVC loci from *P. luminescens* and *P. asymbiotica* demonstrated that they have differing toxicities toward *G. mellonella* upon injection with PVC product derived from the human pathogen *P. asymbiotica* (Gerrard et al. 2004) having greater toxicity for insects than PVC product from the insect pathogen *P. luminescens* TT01 (Duchaud et al. 2003). Although the PVC products showed structural similarity to an antibacterial R-type pyocin, they had no conspicuous antibacterial activity but triggered rapid destruction of insect phagocytes, thus allowing the persistence of recombinant bacteria in wax moth, *G. mellonella* larvae. Comparison of the genomic organizations of PVCs in different *Photorhabdus* species revealed that they have a conserved phage-like structure with a variable number of putative anti-insect effectors encoded at one end. Expression of these putative effectors directly inside cultured cells showed that they are capable of rearranging the actin cytoskeleton (Yang et al. 2006).

Hemolysins or Hemagglutinin-Related Proteins

Hemolysins are extracellular toxic proteins that function as virulence factors and derive their name because of their activity toward red blood cells (Brillard et al. 2002; Cowles and Goodrich-Blair 2005). Hemolysins are produced by a wide spectrum of bacterial species that include the gram-positive (e.g., *Listeria* spp., *Streptococcus* spp.) and the gram-negative (e.g., *E. coli*, *Vibrio* spp., *Photorhabdus* spp., *Xenorhabdus* spp., *Serratia* spp.) bacteria, and these proteins frequently target the immune cells and may aid in evading insect immune responses during infection (Konig et al. 1987; Swihart and Welch 1990; Cowles and Goodrich-Blair 2005). *P. luminescens phlBA* operon, a locus encoding a hemolysin, shows similarities to the pore-forming, calcium-independent hemolysins from *S. marcescens*, and *Proteus mirabilis* type of hemolysins, and belongs to the two-partner secretion (TPS) family of proteins (Brillard et al. 2002). Hemolysins target red blood cells to provide access to iron and may mediate the successful occupation of the different host environments (nematode and insect) it encounters during its life cycle. In case of *X. nematophila*, XhlA (**X**. nematophila **h**aemolysin) was observed to be necessary for full virulence against *M. sexta* larvae (Cowles and Goodrich-Blair 2005).

5.2.2   Insect Virulence Factors Produced by Xenorhabdus Species

Xenorhabdus species are motile gram-negative bacteria of the family Enterobacteriaceae that are mutualistic symbionts of the soil-dwelling nematodes from the family *Steinernematidae*. The life cycle of *Xenorhabdus* is similar to that described for *Photorhabdus*, with the *Steinernema* nematodes playing a key role in vectoring these bacteria from one host to another (Hinchliffe et al. 2010; Castagnola and Stock 2014). Although both types of bacteria are mutualists with nematodes and are entomopathogens, they use distinct, functionally different approaches for these roles (Poinar 1993; Griffin et al. 2001; Goodrich-Blair and Clarke 2007), suggesting that *Xenorhabdus* and *Photorhabdus* underwent divergent evolution that arrived at convergent lifestyles (Chaston et al. 2011). The ingestion of hemolymph was found to trigger the release of *Xenorhabdus* through the anus of nematode host, *S. carpocapsae*. Five species (*Xenorhabdus beddingii*, *Xenorhabdus bovienii*, *Xenorhabdus japonicus*, *Xenorhabdus nematophilus*, and *Xenorhabdus poinarii*) were recognized in the genus *Xenorhabdus* after initial reclassification, and a total of 15 new species have been identified from *Steinernema* nematode collections since then (Lengyel et al. 2005; Somvanshi et al. 2006; Tailliez et al. 2006).

Of these, the best studied nematode-bacterial associations are those of *X. nematophila–S. carpocapsae*, and it has been demonstrated that certain *Steinernema–Xenorhabdus* associations are exclusive and non-cognate pairs will not associate during experimental mixing (Akhurst 1983; Sicard et al. 2004). The complete genomes of *X. nematophila* ATCC 19061 and *X. bovienii* SS-2004 have been sequenced (Chaston et al. 2011).

Xenorhabdus overcomes the insect's defense systems and produces an array of virulence factors (proteases, lipases, hemolysins, immunosuppressants, and toxins) that participate in suppressing insect immunity and killing the host (Forst and Nealson 1996). Two types of hemocytes (the granulocytes and plasmatocytes) comprise greater than 70 % of the cells found in the lepidopteran larval hemolymph (Gillespie et al. 1997). During *X. nematophila* infection, the overall numbers of circulating insect hemocytes are drastically reduced (da Silva et al. 2000). Two factors (C1 and C2) that are produced in liquid cultures and target the insect hemocytes have been identified in *Xenorhabdus* (Brillard et al. 2001). Two cyto-toxins, the αX and Xax, having identical biological effects on insect hemocytes and associated with C1 factor have been characterized from *Xenorhabdus* (Ribeiro et al. 2003; Vigneux et al. 2007). The toxin complex genes (*xpt*) encoding high molecular weight insecticidal proteins have also been observed in *X. nematophila*. *Xenorhabdus xpt* genes exist on a pathogenicity island (PAI) like the *tc* genes of *Photorhabdus*, and it has been demonstrated that the PAIs of strains of *Xenorhabdus* are nearly identical, indicating that the presence of PAIs corresponds to either an evolutionary advantage or increased fitness (Sergeant et al. 2006).

5.2.2.1 The Toxin Complexes (Tc's) of *Xenorhabdus*

X. nematophila contains only a single *tc* locus encoding all three subunits with two [A] genes, *xptA1* and *xptA2*; a single [B] gene, *xptC*; and a single [C] gene, *xptB*. These show the greatest levels of identity to the *P. luminescens* genes tcdA, tcaC, and tccC, respectively (Hinchliffe et al. 2010). The native toxin complex (toxin complex 1) from *Xenorhabdus* is composed of three different proteins XptA2 [284 kDa], XptB1 [110 kDa], and XptC1 [158 kDa], representing class A, B, and C proteins that were found to interact in a 4:1:1 (XptA2:XptB1:XptC1) stoichiometry (Sheets et al. 2011), while *Xenorhabdus tc2* contains XptA1 [287 kDa], in addition to XptB1 and XptC1 where the two separate [A} genes, XptA1 and XptA2, have been shown to be responsible for different host species specificity within the *tc's* (Lee et al. 2007). XptA1 protein confers specificity toward *Pieris brassicae* and *Pieris rapae*, and the XptA2 protein confers specificity toward *H. virescens* (Sergeant et al. 2003). This indicates that these [A] subunits must interact with some kind of specific receptor in order for the complex to cause toxicity. The *tc* gene products can be categorized into toxins and potentiators. The potentiators synergize with their Tc toxin counterpart for full insecticidal activity (Waterfield et al. 2005a). The A component of Tc complexes has toxin activity potentiated by the [BC] components, and this has been demonstrated for the Tc's of *X. nematophila* as well (Sergeant et al. 2003). All three components were found essential for the formation of a biologically active toxin complex. Although the Tc's have often been thought of as being possible pore-forming toxins, Lee et al. (2007) observed that purified XptA1 binds specifically to brush border membrane vesicles (BBMV) from *P. brassicae* and to *Sf*21 cells but does not form pores in the membranes. It is likely that the [A] tetramer alone cannot form pores, whilst a mature complex containing [BC] subunits can, thereby suggesting that [BC] is probably mediating the insertion of [A] into membranes and thus "potentiating" the toxicity of [A] (Hinchliffe et al. 2010).

5.2.2.2 Txp40 Toxin

Brown et al. (2004) described a novel 42 kDa toxin, A24tox, from *X. nematophila* strain A24 that had a lethal with effect on lepidopteran larvae such as *G. mellonella* and *H. armigera* when injected at doses of 30 to 40 ng/g larvae. Injection of the A24tox protein into lepidopteran larvae caused the larvae to cease feeding almost immediately, indicating that the midgut may be the primary site of action for the toxin. Detection, characterization and alignment of the *txp40* gene sequences from several strains of *Xenorhabdus* and *Photorhabdus* highlighted the conserved nature of the gene and its ubiquitous occurrence within this group. (Brown et al. 2006; Castagnola and Stock 2014). The insecticidal activity of the Txp40 toxin and the histopathology of larvae treated with the toxin are similar to those observed with the *Photorhabdus* Txp40 protein and discussed in detail in the Sect. 5.2.1.4.

5.2.2.3 Insecticidal Pilin Protein

All gram-negative bacterial pathogens have been shown to secrete their virulence factors enclosed in outer membrane vesicles (OMVs) (Beveridge 1999). The naturally secreted OMVs of *X. nematophilus* contained a number of proteins and showed larvicidal activity when they were incorporated into the diet of neonatal larvae of *H. armigera* (Khandelwal and Banerjee-Bhatnagar 2003). A 17 kDa pilin subunit protein present in the *X. nematophila* OMV was found to be cytotoxic to the cultured larval hemocytes of *H. armigera*, causing agglutination, and subsequent release of the cytoplasmic enzyme lactate dehydrogenase (Khandelwal et al. 2004). The 17 kDa pilin subunit demonstrated oral toxicity to the fourth or fifth instar larvae of *H. armigera* in a dose-dependent manner, causing the breakdown of the gut epithelial lining, thereby affecting the integrity of the cellular lining, resulting in the sloughing of the cell debris into the lumen (Khandelwal et al. 2004).

5.2.2.4 Insecticidal GroEL Protein

XnGroEL is a ~58 kDa OMV protein secreted by *X. nematophilus* that belongs to a highly conserved family of molecular chaperones and is required for the proper folding of cellular proteins. XnGroEL has chitin-binding property and interacts with the larval peritrophic lining, and oral ingestion of this protein caused inhibition of the growth and development of *H. armigera* larvae. While all three domains (apical, intermediate, and equatorial) of the protein were found to be necessary for optimal insecticidal activity, two surface-exposed residues Thr-347 and Ser-356 in the apical domain were found to be vitally important for binding to the gut epithelium and insect-toxicity (Joshi et al. 2008). The oral toxicity of XnGroEL against *H. armigera*, evaluated by transgenic expression of the protein in tobacco, showed 100 % reduction in the larval survival on transgenic plants (Kumari et al. 2014).

5.2.2.5 *Xenorhabdus* Alpha-Xenorhabdolysin (Xax) Toxin

The XaxAB cytotoxin produced by *X. nematophila* is encoded by two genes *xaxA* and *xaxB* and appears to be the prototype of a new family of binary toxins based on molecular characterization of the locus, gene and amino acid sequences. Xax triggers apoptosis in both insect (*Spodoptera littoralis*) hemocytes and mammalian cells (Vigneux et al. 2007). Active protein was produced when the two genes were expressed in recombinant *E. coli*. However optimum hemolytic activity was observed when these proteins were added to cells in vitro in a specific order (Xax A and then Xax B) and at equal concentrations. *xax* genes A and B were found to be present in the genome sequences from various bacterial pathogens of insects (*Xenorhabdus, Photorhabdus, Pseudomonas entomophila*), plants (*Pseudomonas syringae*), and humans (*Yersinia enterocolitica* and *P. mirabilis*) (Vigneux et al. 2007).

5.2.3 Insect Virulence Factors Produced by Pseudomonas Species

Pseudomonas spp. (Enterobacteriaceae) are metabolically versatile gram-negative bacteria that are ubiquitous in their distribution and have been recovered from a wide variety of ecological niches, including soil and plants (Vodovar et al. 2006). The root-associated bacteria of the genus *Pseudomonas* exhibiting inhibitory activity toward fungal plant pathogens have been extensively used for crop protection. The insecticidal properties of *P. entomophila*, *P. syringae*, and *Pseudomonas fluorescens* were discovered thereafter (Castagnola and Stock 2014).

P. entomophila is highly pathogenic to *Drosophila melanogaster* and also exhibits considerable insecticidal potency against other insects (e.g., *Bombyx mori*, *Anopheles gambiae*) upon ingestion. The precise mechanism of death remains unclear, but the invading bacteria were found to be resistant to the immune response triggered in the insect after oral ingestion (Vodovar et al. 2005). It is speculated that *P. entomophila* evades the insect immune response by making proteases (alkaline protease AprA) and exotoxins (hemolysins and lipases) that punch holes in the cell membranes of the insect phagocytes. *P. entomophila* genome encodes several gut-specific toxin complexes (Tc's) with multiple copies of C-like elements and one copy of a B-like gene that are not organized as a single operon but are scattered across the genome. The genome lacks a gene encoding an A toxin which determines the host range, suggesting that the insecticidal activities of toxin complexes from *P. entomophila* may be fairly restricted. *P. entomophila* genome encodes the apparatus required to produce hydrogen cyanide, the precise role of which in bacterial biology remains to be established (ffrench-Constant and Waterfield 2006). *P. entomophila* has a large repertoire of potential virulence factors such as insecticidal toxins, proteases, putative hemolysins, hydrogen cyanide, and novel secondary metabolites that are envisaged to be important for virulence toward insects. The two-component regulatory system GacS/GacA was observed to play a key role in *P. entomophila* pathogenicity by regulating the expression of many virulence factors (Vallet-Gely et al. 2010).

P. fluorescens produces a proteinaceous insecticidal toxin Fit that demonstrated hemocoel-based toxicity in *M. sexta* and *G. mellonella* (Pechy-Tarr et al. 2008) and caused complete loss of turgor pressure and melanization. Fit has been shown to be orally toxic to *S. littoralis*, *H. virescens*, and *P. xylostella*. *P. chlororaphis* also expresses a Fit toxin and has oral insecticidal activity (Ruffner et al. 2013). Certain strains of *P. fluorescens* contain genes encoding δ-endotoxins, Mcf toxins, *tc* genes encoding B and C components only, lipases, and exotoxins with hemolytic activity. The genome of *P. syringae pv. syringae* contains an intact toxin complex encoding ABC complement, suggesting that this species which was formerly thought to be only plant pathogenic may have an association with insects (ffrench-Constant and Waterfield 2006).

5.2.4 Insect Virulence Factors Produced by Serratia Species

Serratia spp. (Enterobacteriaceae) are commonly isolated from grassland soils, and they often exist as endophytic rhizobacteria possessing antifungal activity. However, several species within the genus *Serratia* are often found associated with insects of many orders (Grimont and Grimont 1978; Lamelas et al. 2011) and nematodes (Rae et al. 2008; Abebe et al. 2011) in a facultative manner. *S. plymuthica*, isolated from the intestine of *Neombius fasciatus* (Steinhaus 1941) caused no infection in the insect host; however, *S. marcescens* and *S. liquefaciens* were regarded as facultative pathogens. *S. marcescens* was found to infect lepidopteran hosts such as poorly reared *H. virescens* (Sikorowski et al. 2001). Contrarily,

S. entomophila and *S. proteamaculans*, the causal agents of amber disease of the New Zealand grass grub, *Costelytra zealandica* (Coleoptera: Scarabaeidae), are considered as true entomopathogenic bacteria (Jackson et al. 2001).

S. entomophila and *S. proteamaculans* colonize and propagate in the grass grub crop; consequently, the infected larvae cease to feed, clear their gut, and become amber-colored (Jackson et al. 1993; Jackson 1995). Eventually the bacteria invade the hemolymph, causing general septicemia and death (Nuñez-Valdez et al. 2008). The pathogenicity determinants in *S. entomophila* are encoded at two regions on a 153-kb plasmid designated as pADAP (amber disease-associated plasmid). The *afp* (anti-feeding prophage) gene cluster, encoding an R-type pyocin structure, mediates the transport of toxins to a target site and causes a cessation of feeding by the grass grub larvae (Hurst et al. 2004; 2007a). The *sep* virulence-associated region which comprises of three genes designated as *sepA* (tcdA-like), *sepB* (tcdB-like), and *sepC* (tccC-like) mediate for the amber disease symptoms of gut clearance and amber coloration of the larvae (Hurst et al. 2000). Both the *sep* genes and *afp* are needed for full virulence of *Serratia* in grass grubs. Amber disease was found to be chronic in nature and despite widespread testing; no other insect species have been shown to be susceptible to *S. entomophila*. The host-specific nature of insecticidal proteins from *Serratia* therefore limits their use in crop protection as these proteins can be deployed against only specific species of pest insects. The *sepABC* genes show homology to the components of the Tc proteins from *P. luminescens* (Bowen et al. 1998) and *xpt* genes identified from *X. nematophilus* (Morgan et al. 2001). However, while Tc toxins have shown cytotoxic effects, direct toxicity of the Sep proteins is unknown. Sep gene orthologues were found to be plasmid-borne in *S. entomophila*, *S. proteamaculans*, *S. liquefaciens*, and *Yersinia frederiksenii*. The *sepA* and *sepB* genes of *S. entomophila* show high nucleotide identity to *Y. frederiksenii* genes thus suggesting a horizontal gene transfer between the two species (Dodd et al. 2006).

5.2.5 Insect Virulence Factors Produced by Yersinia Species

Members of the genus *Yersinia* (Enterobacteriaceae) have undergone extensive diversification during the course of their evolution and are represented by pathogenic species such as *Y. pestis* and *Yersinia ruckeri*, the causative agent of bubonic plague (Perry and Fetherston 1997) and the causative agent of enteric redmouth disease in salmonid fish, respectively (Ewing et al. 1978), while other species (e.g., *Y. aldovae*) have diverged into nonpathogenic organisms (Sulakvelidze 2000). *Y. entomophaga* is a non-sporulating entomopathogenic bacterium that was isolated from diseased larvae of the New Zealand grass grub, *Costelytra zealandica* White (Coleoptera: Scarabaeidae) (Hurst et al. 2011a). The pathogenicity island of *Y. entomophaga*, termed PAI^{Ye96}, is composed of the multi-subunit toxin complex (Yen-Tc) showing homology with toxin complexes produced by *Photorhabdus* spp.

and insecticidal chitinases. The Tc complex includes three protein families termed A (YenA1, Yen A2), B (YenB), and C (YenC1, YenC2) and two chitinases (Chi1 and Chi2) with high endochitinase activity (Hurst et al. 2011b). The 3D structures of the Tc complex showed that subunits YenA1 and YenA2 form the basis of a fivefold symmetric assembly, while subunits B and C form a surface accessible region and are the main toxicity determinants. The structure of the chitinases that adorn the surface of the TcA scaffold has been analyzed and different hypothesis have been proposed to explain their role in mediating Tc toxicity (Landsberg et al. 2011; Busby et al. 2012). Tc protein complex from *Y. entomophaga* exhibits broad host range oral insecticidal activity, causing rapid mortality in many insect pests belonging to the orders Coleoptera, Lepidoptera, and Orthoptera. The culture supernatants of *Y. entomophaga* were found to be toxic to a variety of coleopteran species, including the New Zealand grass grub, *C. zealandica* (Coleoptera: Scarabaeidae); the redheaded cockchafer, *Adoryphorus couloni* (Coleoptera: Scarabaeidae); the blackheaded pasture cockchafer, *Acrossidius tasmaniae* (Coleoptera: Scarabaeidae); and the diamondback moth, *P. xylostella* (Lepidoptera: Plutellidae) (Hurst et al. 2011b; Castagnola and Stock 2014). In *P. xylostella*, initial apical swelling of gut columnar cells occurred after ingestion of purified Tc from *Y. entomophaga*, followed by complete dissolution of the gut lining (Hurst et al. 2011b). The orally active nature and the broad-spectrum insecticidal activity of the Tc protein derived from *Yersinia* species indicates that it may be a potential alternative to *B. thuringiensis* toxins for use in insect control (Bravo and Soberon 2008).

5.3 Toxins Shared by Gram-Negative Bacterial Pathogens of Invertebrates: New Insights from the Comparative Genomics of Entomopathogens

The proliferation of genomic information of invertebrate pathogens (*P. luminescens*, *X. nematophilus*, *P. entomophila*, etc.) and functional analysis of genome data has revealed that the composition of bacterial genomes is dynamic and susceptible to many changes through the process of genome reduction (Moran 2002), gene duplication and divergence (Ohno 1970), vertical inheritance (Woese 1987), and horizontal gene transfer (Ochman et al. 2000) that occur due to multiple pressures, including the environment, mutation, and competition (Chaston et al. 2011). Therefore, comparative genomics will provide an excellent opportunity to elucidate the genetic similarities and differences in different species that will have an impact on the innovations in crop protection technologies, in addition to providing a fundamental understanding of evolutionary relationships and changes contributing to pathogenesis in bacteria. Bacterial genome can be subdivided into "core genome" and "flexible genome." Core genome of bacterial communities is common to all bacterial strains in a defined set of species and contributes to basic cellular functions.

The genomic regions variably present between individual strains constitute "flexible genome" component, and these regions are organized principally into polymorphic strain-specific segments called regions of genome plasticity (RGPs) that play an important role in bacterial adaptation to special growth conditions, such as those involved in the colonization of new ecological niches, symbiosis, host-cell interaction, and pathogenicity. Flexible gene pools act as a site for inter-genomic and intragenomic rearrangements (Frost et al. 2005; Gaudriault et al. 2008). RGPs (underlining the continuous gene transfer among the bacterial genomes), insertion sequences (IS), putative transposons, and the presence of phage remnants are the key indicators of the transfer of genetic elements among different microbes, especially in the bacterial genomes. The flexible genome of the *Xenorhabdus* and *Photorhabdus* genera accounted for 52.6 to 61.5 % of the entire genome, and this region was found to be larger in *Photorhabdus* than in *Xenorhabdus* (Ogier et al. 2010).

A comparison between entomopathogenic bacteria in the genera *Photorhabdus* and *Xenorhabdus* revealed that despite their similar lifestyles the species within the two genera use functionally different approaches to achieve successful host interactions. *Photorhabdus* spp. encode a dedicated type three secretion system (TTSS) which can suppress phagocytosis and nodule formation by injection of effectors, such as LopT and SctC directly into hemocyte cells (Brugirard-Ricaud et al. 2004; Brugirard-Ricaud et al. 2005). *Xenorhabdus* spp. lack a TTSS and associated effectors; however, they do encode several other cytotoxic strategies in order to evade the host cellular responses (Hinchliffe et al. 2010). The TTSS of *Photorhabdus* is highly similar to the plasmid encoded system of *Y. pestis* (Wolters et al. 2013). However, in *Yersinia*, the effector protein YopT is a cytotoxic cysteine protease, whereas the homolog in *P. luminescens*, called LopT, has been shown to prevent phagocytosis (Brugirard-Ricaud et al. 2004).

Insecticidal toxin complexes (Tc's) were first identified in *P. luminescens* and have been studied extensively by independent research groups (Bowen et al. 1998; ffrench-Constant et al. 2000, 2003). Tc toxins were subsequently identified in the genomes of other gram-negative insect pathogens such as *X. nematophila* (Morgan et al. 2001), *S. entomophila* (Hurst et al. 2007b), and *Y. entomophaga* (Hurst et al. 2011b) and have even been reported in the gram-positive *Paenibacillus* (Hinchliffe et al. 2010). The genome of *Y. pestis* contains a locus encoding the Tc protein homologues *yitA* (TcaA-like), *yitB* (TcaB-like), and *yitC* (TcaC-like) and *YipA* and *YipB* (two TccC-like proteins). The *tcaB* and *tcaC* genes of *Y. pestis* contain a frame shift mutation and internal deletion, respectively, which is indicative of a loss of function (Parkhill et al. 2001; Spinner et al. 2012). The nomenclature of the Tc proteins has been revised (ffrench-Constant and Waterfield 2005), and the ABC designation is adopted currently to describe the components of the Tc complex. The Tc-Bs and Tc-Cs are known to make the Tc-As more toxic. The *tc*-like genes identified in *P. luminescens*, *X. nematophila*, and *Y. pestis* are chromosomally borne, while the *sep* genes of *S. entomophila* and *tc*-like genes of *Y. frederiksenii* strain 49 are plasmid-borne. A toxin-encoding operon similar to the *tca* of *P. luminescens* W14 was found to be present in an isolate of *Bt* (*Bt*-IBL200) from the Invasive Insect Biocontrol and Behavior Laboratory (IIBBL, Beltsville, MD, USA) (Blackburn et al. 2011). All three components of *tc* (A, B, and C) were present in the *Bt* IBL200 isolate. The genomic organization and diversity of Tc proteins among different species of insect-pathogenic bacteria increases the likelihood of tc gene transfer between species via plasmids, and suggests that the chromosomally located *tc* genes could also once have been plasmid-borne or mobile in other bacteria (Dodd et al. 2006). Tc toxins are active against different tissues within individual hosts, namely, Tcb against hemocytes and Tcd and Tca against cells of the insect gut. The Tc toxins reside as multiple but dissimilar orthologues throughout the *P. luminescens* TT01 genome with different insecticidal activities attributed to a different Tc cluster (Duchaud et al. 2003; Hey et al. 2006). Plasmid-borne Sep proteins of *S. entomophila* are host-specific. The *xpt* genes of *Xenorhabdus* exist on a PAI like the *tc* genes of *Photorhabdus*; however, PAIs are nearly identical in *Xenorhabdus*. The insecticidal genes *tcdB1*, *yitC*, and *spvB* of *P. luminescens*, *Y. pestis*, and *Salmonella*, respectively, have regions of homology, viz., the N-terminal 367 amino acids of *yitC* are similar to the N-terminus of the putative effector *spvB* of *Salmonella* (Browne et al. 2002; Castagnola and Stock 2014). The *Yersinia* spp. contain islands harboring insecticidal *tc*-like genes; however, insecticidal activity was observed only when a low-growth temperature was used to culture the bacteria and produce a protein extract (Bresolin et al. 2006).

Chitinases having antimycotic activity have been found in both *Xenorhabdus* and *Photorhabdus* (Chen et al. 1996). Interestingly, the assembled Tc toxin of *Y. entomophaga* was found to have endochitinase activity, which was attributed to putative chitinase subunits associated with TcA scaffold. This has not previously been reported in a Tc (Hurst et al. 2011b; Landsberg et al. 2011). *S. marcescens* was found to produce orally active insecticidal toxins and chitinases (Jeong et al. 2010; Brurberg et al. 1996). *S. marcescens* cultures and *Bt* Cry1C toxin exhibited a synergistic insecticidal effect against *S. litura* (Asano et al. 1999).

The apoptotic binary toxin Xax is found in *P. luminescens* and *X. nematophila*. *X. nematophila* XaxA and XaxB showed the strongest similarity to *plu3075* (61 %) and *plu1961* (56 %), respectively, from *P. luminescens*. The putative hemolysin loci, containing two closely linked genes, *xaxA* and *xaxB*, are found together in genome sequences from various bacterial pathogens of plants (*P. syringae*), insects (*Photorhabdus*, *Xenorhabdus*, *P. entomophila*), and humans (*P. mirabilis*, *Y. enterocolitica*). The *xaxAB* homologues in *X. nematophila* are found in a unique genomic context that does not show characteristic features of genome flexibility, such as genomic islands, transposon-related structures, or phages (Vigneux et al. 2007). Interestingly, the *xax* hemolysin locus was found to be present in *Y. enterocolitica* and not in *Y. pestis* even though the latter, like *X. nematophila*, spends part of its life cycle in an insect.

The *mcf* (makes caterpillar floppy) gene, which encodes a large proapoptotic multidomain protein, is present in *P. luminescens*, *P. temperata*, and *P. asymbiotica* (Daborn et al. 2002; Forst and Goodner 2006). Two *mcf* paralogous genes (*mcf1* and *mcf2*) are found in the strains Pl W14 and Pl TT01. Genome sequencing has revealed the presence of other Mcf-like proteins in other bacterial species. The *FitD* gene of *P. fluorescens* encodes a Mcf1-like protein and shows 73.5 % identity to *P. luminescens mcf1* (Pechy-Tarr et al. 2008). The *fitD* locus of *P. fluorescens* is associated with a TolC-family outer membrane efflux protein (*fitE*), two response regulators (*fitF*, *fitH*), and a LysR-like regulator (fitG). However, paralogues of *fitE–H* are not present in *Photorhabdus* and may be involved in the specific regulation of *fitD* in *P. fluorescens* (Hinchliffe et al. 2010).

RTX, the repeats in toxin family possessing different enzymatic activities including cytolytic, protease, or lipase activity are observed to be dramatically expanded in *P. luminescens* (eight *rtxA* genes), but this gene has not been found in *Y. pestis*. Four of the eight *rtxA* genes of *P. luminescens* were disrupted by either insertion sequence (IS) elements or inactivated by frameshift mutations. An RTX-like alkaline protease has recently been isolated from *P. luminescens* W14 and *P. temperata* (Bowen et al. 2003; Forst and Goodner 2006). Phage-related loci are found in both *Serratia* and *Photorhabdus*. The PVCs of *Photorhabdus* are homologous to a prophage-like locus on the pADAP plasmid of *S. entomophila* (Yang et al. 2006). The PVCs have injectable toxicity toward *G. mellonella* hemocytes, whereas the pADAP locus has been associated with anti-feeding effects (Hurst et al. 2004). The *txp40* gene, encoding a 42 kDa protein with injectable cytotoxic activity, was identified in several strains of *Xenorhabdus* and *Photorhabdus*, indicating that it is both highly conserved and widespread among these bacteria (Brown et al. 2006). Proteins similar to the δ-endotoxin from *B. thuringiensis* have also been identified in *P. luminescens*. The *pir* gene of *Photorhabdus* shows a 30 % amino acid sequence similarity to part of the insecticidal crystal protein of *B. thuringiensis* (Li et al. 2009).

The genome of *P. luminescens* was found to be ~1 Mb larger than closely related bacteria such as *Xenorhabdus* and *Yersinia spp*. The average genome size of most members of the Enterobacteriaceae family is approximately 4.6 Mb (Forst and Goodner 2006). The emerging human pathogen *P. asymbiotica* has a smaller genome than that of *P. luminescens* genome. A reduction in the genome size of *P. asymbiotica* was found to coincide with a reduction in different classes of anti-insect virulence factors. Unlike *P.luminescens* strains, the *P. asymbiotica* strains carry a plasmid related to pMT1 from *Y. pestis* that promotes deep tissue invasion, and several PAIs including a novel TTSS; these features suggested that human pathogenicity in *P. asymbiotica* was acquired through the acquisition of pMT1-like plasmid and specific effectors. Despite these molecular changes, the pathogenicity toward insects was found to remain intact in *P. asymbiotica* illustrating a lifestyle superior to *P. luminescens*, which is pathogenic only to insects (Wilkinson et al. 2009).

Comparative genome analysis, gene profiling, functional genomics, and the newly developed genetic approaches like microarrays and software tools like RGP finder will be helpful for several unresolved, mechanistic and evolutionary questions about members of soil bacteria in future. Whole-genome sequencing approaches and comprehensive analysis undertaken by the rapid virulence annotation (RVA) technique (Waterfield et al. 2008) have begun to reveal previously unidentified insecticidal toxins, uncharacterized secondary metabolites possessing toxic activities, putative lipases, and type VI secretion systems with insecticidal activity. Thus, a significant challenge of comparative genomics is to interpret the interrelationships between anti-invertebrate and anti-vertebrate virulence factors and assign biological roles to the candidate virulence factors (Hinchliffe et al. 2010).

5.4 Strategies for Enhancing Transgenic Resistance to Lepidopteran Pests: A Dynamic Landscape

Undoubtedly, *B. thuringiensis* and its insecticidal toxins have been overwhelmingly successful for agronomical pest control for decades. Transgenic crops protected from the damage of lepidopteran and coleopteran insect feeding express insecticidal proteins derived from the entomopathogenic bacteria, *B. thuringiensis* (Huesing and English 2004). The first generation of genetically modified (GM) crops conferring insect protection has been extraordinarily successful, and GM crops are considered as the fastest adopted crop technology in the history of modern agriculture (James 2012). Insect-resistant products expressing *Bt*-derived proteins have been available for corn and cotton since 1996. *Bt* cotton, in particular, has provided effective control of several lepidopteran pest species including tobacco budworm, *H. virescens* F. (Lepidoptera: Noctuidae); pink bollworm, *P. gossypiella* Saunders (Lepidoptera: Gelechiidae); the cotton bollworm, *H. armigera* Hübner (Lepidoptera: Noctuidae); and *Spodoptera* spp. (Lepidoptera: Noctuidae) (Sanahuja et al. 2011; Baum et al. 2012). The second-generation GM varieties generated by stacking and pyramiding resistance genes were also approved for commercialization (Marra et al. 2010). *Bt* cotton planted in more than 18.8 million hectares in 13 countries was the third most dominant crop grown in 2012, which is equivalent to 11 % of the global biotech area (James 2012). Despite the commercial success of the *Bt* technology, there are concerns over the development of resistance by insect species, the problem of efficacy, and narrow spectrum of activity of *Bt* proteins (de Maagd et al. 2001; Pereira et al. 2008; Sayyed et al. 2008). Therefore, efforts are ongoing for the discovery of potent *Bt* strains expressing novel toxins with improved and broad-spectrum activity as well as for the characterization of genes exhibiting biopesticidal properties from other entomopathogens (Christou et al. 2006; Crickmore 2006).

Whole-genome sequencing and screening of genomes of soil-dwelling gram-negative entomopathogenic bacteria and gram-negative bacterial symbionts of soil-dwelling EPNs has yielded a gold mine of potential novel insect toxins that augment a growing list of candidates for use in crop protection (Hinchliffe et al. 2010). The insect virulence factors from gram-negative bacteria that can be used as biocontrol agents against Lepidoptera include a host of insecticidal toxins, proteases, putative hemolysins, and other previously unidentified proteins. Although a surge in the patent applications has been observed with the discovery, functional annotation, and insecticidal efficacy evaluation of novel genes/proteins, data on how far many of these potential candidates have progressed toward commercial field applications is ambiguous (Hinchliffe et al. 2010).

Photorhabdus, *Xenorhabdus*, and other gram-negative entomopathogens have provided novel candidates having both oral and hemolymph-based toxic activity. Proteins that are active upon ingestion and damage the insect midgut are good candidates for developing transgenic crops. The "toxin complexes" are a prime candidate fulfilling these criteria. However, the orally toxic nature of the Tc proteins produced by an insect pathogen that is directly delivered into the insect hemolymph is intriguing. The presence of tc genes in many organisms which are not directly delivered into the insect hemolymph may somewhat explain the oral toxicity of Tc proteins (Castagnola and Stock 2014). The histopathological effects on lepidopteran larvae of toxins that include Tc proteins, the *mcf* gene product, the PirAB binary toxins, the 17 kDa *pilin* subunit, and the Txp40 proteins from *Photorhabdus* and *Xenorhabdus* demonstrated a distinct damage to the midgut. The *tc* genes have been studied extensively by researchers in academia and industry and were considered suitable for commercial product development. One of the Tc proteins, the *tcdA* protein expressed in *A. thaliana* to sufficient levels, was found to be toxic to *M. sexta* (Liu et al. 2003). The demonstration that the large toxin genes such as *tc* can be engineered and expressed in transgenic plants makes them suitable alternative toxins to *B. thuringiensis* (Schnepf et al. 1998). Application of Pir toxins in insect control may be limited, as they were found to be effective only against the diamondback moth, *P. xylostella*. The limited activity spectrum and the possible relation of the Pir toxins to the leptinotarsins may indicate problems with vertebrate toxicity. Toxins like Mcf which act on both the gut and insect immune system represent a promising, yet underexploited avenue for the development of insect-resistant crops in the future (Daborn et al. 2002). It can be speculated that there are many more toxins yet to be functionally characterized from the *Photorhabdus* genome which are responsible for hemolymph-based toxicity.

Stacking virulence factors like Mcf which have hemolymph-based toxicity with conventional orally active toxins would expand their application to IPM. However, only those hemolymph-based toxins which are toxic upon ingestion as well could be used in this case, as these proteins have to be stable in the insect gut. Alternatively, two proteins with different modes of action and diverse targets can also be used in developing novel combinations of genes for pest control, as it has the advantage of reducing the development of insect resistance (Gould 1998). Site-directed mutagenesis and domain-engineering have great potential to alter toxin-encoding genes particularly when sufficient structural information is available, these methods can thus be applied to produce novel recombinant toxins (Gatehouse 2008). Using a combination of insect-toxic genes derived from *B. thuringiensis* and gram-negative entomopathogens can also result in synergistic insecticidal activity (Asano et al. 1999).

The ability of *P. entomophila* to orally infect and kill larvae of insect species belonging to different orders makes it a promising model for the study of host–pathogen interactions and for the development of biocontrol agents against insect pests. *S. entomophila* was developed as a biopesticide and used for 15 years as a commercial product for grass grub control in New Zealand. This microbe was initially developed and applied as a liquid biopesticide Invade®, which is New Zealand's first registered, safety-tested, indigenous biological pesticide and the first microbial control agent in the world to be based on a member of the gram-negative Enterobacteriaceae (Jackson et al. 1992). The pathogenic bacterium has recently been developed for application as a solid granule formulation, Bioshield™ (Young et al. 2010).

The important attributes of an insecticidal protein for successful commercial use include its efficacy and specificity. First and foremost the insecticidal toxin must be highly toxic to a wide range of potential pests, and it has to prevent crop damage by efficiently killing or deterring the insect pests. Delivery of toxic proteins to potential pests is either through their host plants via the expression of the toxin, or toxic subunit/domain, in transgenic crops or developing a pesticide formulation and coating them onto the crops in a stable form (Hinchliffe et al. 2010; Ruiu et al. 2013). Due to specific biological properties and technical reasons such as the specific mode of action (oral or injectable activity), target site of action, and stability, commercially available strains including *Bt* have their restrictions in terms of performance in the field. The toxin protein has to be orally active for developing insect-resistant GM crops. Although "toxin complex" proteins described in different gram-negative entomopathogens are orally active, the Tc's consist of large protein subunits which are inherently difficult to express transgenically in crop plants. Therefore, only individual proteins having limited toxicity can be expressed with the technology available currently. As both toxins and potentiators have to be co-expressed in order to harness the full toxic potential of *tc* genes, either the proteins have to be cropped down to smaller active domains or the transgenic technology has to be improved to allow all subunits to be expressed to achieve the desired level of efficacy (Hinchliffe et al. 2010). Transgenic expression of ingestible insecticidal proteins confers certain degree of pest specificity as only insects actually feeding on the crop will ingest the toxin directly. The toxicity of insecticidal Tc proteins of *P. luminescens* to cultured mammalian cells may attract criticism particularly with regard to their use in crop protection and the associated concerns on biosafety (Waterfield et al. 2005a; Hares et al. 2008). The bioecological compatibility of *P. luminescens* biopesticide against two species of the beneficial insect *Trichogramma* was investigated by Mohan and Sabir (2005). Their study demonstrated that there was a significant reduction of up to 84 % in the emergence of *Trichogramma* adults from the host, *Corcyra cephalonica* eggs, as 65 % of the eggs exposed to either *P. luminescens* cells or their toxins became flaccid. In some species of pathogenic yersiniae viz., *Yersinia pseudotuberculosis* and *Yersinia pestis* the *tc* genes are not insecticidal but have evolved to show mammalian pathogenicity. These data suggest that the biological activity of the toxins against target species and nontarget species should be thoroughly investigated before being considered for crop protection. However, appropriate evaluation of the effects of the toxin on the target pests should be undertaken so that the protein is active toward the target species and does not cause any harm to the non-pest species (beneficial insects, bystander insect species, predatory species which feed off the intoxicated pests, humans) upon exposure.

5.5   Conclusion and Perspectives

Insecticidal toxins are an important option for the biological control of lepidopteran insect pests. Their use in genetic engineering of plants could provide a new generation of insect-resistant crops that can help in maintaining crop yields. A majority of the toxic proteins expressed by gram-negative bacteria have been tested only against model insects (*M. sexta, G. mellonella, S. litura, P. xylostella, H. armigera*), and besides few reports on mammalian toxicity, there is very limited information on the effects of these proteins on beneficial fauna (predators, parasitoids, and pollinators). Therefore, a significant research need is centered on understanding the specific effects of the insecticidal proteins, their activity spectrum, and their effect on nontarget organisms in the ecological sphere.

Insect pest control has entered the genomic era with the recent sequencing and functional analysis of the genomes of agricultural pests and entomopathogenic bacteria. This has enabled the discovery of novel targets in the pests and novel proteins in the entomopathogens and has provided a comprehensive understanding of invertebrate pathology by providing critical insights into evolutionary patterns of bacterial pathogens. Genome analysis has also raised several pertinent questions about the complex life cycles of these pathogens and their association with various invertebrate hosts and vectors. Ultimately a relevant challenge of comparative genomics is to understand the interrelationships between pathogenic mechanisms targeted to invertebrates and vertebrates as such insight may help us understand the evolution and the probable invertebrate origins, of emerging human pathogens, in addition to using the insecticidal genes derived from these bacteria for large-scale, commercial agricultural applications.

References

Abebe E, Abebe-Akele F, Morrison J (2011) An insect pathogenic symbiosis between a *caenorhabditis* and *serratia*. Virulence 2:158–161

Akhurst RJ (1982) Antibiotic activity of *Xenorhabdus* spp, bacteria symbiotically associated with insect pathogenic nematodes of the families *Heterorhabditidae* and Steinernematidae. J Gen Microbiol 128:3061–3065

Akhurst RJ (1983) Taxonomic study of *Xenorhabdus,* a genus of bacteria symbiotically associated with insect pathogenic nematodes. Int J Syst Bacteriol 33:38–45

Asano S, Suzuki K, Hori H (1999) Synergistic effects of the supernatants from *Serratia marcescens* culture on larvicidal activity of *Bacillus thuringiensis* Cry1C toxin against common cutworm, *Spodoptera litura*. J Pestic Sci 24:44–48

Baum JA, Sukuru UR, Penn SR, Meyer SE, Subbarao S, Shi X, Flasinski S, Heck GR, Brown RS, Clark TL (2012) Cotton plants expressing a hemipteran-active *Bacillus thuringiensis* crystal protein impact the development and survival of *Lygus hesperus* (Hemiptera: Miridae) nymphs. J Econ Entomol 105:616–624

Bedding RA, Akhurst RJ, Kaya HK (1993) Nematodes and the biological control of insect pests. CSIRO Publications, Melbourne, p 178

Beveridge TJ (1999) Structures of gram-negative cell walls and their derived membrane vesicles. J Bacteriol 181:4725–4733

Blackburn M, Golubeva E, Bowen D, ffrench-Constant RH (1998) A novel insecticidal toxin from *Photorhabdus luminescens*, Toxin Complex a (Tca), and its histopathological effects on the midgut of *Manduca sexta*. Appl Environ Microbiol 64:3036–3041

Blackburn MB, Farrar RR, Novak NG (2006) Remarkable susceptibility to the diamondback moth (*Plutella xylostella*) to ingestion of Pir toxins from *Photorhabdus luminescens*. Entomol Exp Appl 121:31–37

Blackburn MB, Martin PA, Kuhar D, Farrar RR Jr, Gundersen-Rindal DE (2011) The occurrence of *Photorhabdus*-like toxin complexes in *Bacillus thuringiensis*. PLoS One 6. doi:10.1371/journal.pone.0018122

Bonning BC, Hammock BD (1996) Development of recombinant baculoviruses for insect control. Annu Rev Entomol 41:191–210

Bowen DJ, Ensign JC (1998) Purification and characterization of a high-molecular-weight insecticidal protein complex produced by the entomopathogenic bacterium *Photorhabdus luminescens*. Appl Environ Microbiol 64:3029–3035

Bowen D, Rocheleau TA, Blackburn M, Andreev O, Golubeva E, Bhartia R, ffrench-Constant RH (1998) Insecticidal toxins from the bacterium *Photorhabdus luminescens*. Science 280: 2129–2132

Bowen D, Rocheleau TA, Grutzmacher CK, Meslet L, Valens LM, Marble D, Dowling D, ffrench-Constant RH, Blight MA (2003) Genetic and biochemical characterization of PrtA, an RTX-like metalloprotease from *Photorhabdus*. Microbiology 149:1581–1591

Bravo A, Soberon M (2008) How to cope with insect resistance to Bt toxins? Trends Biotechnol 26:573–579

Bresolin G, Morgan JA, Illgen D (2006) Low temperature-induced insecticidal activity of *Yersinia enterocolitica*. Mol Microbiol 59:503–512

Brillard J, Ribeiro C, Boemare N, Brehélin M, Givaudan A (2001) Two distinct hemolytic activities in *Xenorhabdus nematophila* are active against immunocompetent insect cells. Appl Environ Microbiol 67:2515–2525

Brillard J, Duchaud E, Boemare N, Kunst F, Givaudan A (2002) The PhlA hemolysin from the entomopathogenic bacterium *Photorhabdus luminescens* belongs to the two-partner secretion family of hemolysins. J Bacteriol 184:3871–3878

Brooke E, Hines E (1999) Viral biopesticides for *Heliothine* control-fact of fiction? Today's Life Sci 11:38–45

Brown SE, Cao AT, Hines ER, Akhurst RJ, East PD (2004) A novel secreted protein toxin from the insect pathogenic bacterium *Xenorhabdus nematophila*. J Biol Chem 279:14595–14601

Brown SE, Cao AT, Dobson P, Hines ER, Akhurst RJ, East PD (2006) Txp40, a ubiquitous insecticidal toxin protein from *Xenorhabdus* and *Photorhabdus* bacteria. Appl Environ Microbiol 72:1653 1662

Browne SH, Lesnick ML, Guiney DG (2002) Genetic requirements for salmonella-induced cytopathology in human monocyte-derived macrophages. Infect Immun 70:7126–7135

Brugirard-Ricaud K, Givaudan A, Parkhill J, Boemare N, Zumbihl R, Duchaud E (2004) Variation in the effectors of the *Photorhabdus* type III secretion system among species revealed by genomic analysis. J Bacteriol 186:4376–4381

Brugirard-Ricaud K, Duchaud E, Givaudan A, Girard PA, Kunst F, Boemare N, Brehelin M, Zumbihl R (2005) Site-specific antiphagocytic function of the *Photorhabdus luminescens* type III secretion system during insect colonization. Cell Microbiol 7:363–371

Brurberg MB, Nes IF, Eijsink VG (1996) Comparative studies of chitinases A and B from *Serratia marcescens*. Microbiology 142:1581–1589

Budd RC (2001) Activation-induced cell death. Curr Opin Immunol 13:356–362

Buetow L, Flatau G, Chiu K, Boquet P, Ghosh P (2001) Structure of the Rho-activating domain of *Escherichia coli* cytotoxic necrotizing factor 1. Nat Struct Biol 8:584–588

Busby JN, Landsberg MJ, Simpson R, Jones SA, Hankamer B, Hurst MRH, Lott JS (2012) Structural analysis of Chi1 chitinase from Yen-Tc: The multisubunit insecticidal ABC toxin complex of *Yersinia entomophaga*. J Mol Biol 415:359–371

Carrière Y, Ellers-Kirk C, Sisterson M, Antilla L, Whitlow M, Dennehy TJ, Tabashnik BE (2003) Long-term regional suppression of pink bollworm by *Bacillus thuringiensis* cotton. Proc Natl Acad Sci USA 100:1519–1523

Carson R (1962) Silent spring. Houghton, Mifflin

Castagnola A, Stock P (2014) Common virulence factors and tissue targets of entomopathogenic bacteria for biological control of lepidopteran pests. Insects 5:139–166. doi:10.3390/insects5010139

Chaston JM, Suen G, Tucker SL, Andersen AW, Bhasin A, Bode E, Bode HB, Brachmann AO, Cowles CE, Cowles KN, Darby C, de Léon L, Drace K, Du Z, Givaudan A, Herbert Tran EE, Jewell KA, Knack JJ, Krasomil-Osterfeld KC, Kukor R, Lanois A, Latreille P, Leimgruber NK, Lipke CM, Liu R, Lu X, Martens EC, Marri PR, Médigue C, Menard ML, Miller NM, Morales-Soto N, Norton S, Ogier JC, Orchard SS, Park D, Park Y, Qurollo BA, Sugar DR, Richards GR, Rouy Z, Slominski B, Slominski K, Snyder H, Tjaden BC, van der Hoeven R, Welch RD, Wheeler C, Xiang B, Barbazuk B, Gaudriault S, Goodner B, Slater SC, Forst S, Goldman BS, Goodrich-Blair H (2011) The entomopathogenic bacterial endosymbionts *Xenorhabdus and Photorhabdus*: convergent lifestyles from divergent genomes. PLoS One 6:e27909

Chen G, Zhang Y, Li J (1996) Chitinase activity of *Xenorhabdus* and *Photorhabdus* species, bacterial associates of entomopathogenic nematodes. J Invertebr Pathol 68:101–108

Christou P, Capell T, Kohli A, Gatehouse JA, Gatehouse AM (2006) Recent developments and future prospects in insect pest control in transgenic crops. Trends Plant Sci 11:302–308

Cowles KN, Goodrich-Blair H (2005) Expression and activity of a *Xenorhabdus nematophila* haemolysin required for full virulence towards *Manduca sexta* insects. Cell Microbiol 7:209–219

Crickmore N (2006) Beyond the spore – past and future developments of *Bacillus thuringiensis* as a biopesticide. J Appl Microbiol 101:616–619

Crosland RD, Fitch RW, Hines HB (2005) Characterization of β-leptinotarsin-h and the effects of calcium flux agonists on its activity. Toxicon 45:829–841

da Silva CC, Dunphy GB, Rau ME (2000) Interaction of *Xenorhabdus nematophilus* (Enterobacteriaceae) with the antimicrobial defenses of the house cricket, *Acheta domesticus*. J Invertebr Pathol 76:285–292

Daborn PJ, Waterfield N, Silva CP (2002) A single *Photorhabdus* gene, makes caterpillars floppy (mcf), allows *Escherichia coli* to persist within and kill Insects. Proc Natl Acad Sci USA 99:10742–10747

de Maagd RA, Bravo A, Crickmore N (2001) How *Bacillus thuringiensis* has evolved specific toxins to colonize the insect world. Trends Genet 17:193–199

Dodd SJ, Hurst MR, Glare TR, O'Callaghan M, Ronson CW (2006) Occurrence of sep insecticidal toxin complex genes in *Serratia* spp. and *Yersinia frederiksenii*. Appl Environ Microbiol 72:6584–6592

Dowling AJ, Daborn PJ, Waterfield NR, Wang P, Streuli CH, ffrench-Constant RH (2004) The insecticidal toxin Makes caterpillars floppy (Mcf) promotes apoptosis in mammalian cells. Cell Microbiol 6:345–353

Duchaud E, Rusniok C, Frangeul L, Buchrieser C, Givaudan A, Taourit S, Bocs S, Boursaux-Eude C, Chandler M, Charles JF, Dassa E, Derose R, Derzelle S, Freyssinet G, Gaudriault S, Medigue C, Lanois A, Powell K, Siguier P, Vincent R, Wingate V, Zouine M, Glaser P, Boemare N, Danchin A, Kunst F (2003) The genome sequence of the entomopathogenic bacterium *Photorhabdus luminescens*. Nat Biotechnol 21:1307–1313

Ewing WH, Ross AJ, Brenner DJ, Fanning GR (1978) Yersinia ruckeri sp. nov., the Redmouth (RM) Bacterium. Int J Syst Bacteriol 28:37–44. doi:10.1099/00207713-28-1-37

Federici BA, Siegel JP (2008) Safety assessment of Bacillus thuringiensis and Bt crops used in insect control. In: Hammond BG (ed) Food safety of proteins in agricultural biotechnology. CRC Press Taylor and Francis Group, Chapter 3, pp 45–102

ffrench-Constant RH, Waterfield NR (2005) An ABC guide to the bacterial toxin complexes. Adv Appl Microbiol 58:169–183

ffrench-Constant RH, Waterfield NR (2006) Ground control for insect pests. Nat Biotechnol 24:660–661

ffrench-Constant RH, Waterfield N, Burland V, Perna NT, Daborn PJ, Bowen D, Blattner FR (2000) A Genomic sample sequence of the entomopathogenic bacterium *Photorhabdus luminescens* W14: Potential implications for virulence. Appl Environ Microbiol 66:3310–3329

ffrench-Constant RH, Waterfield NR, Daborn P, Joyce S, Bennet H, Au C, Dowling A, Boundy S, Reynolds S, Clarke D (2003) Photorhabdus: towards a functional genome analysis of a symbiont and pathogen. FEMS Microbiology Rev 26:433–456

Fischer-Le Saux M, Viallard V, Brunel B, Normand P, Boemare N (1999) Polyphasic classification of the genus *Photorhabdus* and proposal of new taxa: *P. luminescens* subsp. *luminescens* subsp. nov, *P. luminescens* subsp. *akhurstii* subsp. nov, *P. luminescens* subsp. *laumondii* subsp. nov, *P. temperata* sp. nov, *P. temperata* subsp. *temperata* subsp. nov. and *P. asymbiotica* sp. nov. Int J Syst Bacteriol 49:1645–1656

Fitt GP (1994) Cotton pest management: part 3; an Australian perspective. Annu Rev Entomol 39: 532–562

Forst S, Goodner B (2006) Comparative bacterial genomics and its use in undergraduate education. Biol Control 38:47–53

Forst S, Nealson K (1996) Molecular biology of the symbiotic-pathogenic bacteria *Xenorhabdus* spp. and *Photorhabdus* spp. Microbiol Rev 60:21–43

Frost LS, Leplae R, Summers AO, Toussaint A (2005) Mobile genetic elements: the agents of open source evolution. Nat Rev Microbiol 3:722–732

Gassmann AJ, Petzold-Maxwell JL, Keweshan RS, Dunbar MW (2011) Field-evolved resistance to Bt maize by western corn rootworm. PLoS One 6:e22629. doi:10.1371/journal.pone. 0022629

Gatehouse JA (2008) Biotechnological prospects for engineering insect-resistant plants. Plant Physiol 146:881–887

Gatehouse AM, Hilder VA, Powell KS, Wang M, Davison GM, Gatehouse LN, Down RE, Edmonds HS, Boulter D, Newell CA et al (1994) Insect-resistant transgenic plants: choosing the gene to do the 'job'. Biochem Soc Trans 22:944–949

Gaudriault S, Pages S, Lanois A, Laroui C, Teyssier C, Jumas-Bilak E, Givaudan A (2008) Plastic architecture of bacterial genome revealed by comparative genomics of Photorhabdus variants. Genome Biol 9:R117. doi:10.1186/gb-2008-9-7-r117

Gerrard J, Waterfield N, Vohra R, ffrench-Constant RH (2004) Human infection with *Photorhabdus asymbiotica*: an emerging bacterial pathogen. Microbes Infect 6:229–237

Gerrard JG, Joyce SA, Waterfield NR (2006) Nematode symbiont for *Photorhabdus asymbiotica*. Emerg Infect Dis 12:1562–1564

Gillespie JP, Kanost MR, Trenczek T (1997) Biological mediators of insect immunity. Ann Rev Entomol 42:611–643

Goodrich-Blair H, Clarke DJ (2007) Mutualism and pathogenesis in *Xenorhabdus* and *Photorhabdu*s: two roads to the same destination. Mol Microbiol 64:260–268

Gould F (1998) Sustainability of transgenic insecticidal cultivars: integrating pest genetics and ecology. Annu Rev Entomol 43:701–726

Griffin CT, O' Callaghan K, Dix I (2001) A self-fertile species of *Steinernema* from Indonesia: further evidence of convergent evolution amongst entomopathogenic nematodes? Parasitology 122:181–186

Grimont PAD, Grimont F (1978) The genus *Serratia*. Annu Rev Microbiol 32:221–248. doi:10. 1146/annurev.mi.32.100178.001253

Hacker J, Kaper JB (2000) Pathogenicity islands and the evolution of microbes. Annu Rev Microbiol 54:641–679

Haq SK, Atif SM, Khan RH (2004) Protein proteinase inhibitor genes in combat against insects, pests, and pathogens: natural and engineered phytoprotection. Arch Biochem Biophys 431:145–159

Hares MC, Hinchliffe SJ, Strong PC, Eleftherianos I, Dowling AJ, ffrench-Constant RH, Waterfield N (2008) The *Yersinia pseudotuberculosis* and *Yersinia pestis* toxin complex is active against cultured mammalian cells. Microbiology 154:3503–3517

Head G, Moar W, Eubanks M, Freeman B, Ruberson J, Hagerty A, Turnipseed S (2005) A multiyear, large-scale comparison of arthropod populations on commercially managed Bt and non-Bt cotton fields. Environ Entomol 34:1257–1266

Hey TD, Meade T, Burton SL, Merlo DJ, Cai Q, Moon HJ, Sheets JJ, Woosley AT (2006) Insecticidal toxin complex fusion proteins. US Patent 2006/0168683

Hinchliffe SJ, Hares MC, Dowling AJ, ffrench-Constant RH (2010) Insecticidal toxins from the *Photorhabdus* and *Xenorhabdus* bacteria. The Open Toxicol J 3:101–118

Hsiao TH, Fraenkel G (1969) Properties of leptinotarsin, a toxic hemolymph protein from the Colorado potato beetle. Toxicon 7:119–130

Huang DF, Zhang J, Song FP, Lang ZH (2007) Microbial control and biotechnology research on *Bacillus thuringiensis* in China. J Invertebr Pathol 95:175–180

Huesing J, English L (2004) The impact of crops on the developing world. AgBio Forum 7:84–95

Hurst MR, Glare TR, Jackson TA, Ronson CW (2000) Plasmid-located pathogenicity determinants of *Serratia entomophila*, the causal agent of amber disease of grass grub, show similarity to the insecticidal toxins of *Photorhabdus luminescens*. J Bacteriol 182:5127–5138

Hurst MR, Glare TR, Jackson TA (2004) Cloning *Serratia entomophila* antifeeding genes-a putative defective prophage active against the grass grub *Costelytra zealandica*. J Bacteriol 186:5116–5128

Hurst MR, Beard SS, Jackson TA, Jones SM (2007a) Purification and characterisation of the *Serratia entomophila* Afp. FEMS Microbiol Lett 270:42–48

Hurst MR, Jones SM, Tan B, Jackson TA (2007b) Induced expression of the *Serratia entomophila* Sep proteins shows activity towards the larvae of the New Zealand grass grub *Costelytra zealandica*. FEMS Microbiol Lett 275:160–167

Hurst MR, Becher SA, Young SD, Nelson TL, Glare TR (2011a) *Yersinia entomophaga* sp nov., isolated from the New Zealand grass grub *Costelytra zealandica*. Int J Syst Evol Microbial 61: 844–849

Hurst MR, Jones SA, Binglin T, Harper LA, Jackson TA, Glare TR (2011b) The main virulence determinant of *Yersinia entomophaga* MH96 is a broad-host-range toxin complex active against insects. J Bacteriol 193:1966–1980

Jackson TA (1995) Amber disease reduces trypsin activity in midgut of *Costelytra zealandica* (Coleoptera, Scarabaeidae) larvae. Research note. J Invertebr Pathol 65:68–69

Jackson TA, Pearson JF, O'Callaghan M, Mahanty HK, Willocks M (1992) Pathogen to product – development of *Serratia entomophila* (Enterobacteriaceae) as a commercial biological control agent for the New Zealand grass grub (*Costelytra zealandica*). In: Glare TR, Jackson TA (eds) Use of pathogens in scarab pest management. Intercept Ltd., Andover, pp 191–198

Jackson TA, Huger AM, Glare TR (1993) Pathology of amber disease in the New Zealand grass grub *Costelytra zealandica* (Coleoptera: Scarabaeidae). J Invertebr Pathol 61:123–130

Jackson TA, Boucias DG, Thaler JO (2001) Pathobiology of amber disease, caused by *Serratia* spp, in the New Zealand grass grub, *Costelytra zealandica*. J Invertebr Pathol 78:232–243

Jackson RE, Bradley JR, Van DWJ (2003) Field performance of transgenic cotton expressing one or two *Bacillus thuringiensis* endotoxins against bollworm, *Helicoverpa zea* (Boddie). J Cotton Sci 7:57–64

James C (2010) Global status of commercialized biotech/GMCrops ISAAA Briefs No. 42. ISAAA, Ithaca

James C (2012) Global status of commercialized biotech/GM Crops: 2012. ISAAA Briefs No. 44. ISAAA, Ithaca

Jeong HU, Mun HY, Oh HK (2010) Evaluation of insecticidal activity of a bacterial strain *Serratia*, sp EML-SE1 against diamondback moth. J Microbiol 48:541–545

Joshi MC, Sharma A, Kant S, Birah A, Gupta GP, Khan SR, Bhatnagar R, Banerjee N (2008) An insecticidal GroEL protein with chitin binding activity from *Xenorhabdus nematophila*. J Biol Chem 283:28287–28296

Kaur S (2000) Molecular approaches towards development of novel *Bacillus thuringiensis* biopesticides. World J Microbiol Biotechnol 16:781–793

Khan AG (2005) Role of soil microbes in the rhizospheres of plants growing on trace metal contaminated soils in phytoremediation. J Trace Elem Med Biol 18:355–364

Khandelwal P, Banerjee-Bhatnagar N (2003) Insecticidal activity associated with the outer membrane vesicles of *Xenorhabdus nematophilus*. Appl Environ Microbiol 69:2032–2037

Khandelwal P, Choudhury D, Birah A, Reddy MK, Gupta GP, Banerjee N (2004) Insecticidal pilin subunit from the insect pathogen *Xenorhabdus nematophila*. J Bacteriol 186:6465–6476

Klein MG (1990) Efficacy against soil-inhabiting insect pests. In: Gaugler R, Kaya HK (eds) Entomopathogenic nematodes in biological control. CRC, Boca Raton, pp 195–231

Konig W, Faltin Y, Scheffer J, Schoffler H, Braun V (1987) Role of cell-bound hemolysin as a pathogenicity factor for *Serratia* infections. Infect Immun 55:2554–2561

Kumari P, Kant S, Zaman S, Mahapatro GK, Banerjee N, Sarin NB (2014) A novel insecticidal GroEL protein from *Xenorhabdus nematophila* confers insect resistance in tobacco. Transgenic Res 23:99–107

Lacey LA, Georgis R (2012) Entomopathogenic nematodes for control of insect pests above and below ground with comments on commercial production. J Nematol 44:218–225

Lacey LA, Kaya HK (2007) Field manual of techniques in invertebrate pathology, 2nd edn. Springer, Dorcrecht

Lacey LA, Frutos R, Kaya HK, Vail P (2001) Insect pathogens as biological control agents: do they have a future? Biol Control 21:230–248. doi:10.1006/bcon.2001.0938

Lamelas A, Gosalbes MJ, Manzano-Marín A (2011) *Serratia symbiotica* from the aphid *Cinara cedri*: A missing link from facultative to obligate insect endosymbiont. PLoS Genet 7:1–11

Landsberg MJ, Jones SA, Busby JN, Rothnagel R, Busby JN, Marshall SDG, Simpson RM, Lott JS, Hankamer B, Hurst MRH (2011) 3D structure of the *Yersinia entomophaga* toxin complex and implications for insecticidal activity. Proc Natl Acad Sci USA 108:20544–20549

Lee SC, Stoilova-McPhie S, Baxter L, Fülöp V, Henderson J, Rodger A, Roper DI, Scott DJ, Smith CJ, Morgan JA (2007) Structural characterization of the insecticidal toxin Xpt A1 reveals a 1.15 MDa tetramer with a cage like structure. J Mol Biol 366:1558–1568

Lengyel K, Lang E, Fodor A, Szállás E, Schumann P, Stackebrandt E (2005) Description of four novel species of *Xenorhabdus*, family Enterobacteriaceae: *Xenorhabdus budapestensis* sp. nov., *Xenorhabdus ehlersii* sp. nov., *Xenorhabdus innexi* sp. nov. and *Xenorhabdus szentirmaii* sp. nov. Syst Appl Microbiol 28:115–122

Li M, Wu G, Liu C, Chen Y, Qiu L, Pang Y (2009) Expression and activity of a probable toxin from *Photorhabdus luminescens*. Mol Biol Rep 36:785–790. doi:10.1007/s11033-008-9246-z

Liu D, Burton S, Glancy T (2003) Insect resistance conferred by 283-kDa *Photorhabdus luminescens* protein TcdA in *Arabidopsis thaliana*. Nat Biotechnol 21:1307–1313

Liu W, Ye W, Wang Z, Wang X, Tian S, Cao H, Lian J (2006) *Photorhabdus luminescens* toxin-induced permeability change in *Manduca sexta* and *Tenebrio molitor* midgut brush border membrane and in unilamellar phospholipid vesicle. Environ Microbiol 8:858–870

Marra MC, Piggott NE, Goodwin BK (2010) The anticipated value of SmartStax™ for US corn growers. AgBio Forum 13:1–12

Mohan S, Sabir N (2005) Biosafety concerns on the use of Photorhabdus luminescens as biopesticide: experimental evidence of mortality in egg parasitoid *Trichogramma* spp. Curr Sci 89:1268–1272

Moran NA (2002) Microbial minimalism: genome reduction in bacterial pathogens. Cell 108: 583–586

Morgan JA, Sergeant M, Ellis D, Ousley M, Jarrett P (2001) Sequence analysis of insecticidal genes from *Xenorhabdus nematophilus* PMFI296. Appl Environ Microbiol 67:2062–2069

Naranjo SE (2005) Long-term assessment of the effects of transgenic Bt cotton on the function of the natural enemy community. Environ Entomol 34:1211–1223

Neale MC (1997) Bio-pesticides – harmonisation of registration requirements within EU directive 91–414. An industry view. Bull Eur Mediterranean Plant Protect Organ 27:89–93

Nuñez-Valdez ME, Calderón MA, Aranda E, Hernández L, Ramírez-Gama RM, Lina L, Rodríguez-Segura Z, Gutiérrez M, Villalobos FJ (2008) Identification of a putative Mexican strain of *Serratia entomophila* pathogenic against root-damaging larvae of Scarabaeidae (Coleoptera). Appl Environ Microbiol 74:802–810

Ochman H, Lawrence JG, Groisman EA (2000) Lateral gene transfer and the nature of bacterial innovation. Nature 405:299–304

Ogier JC, Calteau A, Forst S, Blair HG, Roche D, Rouy Z, Suen G, Zumbihl R, Givaudan A, Tailliez P, Médigue C, Gaudriault S (2010) Units of plasticity in bacterial genomes: new insight from the comparative genomics of two bacteria interacting with invertebrates, *Photorhabdus* and *Xenorhabdus*. BMC Genomics 11:568

Ohno S (1970) Evolution by gene duplication. Springer, New York, p 160

Parkhill J, Wren BW, Thomson NR (2001) Genome sequence of *Yersinia pestis*, the causative agent of plague. Nature 413:523–527

Paul VJ, Frautschy S, Fenical W, Nealson KH (1981) Antibiotics in microbial ecology: isolation and structure assignment of several new antibacterial compounds from the insect-symbiotic bacteria *Xenorhabdus* spp. J Chem Ecol 7:589–597

Pechy-Tarr M, Bruck DJ, Maurhofer M, Fischer E, Vogne C, Henkels MD, Donahuer M, Grunder J, Loper JE, Keel C (2008) Molecular analysis of a novel gene cluster encoding an insect toxin in plant-associated strains of *Pseudomonas fluorescens*. Environ Microbiol 10: 2368–2386

Pereira EJ, Lang BA, Storer NP, Siegfried BD (2008) Selection for Cry1F resistance in the European corn borer and cross-resistance to other Cry toxins. Entomol Exp Appl 126:115–121

Perlak F, Oppenhuizen M, Gustafson K, Voth R, Sivasupramaniam S, Heering D, Carey B, Ihrig RA, Roberts JK (2001) Development and commercial use of Bollgard cotton in the USA-early promises versus today's reality. Plant J 27:489–502

Perry RD, Fetherston JD (1997) *Yersinia pestis* – etiologic agent of plague. Clin Microbiol Rev 14: 35–66

Poinar GO Jr (1979) Nematodes for the biological control of insects. CRC, Boca Raton, p 227

Poinar GO (1993) Origins and phylogenetic relationships of the entomophilic rhabditids, *Heterorhabditis* and *Steinernema*. Fund Appl Nematol 16:333–338

Radake SG, Undirwade RS (1981) Seasonal abundance and insecticidal control of shoot and fruit borer, *Earias* spp. on okra, *Abelmoschus esculentus* (L.). Indian J Entomol 43:283–287

Rae R, Riebesell M, Dinkelacker I (2008) Isolation of naturally associated bacteria of necromenic *Pristionchus nematodes* and fitness consequences. J Exp Biol 211:1927–1936

Reineke A, Karlovsky P, Zebitz CPW (1999) Amplified fragment length polymorphism analysis of different geographic populations of the gypsy moth, *Lymantria dispar* (Lepidoptera: Lymantriidae). Bull Entomol Res 89:79–88

BCC Research (2010) Biopesticides: the global market-BCC Research Report, CHM029C, February 2010

Ribeiro C, Vignes M, Brehelin M (2003) *Xenorhabdus nematophila* (Enterobacteriaceae) secretes a cation selective calcium-independent porin which causes vacuolation of the rough endoplasmic reticulum and cell lysis. J Biol Chem 278:3030–3039

Ruffner B, Péchy-Tarr M, Ryffel F, Hoegger P, Obrist C, Rindlisbacher A, Keel C, Maurhofer M (2013) Oral insecticidal activity of plant-associated *Pseudomonads*. Environ Microbiol 15: 751–763. doi:10.1111/j.1462-2920.2012.02884.x

Ruiu L, Satta A, Floris I (2013) Emerging entomopathogenic bacteria for insect pest management. Bull Insectol 66:181–186

Sanahuja G, Banakar R, Twyman RM, Capell T, Christou P (2011) *Bacillus thuringiensis*: a century of research, development and commercial applications. Plant Biotechnol J 9:283–300

Sanchis V (2011) From microbial sprays to insect-resistant transgenic plants: history of the biospesticide *Bacillus thuringiensis*. A review. Agron Sustain Dev 31:217–231. doi:10.1051/agro/2010027

Saour G (2014) Sterile insect technique and F1 sterility in the European grapevine moth, *Lobesia botrana*. J Insect Sci 14:8

Sayyed AH, Moores G, Crickmore N, Wright DJ (2008) Cross-resistance between a *Bacillus thuringiensis* Cry toxin and non-Bt insecticides in the diamondback moth. Pest Manag Sci 64: 813–819

Schnepf E, Crickmore N, Van Rie J, Lereclus D, Baum J, Feitelson J, Zeigler DR, Dean DH (1998) *Bacillus thuringiensis* and its pesticidal crystal proteins. Microbiol Mol Biol Rev 62:775–806

Sergeant M, Jarrett P, Ousely M, Morgan AW (2003) Interactions of insecticidal toxin gene products from Xenorhabdus nematophila PMF1296. Appl Environ Microbiol 69:3344–3349

Sergeant M, Baxter L, Jarret P (2006) Identification, typing, and insecticidal activity of *xenorhabdus* isolates from entomopathogenic nematodes in United Kingdom soil and characterization of the xpt toxin loci. Appl Environ Microbiol 72:5895–5907

Shahid AA, Rao AQ, Bakhsh A, Husnain T (2012) Entomopathogenic fungi as biological controllers: new insights into their virulence and pathogenicity. Arch Biol Sci Belgrade 64: 21–42

Sheets JJ, Hey TD, Fencil KJ, BurtonS NW, Lang AE, Benz R, Aktories K (2011) Insecticidal toxin complex proteins from *Xenorhabdus nematophilus*: structure and pore formation. J Biol Chem 286:22742–22749

Sicard M, Le Brun N, Pages S, Godelle B, Boemare N, Moulia C (2003) Effect of native *Xenorhabdus* on the fitness of their *Steinernema* hosts: contrasting types of interactions. Parasitol Res 91:520–524

Sicard M, Ferdy JB, Pages S, LeBrun N, Godelle B, Boemare N, Moulia C (2004) When mutualists are pathogens: an experimental study of the symbioses between *Steinernema* (entomopathogenic nematodes) and *Xenorhabdus* (bacteria). J Evol Biol 17:985–993

Sikorowski PP, Lawrence AM, Inglis GD (2001) Effects of *Serratia marcescens* on rearing of the tobacco budworm (Lepidoptera: Noctuidae). Am Entomol 47:51–60

Silva CP, Waterfield NR, Daborn PJ, Dean P, Chilver T, Au CP, Sharma S, Potter U, Reynolds SE, ffrench-Constant RH (2002) Bacterial infection of a model insect: *Photorhabdus luminescens* and *Manduca sexta*. Cell Microbiol 4:329–339

Somvanshi VS, Lang E, Ganguly S, Swiderski J, Saxena AK, Stackebrandt E (2006) A novel species of *Xenorhabdus*, family Enterobacteriaceae: *Xenorhabdus indica* sp. nov, symbiotically associated with entomopathogenic nematode *Steinernema thermophilum*. Syst Appl Microbiol 29:519–525

Spinner JL, Jarrett CO, LaRock DL (2012) Yersinia pestis insecticidal-like toxin complex (Tc) family proteins: characterization of expression, subcellular localization, and potential role in infection of the flea vector. BMC Microbiol 12:1–14

Srinivasan A, Giri A, Gupta V (2006) Structural and functional diversities in Lepidopteran serine proteases. Cell Mol Biol Lett 11:132–154

Steinhaus EA (1941) A study of bacteria associated with thirty species of insects. J Bacteriol 42:757–790

Steinhaus EA (1951) Possible use of *Bacillus thuringiensis* Berliner as an aid in the biological control of the alfalfa caterpillar. Hilgardia 20:350–381

Sulakvelidze A (2000) *Yersinia* other than *Y. enterocolitica, Y. pseudotuberculosis,* and *Y. pestis,* the ignored species. Microbes Infect 2:497–513

Swihart KG, Welch RA (1990) Cytotoxic activity of the Proteus hemolysin HpmA. Infect Immun 58:1861–1869

Tailliez P, Pagès S, Ginibre N, Boemare N (2006) New insight into diversity in the genus *Xenorhabdus*, including the description of ten novel species. Int J Syst Evol Microbiol 56: 2805–2818

Talekar NS, Shelton AM (1993) Biology, ecology, and management of the Diamondback moth. Annu Rev Entomol 38:275–301

Vallet-Gely I, Opota O, Boniface A, Novikov A, Lemaitre B (2010) A secondary metabolite acting as a signalling molecule controls *Pseudomonas entomophila* virulence. Cell Microbiol 12: 1666–1679

Vermunt AMW, Koopmanschap AB, VLak JM, de Kort CAD (1997) Cloning and sequence analysis of cDNA encoding a putative juvenile hormone esterase from the Colorado potato beetle. Insect Biochem and Mol Biol 27:919–928

Vigneux F, Zumbihl R, Jubelin G, Ribeiro C, Poncet J, Baghdiguian S, Givaudan A, Brehélin M (2007) The xaxAB genes encoding a new apoptotic toxin from the insect pathogen *Xenorhabdus nematophila* are present in plant and human pathogens. J Biol Chem 282: 9571–9580

Vodovar N, Vinals M, Liehl P, Basset A, Degrouard J, Spellman P, Boccard F, Lemaitre B (2005) Drosophila host defense after oral infection by an entomopathogenic *Pseudomonas* species. Proc Natl Acad Sci USA 102:11414–11419

Vodovar N, Vallenet D, Cruveiller S, Rouy Z, Barbe V, Acosta C, Cattolico L, Jubin C, Lajus A, Segurens B, Vacherie B, Wincker P, Weissenbach J, Lemaitre B, Médigue C, Boccard F (2006) Complete genome sequence of the entomopathogenic and metabolically versatile soil bacterium *Pseudomonas entomophila*. Nat Biotechnol 24:673–679

Wang JX, Li LY (1987) Entomogenous nematode research in China. Rev Nematol 10:483–489

Waterfield N, Dowling A, Sharma S (2001) Oral toxicity of *Photorhabdus luminescens* W14 toxin complexes in *Escherichia coli*. Appl Environ Microbiol 67:5017–5024

Waterfield NR, Daborn PJ, ffrench-Constant RH (2002) Genomic islands in *Photorhabdus*. Trends Microbiol 10:541–545

Waterfield NR, Daborn PJ, Dowling AJ, Yang G, Hares M, ffrench-Constant RH (2003) The insecticidal toxin makes caterpillars floppy 2 (Mcf2) shows similarity to HrmA, an avirulence protein from a plant pathogen. FEMS Microbiol Lett 229:265–270

Waterfield N, Hares M, Yang G, Dowling A, ffrench-Constant R (2005a) Potentiation and cellular phenotypes of the insecticidal Toxin complexes of *Photorhabdus* bacteria. Cell Microbiol 7:373-382

Waterfield N, Kamita SG, Hammock BD, ffrench-Constant R (2005b) The *Photorhabdus* Pir toxins are similar to a developmentally regulated insect protein but show no juvenile hormone esterase activity. FEMS Microbiol Lett 245:47–52

Waterfield NR, Sanchez-Contreras M, Eleftherianos I, Dowling I, Yang G, Wilkinson P, Parkhill J, Thomson N, Reynolds SE, Bode HB, Dorus S, ffrench-Constant R (2008) Rapid virulence annotation (RVA): identification of virulence factors using a bacterial genome library and multiple invertebrate hosts. Proc Natl Acad Sci USA 105:15967–15972

Waterfield NR, Ciche T, Clarke D (2009) Photorhabdus and a host of hosts. Annu Rev Microbiol 63:557–574

Whitehouse MEA, Wilson LJ, Fitt GP (2005) A comparison of arthropod communities in transgenic Bt and conventional cotton in Australia. Environ Entomol 34:1224–1241

Wilkinson P, Waterfield NR, Crossman L, Corton C, Sanchez-Contreras M, Vlisidou l, Barron A, Bignell A, Clark L, Ormond D, Mayho M, Bason N, Smith F, Simmonds M, Churcher C, Harris D, Thompson NR, Quail M, Parkhill J, ffrench-Constant RH (2009) Comparative genomics of the emerging human Pathogen Photorhabdus asymbiotica with the insect pathogen *Photorhabdus luminescens*. BMC Genomics 10:302–324

Woese CR (1987) Bacterial evolution. Microbiol Rev 51(221):271

Wolters M, Boyle EC, Lardong K (2013) Cytotoxic necrotizing factor-Y boosts *Yersinia* effector translocation by activating Rac. J Biol Chem 288:23543–23553

Wu K, Lu Y, Feng H, Jiang Y, Zhao J (2008) Suppression of Cotton Bollworm in multiple crops in china in areas with Bt toxin-containing cotton. Science 321:1676–1678

Yang G, Dowling AJ, Gerike U, ffrench-Constant RH, Waterfield NR (2006) *Photorhabdus* virulence cassettes confer injectable insecticidal activity against the wax moth. J Bacteriol 188:2254–2261

Young SD, Townsend RJ, Swaminathan J, O'Callaghan M (2010) *Serratia entomophila*-coated seed to improve ryegrass establishment in the presence of grass grubs. N Z Plant Protect 63: 229–234

Chapter 6
The Management of *Spodopteran* Pests Using Fungal Pathogens

Md. Aslam Khan and Wasim Ahmad

6.1 Introduction

An estimated one third of global agricultural production valued at several billion dollars is destroyed annually by insect pests in field and storage (Mariapackiam and Ignacimuthu 2008). As the world population increases, the need to keep insects away from destroying food crops becomes even more urgent. Insects are central to the performance of many ecosystem processes. The class insecta contains nearly one million described species (May 2000) which comprise approximately 67 % of the world described fauna and flora. However, it is in their role as herbivores that conflicts arise with agricultural production due to direct consumption of cultivated crops and indirect damage by plant virus transmission resulting in reduction of potential yield. According to Oerke and Dehne (2004), insect pests are responsible for an estimated 42 % of all losses in crop production.

The last century has been witness to both the keynote augmentation of the chemical pesticides as well as its backlashes. Insecticides of synthetic origin have been used to manage insect pests for more than 50 years (Charnley and Collins 2007). Misuse of chemical pesticides and their negative impacts on soil and water quality, human health, wildlife, and the ecological balance within agroecosystems are increasingly becoming a concern, underlining the need for development of alternative eco-friendly pest control methods (Aktar et al. 2009). In view of growing public demand for food with low or no chemical residues, frantic efforts are being directed towards nonchemical and sustainable plant protection (Flexner and Belnavis 2000).

M.A. Khan
Department of Biology, Faculty of Science, Jazan University, Jazan, Saudi Arabia

W. Ahmad (✉)
Section of Nematology, Department of Zoology, Aligarh Muslim University, Aligarh 202002, India
e-mail: ahmadwasim57@gmail.com

© Springer International Publishing Switzerland 2015

123

K.S. Sree, A. Varma (eds.), *Biocontrol of Lepidopteran Pests*, Soil Biology 43,
DOI 10.1007/978-3-319-14499-3_6

Biological control agents are considered suitable alternatives to the use of chemical pesticides as these organisms are highly specific to host insects, besides being safe to the environment and mankind. Though the bio-management is not as rapid as the chemicals, it is practically more feasible and sustainable than chemical cure (McClintock et al. 2000). Biological control is defined as the use of living organisms to suppress the population density or impact of a specific pest organism, making it less abundant or less damaging than it would otherwise be (Eilenberg et al. 2001). Thus, the aim of biological control is to reduce pest population below the economic threshold. Many biological control agents are being employed for the abatement of pest and vector insects of veterinary and medical importance (Burges 1981; Tanada and Kaya 1993; Lacey and Kaya 2000). Natural enemies such as predators, parasitic wasps, flies, and insect pathogens have long been studied for exploitation in biological control and integrated pest management (IPM) strategies. In recent years, biological control of crop pests with entomopathogenic bacteria, viruses, fungi, and nematodes has been recognized as a valuable tool in pest management (Bhattacharya et al. 2003; Rosell et al. 2008).

Microbial control includes all aspects of utilization of microorganisms or their by-products for the control of pests. Microbial control agents are relatively host specific and do not upset other biotic systems. They are safe to humans, vertebrates, and beneficial organisms; do not cause environmental pollution; and are compatible with most other control methods. They are ideal for both short- and long-term pest suppression. Unlike chemical pesticides, they do not leave chemical residues on crops, are easy and safe to dispose off, and do not contaminate water systems. Compared to other microorganisms, entomopathogenic fungi have received considerable attention as they are exceptionally virulent and function as lethal parasites of insect pests. Fungal diseases are known to cause insect mortality naturally (Roberts and Humber 1981; Vimladevi and Prasad 2001; Gupta 2003). Some entomopathogenic fungi cause regular epizootics (Devi et al. 2003; Rios et al. 2010), resulting in very high levels of mortality in host populations. Entomopathogenic fungi are potential agents for pest control due to their host specificity, broad host range, mode of action, ease of production, and application (Dhaliwal and Koul 2007; Ignacimuthu 2008). This has promoted the evaluation of the entomopathogenic fungi as biocontrol agents in many countries. This chapter outlines the current state of knowledge of insect fungal pathogens, their use as mycoinsecticides, with special reference to spodopteran pest.

6.2 *Spodoptera* spp.

Spodoptera spp. [*Spodoptera litura* (Fabricius), tobacco cutworm; *S. littoralis* (Boisduval), Egyptian cotton leafworm; *S. frugiperda* (Smith), fall armyworm; *S. exigua* (Hübner), beet armyworm; and *S. exempta* (Walker), African armyworm] (Lepidoptera: Noctuidae) are major pests of subtropical and tropical agricultural crops with extensive host range of economically important crops.

Though described way back in the eighteenth century, the importance of *S. litura* Fabricius, 1775, as pest was brought to light only in the beginning of the twentieth century through the accounts of Willcock (1905), Lefroy (1908, 1909), and Fletcher (1914) which marked the beginning of contemporary researches on this obnoxious insect. *S. litura* is widely distributed throughout the world (Anand et al. 2009) and has been reported to feed on 120 cultivated food plants all over the world (http://www.cabi.org/isc/datasheet/44520), of which 40 are grown in India (Basu 1981; Muthukrishnan et al. 2005) including tobacco, cotton, groundnut, jute, lucerne, maize, rice, soybeans, tea, cauliflower, cabbage, and castor (Matsuura and Naito 1997; Sahayaraj and Paulraj 1998; Sharma and Bisht 2008). The moth is dark, with wavy markings on the fore wings and white hind wings and margin having a brown color. The moths have a flight range of 1.5 km during a period of 4 h overnight, facilitating dispersion and oviposition on different host plants (Salama and Shoukry 1972). They can accordingly fly quite long distances.

S. littoralis is similarly one of the most destructive agricultural lepidopteran pests within its subtropical and tropical range. Larvae of *S. littoralis* attack numerous economically important crops belonging to 44 different families including legumes, crucifers, grasses, and deciduous fruit trees, all the year round (Abdel-Megeed 1975). On cotton, the pest may cause considerable damage by feeding on the leaves, fruiting points, flower buds, and, occasionally, also on bolls. In tomatoes, larvae bore into the fruit which is thus rendered unsuitable for consumption. The pest could be active 9 months of a year and complete a generation within 30 days (Gharib 1979).

S. frugiperda is well known as a voracious insect pest of multiple agricultural crops in the Western Hemisphere (Clark et al. 2007; Murua et al. 2009). It is one of the most destructive insect pests attacking corn in the USA.

S. exigua was first discovered in North America around 1876 and reached Florida in 1924. In warm locations such as Florida, all stages can be found throughout the year, although development rate and overall abundance are reduced during the winter months (Tingle and Mitchell 1977). The life cycle can be completed in as few as 24 days, and six generations have been reared during 5 months of summer weather in Florida (Wilson 1934). The beet armyworm has a wide host range, occurring as a serious pest of vegetable, field, and flower crops. Weeds also are suitable for larval development. Larvae feed on both foliage and fruit. In Florida it is regarded as a serious defoliator of flower crops and cotton, though much of the injury is induced by insecticide use that interferes with natural enemy activity. Young larvae feed gregariously and skeletonize foliage. As they mature, larvae become solitary and eat large irregular holes in foliage. They also burrow into the crown or center of the head on lettuce, or on the buds of cole crops. As a leaf feeder, beet armyworm consumes much more cabbage tissue than the diamondback moth, *Plutella xylostella* (Linnaeus), but is less damaging than the cabbage looper, *Trichoplusia ni* (Hübner) (East et al. 1989). Tomato fruit is most susceptible to injury, especially near fruit maturity, but beet armyworm is not considered to be as threatening to tomato as is the corn earworm, *Helicoverpa zea* (Boddie) (Zalom et al. 1986).

6.2.1 Biology

In *S. litura,* female lays 1,000–2,000 eggs in whitish-yellow egg masses of 100–300 on the lower leaf surface of the host plant (Miyahara et al. 1971). The masses are covered by hairlike scales from the end of the insect's abdomen. Fecundity is adversely affected by high temperature and low humidity. The eggs hatch in about 4 days in warm conditions, or up to 11–12 days in winter. The larvae pass through six instars in 15–23 days at 25–26 °C. The young larvae (first to third instar) feed in groups, leaving the opposite epidermis of the leaf intact. Grown-up larva is stout, cylindrical, and pale greenish brown with dark markings, some have transverse and longitudinal gray and yellow bands, and others have rows of dark spots. Later, the (fourth to sixth instar) larvae disperse and spend the day in the ground under the host plant (Fig. 6.1), feeding at night and early in the morning. The pupal period is spent in earthen cells in the soil and lasts about 11–13 days at 25 °C. Longevity of adults is about 4–10 days, being reduced by high temperature and low humidity. Thus, the life cycle generally completes in about 5 weeks, while Saeedeh et al. (2013) reported completion of life cycle between 29 and 35 days at different

Fig. 6.1 Different developmental stages of *Spodoptera litura*. (**a**) Egg mass. (**b**) Larva. (**c**) Pupae. (**d**) Adult moth (photos courtesy: Dr. K. Sowjanya Sree, Amity University UP, India)

temperature and humidity. In the seasonal tropics, several generations develop during the rainy season, while the dry season is survived in the pupal stage. Biological parameters, larval and pupal development, and survival, longevity, and fecundity of adults of *S. litura* are affected by the host plants (Ming et al. 2010).

6.3 Entomopathogenic Fungi

Entomopathogenic fungi, also referred as disease-causing fungi among insects, contribute to the natural regulation of insect populations (Butt et al. 2001; Kleespies et al. 2008). These fungi are among the first microorganisms to be used for the biological control of insect pests. They are cosmopolitan organisms which have been isolated from soils and infected insects from around the world. The natural occurrence of these fungi is also documented by numerous researchers as summarized in the excellent review by Zimmermann (2007a, b, 2008). More than 700 species of fungi from around 90 genera are pathogenic to insects (Thackar 2002; Wraight et al. 2007; Khachatourians and Sohail 2008; Hemasree 2013); however, only few have been thoroughly investigated for their use against insect pests in agriculture.

Modern exploration of entomopathogenic fungi began with the work of Russian entomologist, Metschnikoff, who conducted first systematic experiments on the control of injurious insects by infecting grubs of the grain beetle, *Anisopliae austriaca*, with the green muscardine fungus, *Metarhizium anisopliae* Sorokin. But the real breakthrough in the development of microbial control agents came with the discovery and practical application of the milky disease bacteria, *Bacillus popilliae* Dutky, for the control of Japanese beetle, *Popilliae japonica* Newman, in the USA during the 1940s (Dhaliwal and Koul 2007).

The most widespread insect pathogenic fungal genera are found in the order *Hypocreales*, phylum Ascomycota, viz., *Metarhizium* and *Beauveria*. *M. anisopliae* and *B. bassiana* (Balsamo) Vuillemin, the white muscardine fungus, are the two most recognized species (Vincent et al. 2007). They grow naturally throughout the world and act as parasites of many arthropod species causing white and green muscardine diseases due to the color of their spores (Vincent et al. 2007). There has been overwhelming interest in exploring the potential for *B. bassiana* and *M. anisopliae*.

Entomopathogenic fungi possess added advantage over other microbial control agents as they are capable of attacking all developmental stages of insects including pupal stages (Ferron 1978; Anand et al. 2009). Most importantly, these fungi are the principal pathogens of sucking pests, since these hosts cannot ingest other pathogens like bacteria or viruses that infect through gut wall. None of the entomopathogenic fungi currently in use or under consideration are invasively pathogenic to humans (Kubicek and Druzhinina 2007). Certain ecological niches lend themselves particularly well to the deployment of mycoinsecticides.

B. bassiana, *M. anisopliae*, and *Lecanicillium* (*Verticillium*) *lecanii* (Zimmermann) are intensively studied as common natural enemies and important epizootics of agricultural pests (Milner 1997; Roberts and St. Leger 2004; Wang et al. 2004; Thomas and Read 2007; Li and Sheng 2007). Devi et al. (2003) reported epizootics caused by *Nomuraea rileyi* Farlow (Samson) in south Indian fields during winter. About 33.9 % of the mycoinsecticides are based on *B. bassiana*, followed by *M. anisopliae* (33.9 %), *Isaria fumosorosea* (5.8 %), and *B. brongniartii* (4.1 %) (Faria and Wraight 2007). *B. bassiana* is reported to limit the growth of plant pathogenic fungi in vitro as well as reducing the diseases caused by soilborne plant pathogens like *Pythium*, *Rhizoctonia*, and *Fusarium* (Ownley et al. 2010). Some isolates of *M. anisopliae* are rhizosphere competent and can establish a symbiotic relationship with plant roots (Hu and St. Leger 2002; St. Leger 2008).

6.3.1 Classification/Systematic Position

Entomopathogenic fungi are a very heterogeneous group of insect pathogens, belonging to different systematic groups. Most entomopathogenic fungi are from the fungal divisions Ascomycota and Zygomycota (Samson et al. 1988). The Ascomycete fungi were previously divided into two groups, the Ascomycota and the Deuteromycota. The Fungi Imperfecti of Deuteromycota was known for having no sexual stage. But later, cultural and molecular studies have demonstrated that some of these "imperfect fungi" were in fact anamorphs (asexual forms) of the Ascomycota within the order *Hypocreales* (formerly called Hyphomycetes) (Krasnoff et al. 1995; Fukatzu et al. 1997; Hodge 2003). Within Zygomycota, the most entomopathogenic species are in the order *Entomophthorales* (Roy et al. 2006; Hussain et al. 2012).

The genus *Metarhizium* presently includes several morphologically distinctive species including *M. anisopliae*, *M. album*, *M. flavoviride*, *M. cylindrosporae*, *M. guizhouense*, and *M. pingshaense* (Guo et al. 1986; Rath et al. 1995; Driver et al. 2000) that cause diseases in different insect species. Despite recent interest in the genetic diversity of many groups of entomopathogens, the genus *Beauveria*, in contrast to other genera, has not received critical taxonomic review (Rehner et al. 2011). Neelapu et al. (2009) studied molecular phylogeny of asexual entomopathogenic fungi with special reference to *B. bassiana* and *N. rileyi* using the sequences of partial regions of beta-tubulin and rRNA genes.

6.3.2 Dispersal/Transmission

Entomopathogenic fungi are widely distributed with both restricted and wide host ranges having different biocontrol potentials against insect pest. Different insect pathogenic mycofloras could be found in soil and in the overground environment.

Asexually produced fungal spores or conidia are generally responsible for infection and are dispersed throughout the environment in which the insect hosts are present. Entomopathogenic fungi are constantly present in populations of insect hosts, but when density of the host population is normal, infections occur sporadically (enzootic phase of insect diseases). However, during insects' outbreak, fungi can increase their numbers enough to spread in the environment and contribute to the reduction of insect population (epizootic phase) (Fuxa and Tanada 1987).

Soil is a natural environment for entomopathogenic fungi. Many insects spend at least part of its life in the soil; such natural behavior of insects related to their biology, such as accumulation in the soil or leaf litter to wintering or pupation, is conducive to fungal infections leading to natural reduction of many insect pests. It has been shown experimentally that spores of the lepidopteran pathogen, *N. rileyi*, adhere to leaves of plant seedlings as they emerge through the soil (Ignoffo et al. 1977). Despite the fact that both *B. bassiana* and *M. anisopliae* are common everywhere, *B. bassiana* seems to be very sensitive to disturbance and thus restricted to natural habitats. Entomopathogenic fungi in the insect populations may transmit horizontally (from infected insects on healthy individuals) or by vectors. This second method plays an important role in the transmission of fungi to new habitats (Fuxa and Tanada 1987).

6.3.3   Use of Fungi for Insect Biological Control

Entomopathogenic fungi can be employed for classical biological control, augmentation, and conservation. Views by Jackson et al. (2000) and Butt et al. (2001) provide further information. Classical biological control using fungal entomopathogens can provide a successful and environment friendly avenue for controlling arthropod pests, including the increasing numbers of invasive nonnative species (Hajek and Delalibera 2010).

6.3.3.1   Classical Biological Control

Classical biological control is the intentional introduction of an exotic biological control agent for permanent establishment and long-term pest control (Eilenberg et al. 2001). In the case of microorganisms widely distributed in nature, the term exotic means the use of a particular strain or biotype, which is not native to the area where the pest is controlled. Introduced species to induce long-term effect has to acclimate to the area under certain climatic conditions, multiply, and spread. *Entomophaga maimaiga* was introduced from Japan into Eastern North America in 1910–1911 to control the gypsy moth *Lymantria dispar*.

6.3.3.2 Augmentation

In many situations, natural enemies are present in indigenous pest populations, but they are either too few or active too late to limit crop damage. In these cases, the natural enemies can be augmented. There are two approaches to augmentation: inoculation and inundation. In an inoculation approach, the fungus is applied, often in small amounts, early in the season of the crop, with the expectation that it will establish epizootics in pest populations and thereby maintaining the pest population below the economic threshold. Inundative augmentation involves applying the fungus, often in large amounts, for rapid short-term control with no expectation for secondary infection (Eilenberg et al. 2001). In this way, the fungus is used in a similar way to a chemical insecticide. The term "mycopesticide" or "mycoinsecticide" have been used to describe this approach. Anamorphic entomopathogenic fungi, such as *B. bassiana* and *M. anisopliae*, are usually developed as inundative control agents which are applied in mass to a pest population, and there is little expectation that they will persist and reproduce within the biotic environment.

6.3.3.3 Conservation

Conservation biological control is modification of the environment or existing practices to protect and enhance specific natural enemies or other organisms to reduce the effect of pests (Fuxa 1998; Eilenberg et al. 2001). Fungal pathogens act generally in a density-dependent fashion against their hosts and have relatively slow kill. Therefore, epizootics of fungal pathogens on crop pests often occur too late to be of economic value. One key advantage over chemical pesticides is lost if an epizootic occurs too late or does not occur at all. Epizootics of nuclear poly-hedrosis virus (NPV) are as common as fungal infections, but NPV requires high-density populations for development of epizootics, whereas entomopathogenic fungi can cause high level of infection in low-density as well as high-density populations (Hajek et al. 1995). Mixed infections of entomopathogenic fungi and viruses are also common in other pathogen-host systems. Specificity can be a problem when there is a pest complex and no single pathogen can give control (Powell and Jutsum 1993; Ravensberg 1994). In Columbia, there is a product named Microbiol Completo, which is a mixture of *B. bassiana*, *M. anisopliae*, *N. rileyi*, *P. fumosoroseus*, and the bacterium *B. thuringiensis* at 1×10^9 spores of each pathogen per gram of product, for control of larvae and adults of Lepidoptera, Coleoptera, Hemiptera, Diptera, and mites (Alves et al. 2003). An alternative is to use two treatments sequentially. Early-season use of *B. bassiana* followed by insecticides gave good control of beetle and caterpillar pests on crucifer crops (Vandenberg et al. 1998).

6.3.4 Genetic Manipulation

The potential of entomopathogenic fungi often varies in different ecosystems and among fungal species and strains. Natural strains of these fungi often lack sufficient virulence or tolerance to adversity (Rangel et al. 2005). Therefore, genetic manipulation is necessary to improve their efficacy and ecological fitness (Fang et al. 2005; St. Leger and Wang 2009). In the process to improve the virulence of entomopathogenic fungi, the extensive transcriptomic and genetic study of fungal infection process revealed that a number of different genes were involved in the pathogenicity (Freimoser et al. 2003; Wang et al. 2005; Cho et al. 2006a, b, 2007). Therefore, highly virulent fungal genotypes against particular insect pest were manipulated using molecular techniques. Development of recombinant DNA techniques has made it possible to significantly improve the insecticidal efficacy of fungi, bacteria, and viruses (Inceoglu et al. 2006; Wang and St. Leger 2007a, b, c). These new types of biological insecticides offer a range of environment friendly options for cost-effective control of insect pests (Federici et al. 2008). Integration of insecticidal protein vip3Aa1 into *B. bassiana* enhances fungal virulence against *S. litura* larvae (Yi et al. 2010). Fungal infection of aphids was accelerated by overexpressing a silkworm chitinase or a hybrid chitinase in *B. bassiana* (Fang et al. 2005; Fan et al. 2007). The realization of the economic potential of mycoinsecticides would benefit from advances in biotechnology (Miranpuri and Khachatourians 1995). Review by Raymond and Chengshu (2010) provides comprehensive information for genetic engineering of fungal biocontrol agents.

6.3.5 General Biology of Entomopathogenic Fungi

The life cycle of most entomopathogenic fungi consists of two phases: a normal mycelia growth phase mostly outside the host body and a yeastlike budding phase mostly in the hemocoel of the host. When a spore adheres to the cuticle of insects, a germ tube is generated and passed through the integument by mechanical and enzymatic (e.g., chitinases, proteases, and lipases) process. When it reaches the insect's hemocoel, it produces blastospores which are the final pathogenic parts for host infection (Vincent et al. 2007). Most, if not all, entomopathogenic fungi have life cycles which synchronize with insect host stages and environmental conditions. Complete developmental cycle of *N. rileyi* in *S. litura* lasted approximately 8–9 days (Srisukchayakul et al. 2005). Species, and sometimes isolates within a species, can behave differently. For example, insect host range, infection levels, germination rates, and temperature optima can vary between species and isolates (Sierotzki et al. 2000; Pell et al. 2001; Shaw et al. 2002). Members of the *Hypocreales* such as *Beauveria* and *Metarhizium* spp. are opportunistic hemibiotrophs with a parasitic phase in the live host and saprotrophic phase during postmortem growth on the cadaver. Entomopathogenic fungi can invade their hosts directly through the exoskeleton or cuticle; therefore, they can infect nonfeeding stages such as eggs and pupae. Information on the biology and ecology of entomopathogenic fungi can be obtained from Steinhaus (1949, 1964), Evans (1989), and Balazy (1993). Uribe and Khachatourians (2008) have also described the life cycle of *M. anisopliae* under liquid culture conditions.

6.3.6 Isolation of Entomopathogenic Fungi

6.3.6.1 From Insects

Entomopathogenic fungi have been isolated from infected insects and soil from around the world. Several entomopathogenic fungal pathogens such as *B. bassiana*, *M. anisopliae*, and *Lecanicillium* spp. with good potential for pest control had been isolated from a variety of insects (Wraight et al. 2000; Faria and Wraight 2007). Ramiro et al. (2013) isolated entomopathogenic fungus *N. rileyi* from dead *S. frugiperda* larvae on corn. Satti and Gorashi (2013) isolated *B. bassiana* and *Paecilomyces* sp. from infected dead beetle in Northern Sudan. Mudroncekova et al. (2013) isolated entomopathogenic fungi from infected individuals of *Ips typographus*. Before attempting to isolate a fungal pathogen, it is important to keep the specimens fairly dry to avoid further deterioration by growth of saprophytic fungi and bacteria. The following protocol is quite effective:

- Surface sterilize the insect by immersing it in alcohol or 5 % solution of Na–HCl, $HgCl_2$, or other suitable germicides for several minutes, and then rinse it in three changes of sterile water.
- In a sterile dish, open the specimen and transfer a small portion of infected tissue to a sterile culture plate. Saboraud's dextrose agar with yeast extract produces quick growth for entomogenous fungi, and the acid reaction (pH 5.6) retards bacterial growth. The larva to be used for isolation of the fungus should not be exposed to air once the sporulation is initiated. After sporulation of the fungus, the Petri plate should be sealed with a parafilm and transferred to the refrigerator to avoid any contaminants in isolation.
- The spore from the surface of the cadavers is streaked aseptically on the surface of SMAY slopes and incubated as usual in the dark at 25 °C. Mycelium and spores can also be removed from a fresh specimen and placed directly on the medium; however, this isolate should be compared with that obtained from the infected tissues, since the chances of encountering a saprophytic fungus from the surface of the specimen is much greater than from the internal tissue.
- The cultures can be placed in a moist incubator at 25 °C and examined daily. After growth and sporulation, fungi were identified based on the morphological characters as per Humber (1997).

- After sporulation is complete, the slants are transferred to the refrigerator and stored. For several fungi, sporulation continues for a week after transfer to the refrigerator (Vimladevi and Prasad 2008).

6.3.6.2   From Soil

Soil is considered as excellent habitat for insect pathogenic fungi and other microorganisms since it is protected from UV radiation and buffered against extreme biotic and abiotic influences (Keller and Zimmerman 1989). Fungal epizootics in soil insect populations are also well documented (Samson et al. 1988; Keller and Zimmerman 1989; Klingen and Haukeland 2006). Insect pathogenic fungi in the genera *Beauveria*, *Metarhizium*, and *Paecilomyces* are all commonly found in the soil (Domsch et al. 1980); David et al. (2003); Neuman and Shields (2004); Torasco and poliseno (2005). Serigo et al. (2010) collected entomopathogenic fungi *B. bassiana* and *M. anisopliae* from soil using bait method. Greater wax moth, *Galleria mellonella* (L.) larvae were used as bait insects. The "*Galleria* bait method" was first introduced by Zimmermann (1986) as a sensitive method to detect a broad spectrum of insect pathogenic fungi in soil samples. Use of insect bait is a very sensitive detection method, and entomopathogenic fungi can be selectively isolated. Insect-associated fungi were detected in 55.5 % of the 425 soil samples collected from different field crops and orchards in China by Sun and Liu (2008) using *Galleria* bait method. They also reported that diversity of insect pathogenic fungi was greatest in field crop soil than in orchard. Sookar et al. (2008) also reported three isolates of entomopathogenic fungi, *B. bassiana*, *M. anisopliae*, and *Paecilomyces fumosoroseus*, from soil samples. Mudroncekova et al. (2013) also isolated entomopathogenic fungi from soil samples. Further a critical review by Hemasree (2013) on the natural occurrence of entomopathogenic fungi in agricultural ecosystem provides complete information.

6.3.7   Mode of Action of Entomopathogenic Fungi

Unlike bacteria and viruses, which must be consumed, toxicity from entomopathogenic fungi most often occurs from contact of the fungal conidia with the host cuticle. Insect cuticle is composed of chitin fibrils embedded in a matrix of proteins, lipids, pigments, and N-acylcatecholamines. Entomopathogenic fungi secrete extracellular enzymes, proteases, chitinases, and lipases to degrade the major constituents of the cuticle and allow hyphal penetration (Wang et al. 2005; Cho et al. 2006b). The success of invasion is directly proportional to secretion of exoenzymes (Khachatourians 1996). However, the ability of fungus to withstand antifungal compounds such as short-chain fatty acids, in the insect cuticle, is a prerequisite for successful invasion (Boucias and Pendland 1991). Numerous light and electron microscope studies on the invasion of host cuticle by entomopathogenic fungi are consistent with the involvement of both enzymes and mechanical pressure. Besides exoenzymes, entomopathogenic fungi are reported to produce toxin proteins (Roberts 1981; Gillespie et al. 2000; Strasser et al. 2000; Freimoser et al. 2003). Toxin production is reported for *B. bassiana*, *M. anisopliae*, and *V. lecanii*. However, *B. bassiana* is reported to produce low molecular weight cyclic peptides and cyclosporins A and C with insecticidal properties (Roberts 1981; Vey et al. 2001). Some of the strains of *B. bassiana* are reported to produce high molecular weight compounds with toxic activity against insect pest (Enrique and Alain 2004). Maoye et al. (2014) mentioned great differences in chitinase activities of *Beauveria* isolates.

The development of fungal infections in terrestrial insects is largely influenced by environmental conditions. High humidity is vital for germination of fungal spores and transmission of the pathogens from one insect to another. With most entomopathogenic fungi, disease development involves the following steps:

Infection process starts with adhesion of conidia on the insect cuticle (Srisukchayakul et al. 2005). Once the fungus breaks through the cuticle and underlying epidermis, it may grow profusely in insect hemolymph. Fungus invade host cuticle through the body wall and spiracles primarily and also through the mouth parts (Hajek and St Leger 1994; Clarkson and Charnley 1996). The conidia germinate on the insect cuticle by producing germ tubes which penetrate the body wall. Growth of the fungus after it reaches the hemocoel is by budding which produces hyphal bodies. These are transported throughout the hemocoel and give rise to localized concentration of mycelia. A heavy growth of intertwining mycelia develops in the hemocoel 1–2 days later. Insect death is probably the result of starvation or physiological/biochemical disruption brought about by the fungus (Feng et al. 1994). At the end of the infection cycle, mycelia emerge from the cuticle and produce conidiophores (Srisukchayakul et al. 2005; Joseph et al. 2010). Under suitable conditions, particularly high relative humidity, external sporulation helps to spread the fungus and establish an epizootic.

6.3.8 Insect Immune Responses to Entomopathogenic Fungi

Insects have effective immune systems composed of both cellular and humoral responses for fighting back against different kinds of microbial agents (Boman and Hultmark 1987). When cuticular defense mechanism fails to overcome the invading microorganisms, cellular defense regulated via interaction of different hemocytes is initiated (Gupta et al. 2005). Hemolymph of insects is a medium for several physiological processes like immune responses and intermediary metabolism. When an invader enters the hemocoel of insects, hemocytes get engaged to remove non-self-target by phagocytosis, nodule formation, encapsulation, synthesis of antimicrobial peptides, and reactive metabolites (Beckage 2008). Meshrif et al. (2011) reported a time-dependent decrease and increase in the phagocytosis activity after injection of *B. bassiana* and *N. rileyi*, respectively, against *S. littoralis*.

Both fungi provoked a decrease of the total number of hemocytes at 48 h followed by an increase at 72 h postinjection. Phenoloxidase enzyme is an important key enzyme that triggered immune response in *S. litura* larvae against the invasion of *B. bassiana*. Significant increase in PO level was observed at 4.0×10^6–2.0×10^7 spores/ml after 24 h of infection in 3rd and 4th instar larvae of *S. litura* (Gurmeet and Sanehdeep 2013). Study by Liu et al. (2014) gives an insight into better understanding of the molecular mechanisms of innate immune processes in Asian corn borer *Ostrinia furnacalis* larvae against *B. bassiana*. An excellent review on entomopathogenic fungi versus insect cuticle by Almudena and Nemat (2013) provides complete information on advances in broad-host-range entomopathogenic fungi.

6.3.9 Compatibility of Entomopathogenic Fungi with Other Crop Protection Techniques

The integration of biological and chemical control approaches is important for successful IPM programs. Entomopathogenic fungi compatibility with other crop protection techniques such as insecticides and commercial botanicals necessitates a proper understanding of the development and reproduction of pathogen (Malo 1993). For successful biological control, the biology of the control agent and its compatibility with other agents is important (Cuthbertson and Murchie 2007). Fungal biological control agents and selective insecticide may act synergistically increasing the efficiency of the control (Ambethgar 2009). By contrast, use of incompatible insecticides may inhibit growth and reproduction of the pathogens. Therefore, fungal genotypes compatible to particular pesticides can be identified and manipulated. Approach is based on the assumption that, weakened by another stressor, the insect will succumb more readily to mycosis. Beneficial effects were observed in using the two fungi together compared to the separate application. In China, combinations of *B. bassiana* and certain insecticides have been recommended for application against crop and forest pests (Feng et al. 1994). Synergy occurred between *B. bassiana* and imidacloprid against caterpillar *Spilarctia obliqua* (Walker) (Purwar and Sachan 2006). Efficacy of *B. bassiana* (Bb-L-2) and *M. anisopliae* (Ma-L-1) strains enhanced in combination with insecticide, spinosad 45SC at 0.009 %, neem oil 5 %, and NSKE 5 % against *S. litura* larvae (Udayababu et al. 2012).

Gopalkrishnan and Mohan (2000) tested seven insecticides and seven fungicides which are commonly used for the control of pests and diseases of tomato for their inhibitory effect on germination of conidia of *N. rileyi* at three (low, normal, and high) concentrations in vitro. They concluded that monocrotophos, phosphomedon, and dimethoate were safe to the mycopathogen at all the three concentrations. Among the fungicides, captafol, zineb, chlorothalonil, fosetyl-Al, and ziram were safe to the fungus at all the concentrations evaluated. Captan and sulfur on the other hand, though allowed conidial germination at low concentrations but normal concentrations, caused total inhibition at higher concentrations. No synergistic effect was observed on pathogenicity in the combination treatments of *B. bassiana* with *L. muscarium* (Naser et al. 2013). Asi et al. (2010) reported that insecticides significantly inhibit mycelial growth and conidial germination of *M. anisopliae* and *P. fumosoroseus*. Complete growth inhibition (100 %) of new isolate IOF1 of *N. rileyi* was reported by Patil et al. (2013) in compatibility with fungicides, carbendazim, and mancozeb at recommended doses, but insecticides endosulfan and dichlorvos inhibit growth by 52.8 and 50.7 % respectively. Golshan et al. (2013) reported highest adverse effect of fungicides on the germination of *B. bassiana* isolates in the rice fields; however, they recorded highest germination of fungal isolates in the herbicide treatments. Inhibitory effects of pesticides on germination and mycelial growth of entomopathogenic fungi often vary among fungal species and strains (Vanninen and Hokkanen 1988; Anderson et al. 1989). Commercial botanicals and ethanol extracts of seeds and leaves also reduce the mycelial growth of *B. bassiana*, *I. fumosorosea*, and *L. lecanii* (Sahayaraj et al. 2011). Plant extracts, *Annona squamosa* and *Polyalthia longifolia*, showed 55.8 % growth inhibition of *N. rileyi* isolate (Patil et al. 2013).

6.4 Control Measures of *Spodoptera* spp.

6.4.1 Prophylactic Measures

Cultural practices like deep plowing of infested fields so as to expose the pupae to the sun rays and insectivorous birds, destruction of weed and wild host plants, burning of crop residues, growing of castor as ovipositional trap crop, and collection and destruction of egg masses and early instars had been in practice for a very long time. Jayaraj (1978) added a new dimension to the above practices by advocating the cultivation of less susceptible cotton varieties in Tamil Nadu. He further suggested that rationing of cotton crop be stopped as it aggravates the pest problem. Chari and Patel (1983) added that cultivation of summer crops like bhindi (*Abelmoschus esculentus*) and lucerne (*Medicago sativa*) with irrigation be avoided to keep the population of pest under check.

Light traps had been in use for a considerable long time as one of the important tools of limiting the pest population. Pandey (1970) operated light traps and found that early hours of the night provided better catches if ecological factors like wind, rainfall, temperature, and humidity did not affect the pest adversely. Jayaraj (1978) suggested that light traps may be operated between 7 p.m. and 11 p.m. for better catches. Chari and Patel (1983) recommended blacklight traps for this pest. Tucker (1983) studied the association between weather and light-trap catches of *Spodoptera* and reported that high densities of flying moths, which may give rise to caterpillar outbreaks early in the year, often occur where there is rather infrequent early-season rainfall. Field trials on the population dynamics of *S. litura* using light and pheromone traps indicated its activity throughout the year (Nandihalli et al. 1989). Pheromone traps can be a viable alternative in integrated management of tobacco cutworm infestation. It can help farmers to get higher benefits by lowering costs and increasing yields than mere chemical control. Adoption of sex pheromone trapping can effectively suppress adults, largely decrease larvae or egg mass density, and damage rate of *S. litura* (Yang et al. 2009).

6.4.2 Chemical Insecticidal Control

Management of obnoxious lepidopteran pests relies mainly on the use of chemical insecticides including carbamates, pyrethroids, and organophosphates (Liburd et al. 2000). Chemical control of *Spodoptera* has been extensively reported on various crops in India. Two common insecticides, DDT and BHC, which were regarded as effective checks for all pest problems in the early parts of fifties and sixties in India, were used against *S. litura* (Kurup and Joshi 1959). Later, researchers recommended the use of strong insecticides like endrin and endosulfan either singly or in combination with other insecticides for the control of *S. litura* (Jotwani et al. 1961; Sarup and Singh 1974). Besides the chlorinated insecticides, some very promising organophosphatic insecticides were also recommended for the control of *S. litura*. Malathion, dimethoate, etc. were the common organophosphatic insecticides recommended by various workers (Sarup 1970; Saad et al. 1975). Yet another landmark in the insecticidal world was made by the introduction of carbamates. Carbaryl, a potent carbamate insecticide, was recommended for the control of *S. litura* (Yathom and Rivnay 1960; Sarup 1970). Hazards and harmful effects of insecticides as chemical control especially the wide application of conventional insecticides necessitate the new chemistry insecticides which are more effective, safer for humans, and much less toxic to our ecosystem (Korrat et al. 2012). Pyrethroids were least effective due to their high LC_{50} values as compared to new chemistry insecticides (Bhatti et al 2013).

Resistance to insecticides is a major problem associated with the chemical control of insect pests, which is characterized by rapid evolution under strong selection of gene(s) that confers survival to insecticides (Ahmad et al. 2008). The pests have developed high resistance against a wide variety of insecticides including organophosphate, carbamate, and pyrethroids, resulting in their sporadic outbreaks and failure of crops (Armes et al. 1997; Kranthi et al. 2001; Ahmad et al. 2007a, b, 2008; Saleem et al. 2008). The management of the pest has therefore become increasingly difficult all over the world. Shankarganesh et al. (2012) observed resistance against conventional as well as new chemistry insecticides such as quinalphos, monocrotophos, lindane, endosulfan, benzene hexachloride, avermectins, spinosad, fipronil, indoxacarb, and chitin synthesis inhibitors. Tong et al. (2013) reported field resistance of *S. litura* to carbamates (thiodicarb or methomyl) was significantly higher than that of organophosphates and pyrethroids.

6.4.3 Plant Products/Biopesticides

Replacement of synthetic insecticides by bio-rational insecticide is universally acceptable and is a practicable approach worldwide. Throughout history, plant products have been successfully exploited as insecticides. Recent plant protection research, particularly of the last decade, revealed the importance of plant products that disrupt normal insect growth and development. Plants are rich sources of natural substances that can be utilized in the development of environmentally safe methods for insect control (Sadek 2003). Plants contain secondary metabolites that are deleterious to insect pests in diverse ways; through acute toxicity, enzyme inhibition and interference with the consumption and/or utilization of food (Wheeler and Isman 2001) and the biomolecules present in phytopesticides act as a feeding deterrent, ovicide, oviposition deterrent, and growth inhibitor against field insect pests (Baskar et al. 2011; Baskar and Ignacimuthu 2012a, b; Arivoli and Tennyson 2013). Several plant products have been screened and tested against *S. litura* and some promising plants have been reported by Arivoli and Samuel (2012, 2013). However, screening of plant extracts against this pest is still continuing throughout the world to find eco-friendly biopesticides.

6.4.4 Biological Control

Biological control represents an important strategy to reestablish the biodiversity of agroecosystems, especially with the introduction of entomophagous organisms with classic innoculative techniques or the increase of natural populations of predators, parasitoids, and pathogens (Silva 2000).

6.4.4.1 Predators and Parasitoids

A number of predators [*Andrallus spinidens* and *Harpactos costalis* (Hemiptera: Pentatomidae), *Agrypnus fuscipes* and *Broscus punctatus* (Coleoptera: Elateridae) (Chari and patel 1983), and generalist predator *Podisus nigrispinus* (Heteroptera: Pentatomidae)], widespread in the Neotropical region (Thomas 1992), have been reported to feed on *Spodopteran* larvae (Batalha et al. 1997; Zanuncio et al. 1998; 2008).

Parasitoids [*Apanteles flavipes*, *A. ruficrus*, *A. vitripennis*, *Chelonus heliope*, *C. formosanus*, *Microphilis sp.* (Hymenoptera: Braconidae); *Blepharella lateralis*, *Peribaea orbata*, *Strobliomyia egyptia* (Diptera: Tachinidae)] limit the population of *Spodopteran* pest (Chari and patel 1983). Sertkaya et al. (2004) reported braconid larval parasitoids *Microplitis rufiventris* (Kokujev), *M. tuberculifer* (Wesmael), *Meteorus ictericus* (Nees), *Chelonus obscuratus* (Herrich-Schaffer) (an egg-larval parasitoid), and *Apanteles ruficrus* (Halliday); the ichneumonid larval parasitoids

Hyposoter didymator (Thunberg) and *Sinophorus xanthostomus* (Gravenhorst); and the egg parasitoid *Trichogramma evanescens* (Westwood) as natural enemies attacking *S. exigua* in Turkey.

6.4.4.2   *Bacillus thuringiensis*

Owing to its insecticidal activity, *B. thuringiensis* acquired tremendous significance in agriculture. Many lepidopteran insect pests can effectively be controlled by *B. thuringiensis* (Nethravathi et al. 2010). Literature revealed that thousands of toxicogenic strains of *B. thuringiensis* exist (Lereclus et al. 1993). The efficacy of biopesticides is dependent on the dose of the active ingredient (Prabagaran et al. 2003; Knowles 2005). By restricting the amount of damage caused by lepidopteran pests to the infested crop, *B. thuringiensis*-based biopesticides have contributed to yield increases in different crops like rice (Kandibane et al. 2010), cauliflower (Justin et al. 2003), and corn (Tamez-Guerra et al. 1998). Combination of insecticides and *B. thuringiensis* had also been tried to limit the population of lepidopteran pests (Khan et al. 2010).

6.4.4.3   Polyhedrosis Virus

Insect viruses, in particular nuclear polyhedrosis viruses (NPVs), however, seemed useful biological control agents because of their virulence and host specificity (Payne 1986; Bhutia et al. 2012). NPV of *S. litura* (spltNPV) has emerged as an alternative to chemical insecticides for the management of *S. litura* (Ramakrishna and Tiwari 1969). Several highly virulent NPVs have been isolated from *S. exigua* larvae (Vlak and Groner 1980; Vlak et al. 1982; Gelernter and Federici 1986). However, NPVs have both advantages as well as disadvantages. The narrow host range of NPVs generally restricts their effectiveness against the complex of insect pests in fields. The slow speed of their action against target insect is another disadvantage of NPVs (Muralibaskaran et al. 1997). Only the in vivo production of the baculoviruses has so far been economically viable due to the high costs involved in the in vitro production systems (Shieh 1989; Kumar and Singh 2009).

6.4.4.4   Entomopathogenic Nematodes

Entomopathogenic nematodes (Rhabditida: Steinernematidae and Heterorhabditidae) are insect pathogens with great potential for biological control (Kaya and Gaugler 1993; Kaya 2006). Kaya (1985) reported that all larval instars of *S. exigua* were susceptible to *Steinernema feltiae* (*S. carpocapsae*); however, neonate larvae were significantly less susceptible to nematode infection than 3- and 8-day-old larvae. The nematodes usually kill the host insects within 1–3 days (Kaya 1985; Gothama et al. 1995). Entomopathogenic nematode, *Heterorhabditis indica*, has been extensively studied for use in the control of various lepidopteran pests damaging glasshouse and nursery crops (Gouge and Hague 1993; Grewal and Richardson 1993; Elanchezhyan 2006; Divyaa et al. 2010).

6.5 Entomopathogenic Fungi

Insect pathogenic fungi naturally restrict the buildup of insect pests without any interference. This part of the "law of natural balance" is being inadvertently destroyed by the indiscriminate use of chemical pesticides. This has led to the notion that all forms of pest control should be integrated and environmentally acceptable to the agricultural ecosystem. Entomogenous fungi are potentially the most versatile biological control agents, due to their wide host range that often results in natural epizootics. Mycopathogens infect by direct penetration of the cuticle (Wraight et al. 1998; 2000; Srisukchayakul et al. 2005) and are, therefore, the principle pathogens among sucking insects which cannot ingest other pathogens. Fungi often cause spectacular epizootics with large number of pathogenic insects showing visible fungal outgrowth (Hall 1982). Fungal biological control agents have demonstrated efficacy against a wide range of insect pests including *Spodoptera* spp. (Purwar and Sachan 2005; Lin et al. 2007; Amer et al. 2008). *M. anisopliae* is known to attack over 200 species of insects belonging to orders Lepidoptera, Coleoptera, Orthoptera, Homoptera, and Dermoptera (Moore et al. 1996). However, *V. lecanii* popularly called the "white holo" is known to cause mycosis in a number of insects belonging to the insect orders Lepidoptera, Coleoptera, and Homoptera. It is certain that entomogenous fungi will continue to increase their share very rapidly in the IPM. There is a general feeling that the development and spread of biological control will empower the resource of poor farmers to manage their pest problems in an eco-friendly way.

6.5.1 Entomopathogenic Fungus Beauveria

B. bassiana, the most common and ubiquitous fungal entomopathogen, is known to be highly potent for the control of insects belonging to various orders. Wraight et al. (2010) reported that all lepidopteran species of vegetable crop pests, viz., fall armyworm (*S. frugiperda*), beet armyworm (*S. exigua*), diamondback moth (*Plutella xylostella*), European corn borer (*Ostrinia nubilalis*), corn earworm (*Helicoverpa zea*), black cutworm (*Agrotis ipsilon*), cabbage worm (*Pieris rapae*), and cabbage looper (*Trichoplusia ni*), were susceptible to *B. bassiana* isolates. However, *S. exigua* and *H. zea* were most susceptible to fungal infection and *S. frugiperda* was least susceptible. Fungus negatively effects the biology of *S. litura*. Kaur et al. (2011) observed significantly high larval mortality, reduced reproductive potential and adult life span, and a decrease in egg hatching and larval period along with pupal and adult deformities at sublethal doses of *B. bassiana* in *S. litura*. Karthikeyan and Selvanarayanan (2011) reported 86.67, 86.67, and 73.33 % larval mortality of *S. litura*, *H. armigera*, and *E. vittella*, respectively, caused by *B. bassiana*. Yet in another study, Malarvannan et al. (2010) noticed adverse effects of *B. bassiana* on *S. litura* pupae, formed from treated larvae along with adult malformation. Alice and Nadarajan (2003) reported natural occurrence of *B. bassiana* on rice leaf folder *Cnaphalocrocis medinalis* Guenee, while red hairy caterpillar *Amsacta albistriga* Walker was recorded infected with *B. bassiana* in groundnut crop (Veena et al. 2006). Virulence test among five species of entomogenic fungi, *B. bassiana*, *B. brongniartii*, *M. anisopliae*, *N. rileyi*, and *P. fumosoroseus*, showed that *B. brongniartii* and *N. rileyi* had evident pathogenic effects against 2nd instars of *S. litura*. The virulence of *B. brongniartii* and *N. rileyi* to the 3rd instars was lower than that of 2nd instars (Lin et al. 2007).

The potential of entomopathogenic fungi often varies in different ecosystems and among fungal species and strains. Larvicidal and growth inhibitory activities of ten different isolates of *B. bassiana* from Pulney hills of Western Ghats of Tamil Nadu, India, were evaluated against third instar larvae of *S. litura*. *B. bassiana* isolate (Bb10) showed maximum larvicidal activity of 68.06 %, minimum pupal weight of 183 mg, low number of adult emergence (22.91 %), and 100 % abnormalities at 1×10^8 spore/ml concentration (Baskar et al. 2012). They also noticed dose-dependent activities in all the tested isolates. Godonou et al. (2009) screened indigenous isolates of *B. bassiana* for virulence against *P. xylostella* larvae and reported 94 % mortality caused by *B. bassiana* isolate Bba5653. They further reported approximately threefold higher yield compared to plots treated with the insecticide, bifenthrin, or in untreated plots. Pathogenicity of the isolates was greater when pathogens were obtained directly from host insect. *B. bassiana* isolate Bb42 obtained from *S. frugiperda* larvae showed the highest virulence, (96.6 % mortality) against second instar larvae of the same insect (Garcia et al. 2011). In contrary to above results, Hung et al. (1993) reported that insect mycopathogen *B. bassiana* and *Candida albicans* were rapidly phagocytized by circulating hemocytes in *S. exigua*, resulting in the protection of challenged larva. Bioassay of *Beauveria* isolates, under the concentrated standard spray of 1,000 conidia/mm² against economically important rice pest *Nilaparvata lugens*, resulted in 17.2–79.1 % adult mortality, 10 days after inoculation (Maoye et al. 2014).

6.5.2 *Entomopathogenic Fungus Metarhizium*

Metarhizium, commonly known as green muscardine fungus, contains a number of morphologically distinct species, *M. anisopliae*, *M. album*, *M. cylindrosporae*, *M. flavoviride*, *M. guizhouense*, and *M. pingshaense* (Driver et al. 2000), which germinate at low temperature (Meyling and Eilenberg 2007). Krutmuang et al. (2008) reported *M. anisopliae* isolate BCC4849 to be highly effective against third instar larvae of *S. litura* at 6×10^8 spores/ml concentration. However,

M. anisopliae isolate Ma002 showed 87.5 and 81.25 % larval mortality of *S. exigua* under laboratory and semi-field conditions (Freed et al. 2012b). Udayababu et al. (2012) estimated LC_{50} value (16.52×10^5 conidia/ml) of *M. anisopliae* against third instar larvae of *S. litura*. Moraga et al. (2006) reported significant mortality of *S. littoralis* larvae by crude protein extracts of *M. anisopliae* and *B. bassiana* isolates. Quesada-Moraga et al. (2006) also successfully used the crude protein extracts of *M. anisopliae* for the control of *S. litura*. Yet in another study, Contreras et al. (2014) evaluated liquid formulation of *M. anisopliae* on different populations of tomato borer *T. absoluta* and reported the potential of *M. anisopliae* to control pupae of the lepidopteran borer at the recommended rate, in IPM programs. Moreover, a notably lower dose was also sufficiently effective to control the tomato borer populations. Entomopathogenic fungi secrete extracellular enzymes to degrade the major constituents of the cuticle (Wang et al. 2005; Cho et al. 2006b), and the success of invasion is directly proportional to secretion of exoenzymes (Khachatourians 1996). Petlamul and Prasertsan (2012) reported that in comparison with *B. bassiana*, *M. anisopliae* possessed the highest enzyme activities against *S. litura*. *M. anisopliae* strain M6 and M8 possessed the highest protease activity (145.00 mU/ml) and highest chitinase activity (20.00 mU/ml) respectively during 96–144 h cultivation. Amer et al. (2008) reported higher mortality among *S. littoralis* larvae treated with *M. anisopliae* isolates compared to *B. bassiana* isolates. *M. anisopliae* formulations are also highly specific against locusts and grasshoppers. As a result of *M. anisopliae* infection, physiological and biochemical changes in the locusts are reported. Shereen et al. (2012) noticed a decline in total insect proteins, carbohydrates, and lipid contents. The highest physiological phenomena like immune responses and intermediary metabolisms 12 h postinjection by spores of *M. anisopliae* and *B. bassiana* on fifth larval instars of *S. littoralis* (Mirhaghparast et al. 2013) as well as high pathogenicity of *M. anisopliae* against European spruce bark beetle *Ips typographus* (Mudroncekova et al. 2013) have been reported.

Since fungi infect by direct penetration of the insect cuticle, they are capable of infecting even nonfeeding stages, eggs (Ujian and Shahzad 2007; Anand and Tiwary 2009), and pupae of insects (Nguyen et al. 2007; Anand et al. 2009). Asi et al. (2012) tested potential of strains of *M. anisopliae* along with *B. bassiana*, *I. fumosorosea*, and *L. Lecanii* against eggs of *S. litura* and found significant variation (37.50–78.00 %) in hatching of fungal-treated eggs. Entomopathogenic fungi not only affect egg hatching but also infect larvae of *S. litura* after hatching. Rajesh et al. (2009) reported the biocontrol potential of *M. anisopliae*, *L. muscarium*, and *C. cardinalis* against pupae of polyphagous pest *S. litura*. Fungi can also target sucking insects such as aphids, thrips, whiteflies, and mosquitoes. Jandricic et al. (2014) reported high virulence of entomopathogenic fungi *Metarhizium* and *Beauveria* isolates against nymphal stage of aphids *Myzus persicae* and *Aphis gossypii*. Nguya and Sunday (2013) reviewed the potential of entomopathogenic fungi for the control of insect vector, tsetse flies.

Susceptibility of the insect to entomopathogenic fungi decreases with the advancement in age of larvae (Asi et al. 2013); however, the larval mortality increases with increase in conidial concentrations and time elapsed after treatment. Pandey and Hasan (2009) also noticed that susceptibility of *S. litura* larvae against *M. anisopliae* decreases with increase in age of larvae, in terms of both LC_{50} and LT_{50}.

6.5.3 Entomopathogenic Fungus Nomuraea

It is well known that *N. rileyi* induces extensive epizootics among caterpillar and it has potential to be a microbial control agent for lepidopteran pests (Li-chang and Roger 2004). Ingle et al. (2004) observed severe infections of *N. rileyi* in *S. litura* and *H. armigera* on green gram along with lower incidence of the pathogen on Bihar hairy caterpillar *Spilarctia obliqua* (Walker) in soybean. Vimaladevi and Prasad (1997) reported epizootics of *N. rileyi* on *S. litura* and *H. armigera* in kharif groundnut when relative humidity and temperature ranged between 70–92 % and 21–27 °C. Manjula et al. (2003) in a survey for incidence of *N. rileyi* on *S. litura* and *H. armigera* on tomato, cotton, black gram, red gram, and groundnut recorded 100 % mycosed larvae of *S. litura* and *H. armigera* on cotton and groundnut. Sreedhar and Devaprasad (1995) reported higher percent of mycosis caused by *N. rileyi* in *S. litura* larval population, when groundnut was used as food plant than chillies, black gram, and tomato. Vimaladevi et al. (1996) reported epizootics of *N. rileyi* on lepidopteran pests of oil seed crops. They observed more infected larvae of *S. litura* in castor followed by groundnut and pigeon pea. Patil et al. (2013) isolated *N. rileyi* from the cadavers of *S. litura* collected from groundnut field and observed that early instars were highly susceptible with a mortality of 70.17 %, which decreased significantly as the age of the larvae advanced. Lethal time for 1st to 5th instars of *S. litura* larvae was 130.71, 137.77, 148.04, 235.65, and 263.10 h, respectively. During rainy and summer seasons, Rachappa et al. (2007) recorded ten entomopathogenic fungi on lepidopteran caterpillars in different crops.

Vimaladevi (1994) tested efficacy of *N. rileyi* against *S. litura* in a lab, net house, and field and reported 2×10^{11} conidia/l of spray solution brings about effective control of late second to early third instar *S. litura* larvae on castor crop. In the field, larval mortality of 52–60 % was observed at 12 days after spraying with *N. rileyi* conidia. The highest cumulative mortality, 88–97 %, was observed by 19 days. Kulkarni and Lingappa (2002) conducted field experiments from 1996 to 1998 to evaluate the bio-efficacy of *N. rileyi* on defoliator *S. litura* in soybean and reported that pathogen achieved significantly higher reduction of *Spodopteran* larvae at higher concentration (1.2×10^{12} conidia/l) at 14 days.

S. frugiperda has been reported to be susceptible to more than 20 species of entomopathogenic fungi (Sanchez-Pena 2000); one of these is *N. rileyi* (Lezama et al. 2001). Rios et al. (2010) reported natural epizootic of the entomopathogenic fungus, *N. rileyi* infecting *S. frugiperda*. In laboratory bioassays, Domenico et al. (2009) observed *N. rileyi* killing 80 % of *S. frugiperda* larvae. Entomopathogenic fungal isolates from different geographical locations and different hosts vary in their virulence and specificity (Tigano et al. 1995). Vimaladevi et al. (2003) studied eleven geographical isolates of *N. rileyi* and reported best traits to be the Karimnagar isolate of *S. litura* origin. Ramiro et al. (2013) recorded the occurrence of native entomopathogens in *S. frugiperda* larvae, collected from 22 localities of central Mexico, during 2009; *N. rileyi* was recovered from 38 larvae, whereas unidentified microsporidia was found infecting 19 larvae.

6.5.4 Entomopathogenic Fungus Isaria

The entomopathogenic fungi *I. farinosa* and *I. fumosorosea* were known as *P. farinosus* and *P. fumosoroseus*, for more than 30 years. Both fungi have worldwide distribution and a relatively wide host range. While *I. farinosa* currently is of minor importance in research and as biocontrol agent, *I. fumosorosea* is regarded as a species complex, and various strains are successfully used for biocontrol of several economically important insect pests (Zimmermann 2008). Freed et al. (2012a) reported insecticidal and antifeedant activities of crude proteins produced by *I. fumosorosea* against diamond back moth, *P. xylostella*. A significant level of increase in the antifeedant index was recorded with the increase of concentrations and time duration. Zemek et al. (2012) evaluated efficacy of two strains of *I. fumosorosea*, Apopka 97 (an active ingredient of commercial bio-pesticide, Biobest) and CCM 8367 (isolated from *Cameraria ohridella*), against larvae and pupae of *S. littoralis* using standard dip test. Obtained results revealed higher virulence of CCM 8367 to the last instar larvae of *S. littoralis* (93.1 % mortality) compared to Apopka 97 (65.5 % mortality) on the 7th day. More obvious difference was found in pupae, where mortality of CCM 8367-treated pupae was 80.0 %, while mortality in Apopka 97-treated pupae reached only 3.3 % on the 8th day after the treatment. Hussein et al. (2013) also reported that *I. fumosorosea* strain CCM 8367 has strong insecticidal effects against *S. littoralis* and has the potential to be implemented as a novel biocontrol agent. Asi et al. (2012) tested various strains of *I. fumosorosea*, *B. bassiana*, *M. anisopliae*, and *L. lecanii* against third instar larvae of *S. litura*, using larval immersion method, and recorded more profound effects of *I. fumosorosea* and *B. bassiana*. Strains with high virulence potential also showed greater mycosis and sporulation on cadavers.

Three entomopathogenic fungi, *I. fumosorosea*, *B. bassiana*, and *N. rileyi*, were compared against *H. armigera* in laboratory bioassays employing topical versus per os inoculation techniques. *N. rileyi* outperformed both *I. fumosorosea* and *B. bassiana*, causing a mean 87 % mortality. However, no difference was detected between the two inoculation techniques employed. A general trend of accelerated pupation following treatment with *N. rileyi* was noted with both inoculation techniques (Hatting 2012). Among most of the cases, concentration is a critical parameter that determined the "speed of kill" of the exposed insect species. Kavallieratos et al. (2014) tested *I. fumosorosea*, *B. bassiana*, and *M. anisopliae* against the stored-grain pest *Sitophilus oryzae*. The mortality of *S. oryzae* adults during the overall exposure period for the lowest, as well as for the highest, concentrations of tested entomopathogenic fungi ranged from 0 to 100 %. Further they reported that adult mortality of *S. oryzae* was higher when the entomopathogenic fungi were directly applied on adults than fungus applied on food.

6.5.5 Entomopathogenic Fungus Lecanicillium

Species of *Lecanicillium* have a wide host range and have been isolated from a variety of insect orders (Zare and Gams 2001; North et al. 2006; Anand et al. 2009). Wang et al. (2007) tested crude toxins extracted from *Lecanicillium* (*Verticillium*) *lecanii* (Zimmermann) strain V3450 and Vp28 for contact toxicity, feeding deterrence, and repellent activity against the sweet potato whitefly, *Bemisia tabaci* (Gennadius). They reported that both toxins showed ovicidal activity. However, nymphs of *B. tabaci* were the most susceptible stages and adults were the second most susceptible stage. Both toxins exhibited repellent activity at low concentration and antifeedant activity at high concentration. ChunLi et al. (2010) determined the lethal concentration of *L. lecanii* strain MZ041024 against 2nd instar larvae of *Laphygma exigua* and reported strong pathogenicity against pest. 100 % mortality caused by *L. lecanii* was reported by Karthikeyan and Selvanarayanan (2011) against cotton pests, *Aphis gossypii* and *Bemisia tabaci*.

Bio-Catch is a biological insecticide based on a selective strain of *L. lecanii*. Product contains spores and mycelial fragments of *L. lecanii* and is available in liquid (1×10^9 CFU/ml) and powder (1×10^8 CFU/gm) formulation. *L. muscarium* has been commercialized as biopesticides, Mycotal against whiteflies and thrips and Verticillin against whiteflies, aphids, and mites (Faria and Wraight 2007). El-Hawary and Abd-El-Salam (2009) evaluated the efficacy of entomopathogenic fungal commercial products, Bio-Power (*B. bassiana*), Bio-Catch (*L. lecanii*), and Priority (*Paecilomyces fumosoroseus*), against *S. littoralis* larvae and found Bio-Power to be the most effective product followed by Bio-Catch and Priority with 87.5, 72.5, and 67.5 % larval mortality, respectively, at 1×10^9 spores/ml. Oil-based formulation of mycoinsecticides could be an important biocontrol agent in the management of lepidopteran pest. Sahayaraj and Borgio (2012) reported that oil-based conidial formulations of *L. lecanii*, *B. bassiana*, and *Paecilomyces fumosoroseus* were pathogenic to lepidopteran pest *Pericallia ricini* at all tested concentrations. *L. lecanii* caused highest larval mortality followed by *B. bassiana*. However, *P. fumosoroseus* caused least larval mortality.

6.5.6 Entomopathogenic Fungus Fusarium

The genus *Fusarium* comprises a large group of species of filamentous fungi widely distributed in soil usually in association with plants. Most species are saprotrophic and relatively abundant members of the soil microbiota (Leslie and Summerell 2006). Many *Fusarium* species are well known as pathogens of insects and plants. More than 13 *Fusarium* species are pathogenic to insects, and the genus has a host range that includes Lepidoptera, Coleoptera, Hemiptera, and Diptera (Teetor-Barsch and Roberts 1983; Humber 1992). *Fusarium* includes various species/strains that are able to produce potent secondary metabolites, such as trichothecenes (Kilpatrick 1961; Kuno and Ferrer 1973), and fumonisins (Kuruvilla and Jacob 1979).

Ameen (2012) tested *Fusarium* isolates, *F. chlamydosporum*, *F. equiseti*, *F. graminearum*, *F. moniliforme*, *F. oxysporum*, *F. poae*, *F. semitectum*, *F. sacchari*, and *F. solani* isolated from soil samples of different locations and reported pathogenicity of all *Fusarium* isolates against *G. mellonella* larvae. Rajesh and Bhupendra (2009) reported 100 % mortality in *S. litura* unscaled eggs by the infection of *F. lateritium*, *M. anisopliae*, and *C. cardinalis* at 10^6 conidia/ml. The larvae were found susceptible to entomopathogenic fungi in a dose-dependent manner. Further they noticed that when both larvae and the leaves (provided as food) were treated with fungal conidia, mortality further increased. Zhang (2001) isolated *F. lateritium* from dead insect bodies of citrus aphid. Bioassay results indicated higher fungus virulence to the larvae of citrus aphid. Pelizza et al. (2011) reported natural infection caused by *F. verticillioides* in grasshoppers. These results strongly indicate the control potential of entomopathogenic fungi in managing *Spodopteran* species.

6.6 Conclusions

It is well known that entomopathogenic fungi induced extensive epizootics among caterpillar. These fungi have a wide host range and have been isolated from a variety of insect orders. The potential of entomopathogenic fungi often vary in different ecosystems and among fungal species and strains. Highly virulent fungal genotypes against particular insect pest can be manipulated to develop new natural insecticides. Current research efforts are directed at selective native entomopathogenic fungi, their characterization, assessment of virulence, and developing a formulation for them. Finally, entomopathogenic fungi have potential to be implemented as a novel biocontrol agent against lepidopteran pests.

References

Abdel-Megeed MI (1975) Field observations on the vertical distribution of the cotton leafworm, *Spodoptera littoralis* on cotton plants. Z Angew Entomol 78:597–662

Ahmad M, Arif MI, Ahmad M (2007a) Occurrence of insecticide resistance in field populations of *Spodoptera litura* (Lepidoptera: Noctuidae) in Pakistan. Crop Prot 26:809–817

Ahmad M, Sayyed AH, Crickmore N, Saleem MA (2007b) Genetics and mechanism of resistance to deltamethrin in a field population of *Spodoptera litura* (Lepidoptera: Noctuidae). Pest Manag Sci 63:1002–1010

Ahmad M, Sayyed AH, Saleem MA (2008) Evidence for field evolved resistance to newer insecticides in *Spodoptera litura* (Lepidoptera: Noctuidae) from Pakistan. Crop Prot 27: 1367–1372

Aktar MW, Sengupta D, Chowdhury A (2009) Impact of pesticide use in agriculture: their benefits and hazards. Interdisc Toxicol 2:1–12

Alice SJ, Nadarajan L (2003) Occurrence of entomogenous fungi on rice pests on Karaikal region. Insect Environ 9:192

Almudena OU, Nemat OK (2013) Action on the surface: entomopathogenic fungi versus the insect cuticle. Insects 4:357–374

Alves SB, Pereira RM, Lopes RB, Tamai MA (2003) Use of entomopathogenic fungi in Latin America. In: Upadhyay RK (ed) Advances in microbial control of insect pests. Kluwer, New York, pp 193–212

Ambethgar V (2009) Potential of entomopathogenic fungi in insecticide resistance management (IRM): A review. J Biopestic 2:177–193

Ameen MKM (2012) Screening of *Fusarium* isolates pathogenicity in vitro by using the larvae of *Galleria Mellonella* L. J Basrah Res 38:19–28

Amer MM, El-Sayed TI, Bakheit HK, Moustafa SA, El-Sayed YA (2008) Pathogenicity and genetic variability of five entomopathogenic fungi against *Spodoptera littoralis*. Res J Agric Biol Sci 4:354–367

Anand R, Tiwary BN (2009) Pathogenicity of entomopathogenic fungi to eggs and larvae of *Spodoptera litura*, the common cutworm. Biocontrol Sci Technol 19:919–929

Anand R, Prasad B, Tiwary BN (2009) Relative susceptibility of *Spodoptera litura* pupae to selected entomopathogenic fungi. BioControl 54:85–92

Anderson TE, Hajek AE, Roberts DW, Preisler K, Robertson JL (1989) Colorado potato beetle (Coleoptera: Chrysomelidae): Effects of combinations of *Beauveria bassiana* with insecticides. J Econ Entomol 82:83–89

Arivoli S, Samuel T (2012) Antifeedant activity of plant extracts against *Spodoptera litura* (Fab.) (Lepidoptera: Noctiduae). Am-Euras J Agric Environ Sci 12:764–768

Arivoli S, Samuel T (2013) Antifeedant activity, developmental indices and morphogenetic variations of plant extracts against *Spodoptera litura* (Fab) (Lepidoptera: Noctuidae). J Entomol Zool Stud 1:87–96

Arivoli S, Tennyson S (2013) Ovicidal activity of plant extracts against *Spodoptera litura* (Fab) (Lepidoptera: Noctuidae). Bull Environ Pharmacol Life Sci 2:140–145

Armes NJ, Wightman JA, Jadhav DR, Ranga RGV (1997) Status of insecticide resistance in *Spodoptera litura* in Andhra Pradesh, India. Pesti Sci 50:240–248

Asi MR, Bashir MH, Afzal M, Ashfaq M, Sahi ST (2010) Compatibility of entomopathogenic fungi, *Metarhizium anisopliae* and *Paecilomyces fumosoroseus* with selective insecticides. Pak J Bot 42:4207–4214

Asi MR, Bashir MH, Afzal M, Khan BS, Khan MA, Gogi MD, Zia K, Arshad M (2012) Potential of entomopathogenic fungi against larvae and eggs of *Spodoptera litura* (Lepidoptera: Noctuidae). Pak Entomol 34:151–156

Asi MR, Bashir MH, Afzal M, Zia K, Akram M (2013) Potential of entomopathogenic fungi for biocontrol of *Spodoptera litura* fabricius (lepidoptera: noctuidae). J Anim Plant Sci 23: 913–918

Balazy S (1993) Entomophthorales. In: Flora of Poland, fungi (Mycota), vol XXIV. Instytut Botaniki, Krakow

Baskar K, Ignacimuthu S (2012a) Ovicidal activity of *Atalantia monophylla* (L) Correa against *Helicoverpa armigera* Hubner (Lepidoptera: Noctuidae). J Agric Technol 8:861–868

Baskar K, Ignacimuthu S (2012b) Antifeedant, larvicidal and growth inhibitory effect of ononitol monohydrate isolated from *Cassia tora* L. against *Helicoverpa armigera* (Hub) and *Spodoptera litura* (Fab.) (Lepidoptera: Noctuidae). Chemosphere 88:384–388

Baskar K, Maheshwaran R, Kingsley S, Ignacinuthu S (2011) Bioefficacy of plant extracts against Asian army worm *Spodoptera litura* Fab. (Lepidoptera: Noctuidae). J Agric Technol 7: 123–131

Baskar K, Raj GA, Mohan PM, Lingathurai S, Ambrose T, Muthu C (2012) Larvicidal and growth inhibitory activities of entomopathogenic fungus, *Beauveria bassiana* against Asian Army worm, *Spodoptera litura* Fab. (Lepidoptera: Noctuidae). J Entomol 9:155

Basu AC (1981) Effect of different foods on the larval and post larval development of moth of *Prodenia litura* (Fab.). J Bombay Nat Hist Soc 44:275–288

Batalha VC, Zanuncio JC, Picanco M, Guedes RNC (1997) Selectivity of insecticides to *Podisus nigrispinus* (Heteroptera: Pentatomidae) and its prey *Spodoptera frugiperda* (Lepidoptera: Noctuidae). Ceiba 38:19–22

Beckage NE (2008) Insect immunology. Academic Press/Elsevier, San Diego, p 348

Bhattacharya AK, Mondal P, Ramamurthy VV, Srivastava RP (2003) *Beauveria bassiana*: a potential bioagent for innovative integrated pest management programme. In: Srivastava RP (ed) Biopesticides and bioagents in integrated pest management of agricultural crops. International Book Distributing Co, Lucknow, p 860

Bhatti SS, Ahmad M, Yousaf K, Naeem M (2013) Pyrethroids and new chemistry insecticides mixtures against *Spodoptera litura* (Noctuidae: Lepidoptera) under laboratory conditions. Asian J Agric Biol 1:45–50

Bhutia KC, Chakravarthy AK, Doddabasappa B, Narabenchi GB, Lingaraj VK (2012) Evaluation and production of improved formulation of nucleopolyhedrosis virus of *Spodoptera litura*. Bull Insectol 65:247–256

Boman HG, Hultmark D (1987) Cell-free immunity in insects. Annu Rev Microbiol 41:103–126

Boucias DG, Pendland JC (1991) Attachment of mycopathogens to cuticle: the initial event of mycosis in arthropod hosts. In: Cole GT, Hoch HC (eds) The fungal spore and disease initiation in plants and animals. Plenum, New York, pp 101–128

Burges HD (1981) Microbial control of pests and plant diseases 1970-1980. Academic, London. ISBN 9780121433604

Butt TM, Jackson CW, Magan N (2001) Fungi as biocontrol agents: progress, problems and potential. CAB International, Wallingford

Chari MS, Patel NG (1983) Cotton leaf worm *Spodoptera litura* (Fabr.): its biology and integrated control measures. Cotton Dev 13:7–8

Charnley AK, Collins SA (2007) Entomopathogenic fungi and their role in pest control. In: Kubicek CP, Druzhinina IS (eds) Environment and microbial relationships, 2nd edn. The Mycota IV. Springer, Berlin, pp 159–185

Cho EM, Boucias D, Keyhani NO (2006a) EST analysis of cDNA libraries from the entomopathogenic fungus *Beauveria* (Cordyceps) *bassiana*. II. Fungal cells sporulating on chitin and producing oosporein. Microbiology 152:2855–2864

Cho EM, Liu L, Farmerie W, Keyhani NO (2006b) EST analysis of cDNA libraries from the entomopathogenic fungus *Beauveria* (Cordyceps) *bassiana*. I. Evidence for stage-specific gene expression in aerial conidia, in vitro blastospores and submerged conidia. Microbiology 152(2843):2854

Cho EM, Kirkland BH, Holder DJ, Keyhani NO (2007) Phage display cDNA cloning and expression analysis of hydrophobins from the entomopathogenic fungus *Beauveria* (Cordyceps) *bassiana*. Microbiology 153:3438–3447

ChunLi X, ShengYong Y, Qiong K, Yali G, HongRui Z, ZengYue L (2010) Median lethal concentration determination of *Verticillium lecanii* MZ041024 strain against 2nd instar larvae of *Laphygma exigua*. Plant Dis pests 1:57–59

Clark PL, Molina O, Martinelli J, Skoda S, Isenhour SR, Lee DJ, Krumm JT, Foster JE (2007) Population variation of the fall armyworm, *Spodoptera frugiperda,* in the Western hemisphere. J Insect Sci 7:1–10

Clarkson JM, Charnley AK (1996) New insights into the mechanisms of fungal pathogenesis in insects. Trends Microbiol 4:197–203

Contreras J, Mendoza JE, Martínez-Aguirre MR, García-Vidal L, Izquierdo J, Bielza P (2014) Efficacy of enthomopathogenic fungus *Metarhizium anisopliae* against *Tuta absoluta* (Lepidoptera: Gelechiidae). J Econ Entomol 107:121–124

Cuthbertson AGS, Murchie AK (2007) A review of the predatory mite *Anystis baccarum* and its role in apple orchard pest management schemes in Northern Ireland. J Entomol 4:275–278

David I, James RF, Bruce WW, Khuong BN, Byson JA, Richard AH, Michael JH (2003) Survey of entomopathogenic nematodes and fungi in pecan orchards of the southeastern United States and their virulence to pecan weevil. Environ Entomol 5:187–195

Devi KU, Murali Mohan CH, Padmavathi J, Ramesh K (2003) Susceptibility to fungi of cotton boll worms before and after a natural epizootic of the entomopathogenic fungus *Nomuraea rileyi* (Hyphomycetes). Biocontrol Sci Technol 13:367–371

Dhaliwal GS, Koul O (2007) Biopesticide and pest management: conventional and biotechnological approaches. Kalyani Publishers, New Delhi, p 455

Divyaa K, Sankarb M, Marulasiddeshac KN (2010) Efficacy of entomopathogenic nematode, *Heterorhabditis indica* against three lepidopteran insect pests. Asian J Exp Biol Sci 1:183–188

Domenico P, Mayri D, Lesbia T, Blas D (2009) A granular formulation of *Nomuraea rileyi* farlow (samson) for the control of *Spodoptera frugiperda* (lepidoptera: noctuidae). Interciencia 34:130–134

Domsch KH, Gams W, Anderson TH (1980) Compendium of soil fungi, vol 1. Academic, London, p 893

Driver F, Milner RJ, Trueman WH (2000) A taxonomic revision of *Metarhizium* based on a phylogenetic analysis of rDNA sequence data. Mycol Res 104:134–150

East DA, Edelson JV, Cartwright B (1989) Relative cabbage consumption by the cabbage looper (Lepidoptera:Noctuidae), beet armyworm (Lepidoptera: Noctuidae), and diamond back moth (Lepidoptera: Plutellidae). J Econ Entomol 82:1367–1369

Eilenberg J, Hajek A, Lomer C (2001) Suggestions for unifying the terminology in biological control. BioControl 46:387–400

Elanchezhyan K (2006) Compatibility of *Beauveria bassiana* and *Nomuraea rileyi* with EPN for control of *Helicoverpa armigera*. Ann Plant Prot Sci 14:64–68

El-Hawary FM, Abd El-Salam AME (2009) Laboratory bioassay of some entomopathogenic fungi on *Spodoptera littoralis* (Boisd.) and *Agrotis ipsilon* (Hufn.) larvae (Lepidoptera: Noctuidae). Egyptian Acad J Biol Sci 2:1–4

Enrique Q, Alain VEY (2004) Bassiacridin, a protein toxic for locusts secreted by the entomopathogenic fungus *Beauveria bassiana*. Mycol Res 108:441–452

Evans HC (1989) Mycopathogens of insects of epigeal and aerial habitats. In: Wilding N, Collins NM, Hammond PM, Weber JF (eds) Insect fungus interactions. Academic, London, pp 205–238

Fan Y, Fang W, Guo S, Pei X, Zhang Y, Xiao Y, Li D, Jin K, Bidochka MJ, Pei Y (2007) Increased insect virulence in *Beauveria bassiana* strains overexpressing an engineered chitinase. Appl Environ Microbiol 73:295–302

Fang WG, Leng B, Xiao YH, Jin K, Ma JC, Fan YH, Feng J, Yang XY, Zhang YJ, Pei Y (2005) Cloning of *Beauveria bassiana* chitinase gene *Bbchit1* and its application to improve fungal strain virulence. Appl Environ Microbiol 71:363–370

Faria MR, Wraight SP (2007) Mycoinsecticides and Mycoacaricides: a comprehensive list with worldwide coverage and international classification of formulation types. Biol Control 43: 237–256

Federici BA, Bonning BC, St. Leger RJ (2008) Improvement of insect pathogens as insecticides through genetic engineering. In: Hill C, Sleator R (eds) Pathobiotechnology. Landes Bioscience, Austin, pp 15–40

Feng MG, Poprawski TJ, Khachatourians GG (1994) Production, formulation and application of the entomopathogenic fungus *Beauveria bassiana* for insect control: current status. Biocontrol Sci Technol 4:3–34

Ferron P (1978) Biological control of insect pests by entomopathogenic fungi. Annu Rev Entomol 23:409–442

Fletcher TB (1914) Some South Indian insects and other animals of importance considered especially from an economic point of view. Government Press, Madras (presently Chennai), p 565

Flexner JL, Belnavis DL (2000) Microbial insecticides. In: Rechcigl JE, Rechcigl NA (eds) Biological and biotechnological control of insect pests, vol 1. Lewis Publishers Limited, Boca Raton, pp 35–62

Freed S, Feng-Liang J, Naeem M, Shun-Xiang R, Hussian M (2012a) Toxicity of proteins secreted by entomopathogenic fungi against *Plutella xylostella* (Lepidoptera: Plutellidae). Int J Agric Biol 14:291–295

Freed S, Saleem MA, Khan MB, Naeem M (2012b) Prevalence and effectiveness of *Metarhizium anisopliae* against *Spodoptera exigua* (Lepidoptera: Noctuidae) in Southern Punjab, Pakistan. Pak J Zool 44:753–758

Freimoser FM, Screen S, Bagga S, Hu G, St. Leger RJ (2003) Expressed sequence tag (EST) analysis of two subspecies of *Metarhizium anisopliae* reveals a plethora of secreted proteins with potential activity in insect hosts. Microbiology 149:1–9

Fukatzu T, Sato H, Kuriyama H (1997) Isolation, inoculation to insect host, and molecular phylogeny of an entomogenous fungus *Paecilomyces tenuipes*. J Invertebr Pathol 70:203–208

Fuxa JR (1998) Environmental manipulation for microbial control of insects. In: Babbosa P (ed) Conservation biological control. Academic, San Diego, pp 255–289

Fuxa JR, Tanada Y (1987) Epizootiology of insect diseases. Wiley-Interscience, New York. ISBN 047187812X

Garcia GC, Berenice GMM, Nestor BM (2011) Pathogenicity of isolates of entomopathogenic fungi against *Spodoptera frugiperda* (Lepidoptera: Noctuidae) and *Epilachna varivestis* (Coleoptera: Coccinellidae). Rev Colomb Entomol 37:217–222

Gelernter UD, Federici BA (1986) Isolation, identification and determination of virulence of a nuclear polyhedrosis virus from the beet armyworm, *Spodoptera exiqua* (Lepidoptera: Noctuidae). Environ Entomol 15:240–245

Gharib A (1979) Rahe pest in Khozestan. J Pestic Plant Pathol 47:161–178

Gillespie JP, Bailey AM, Cobb B, Vilcinskas A (2000) Fungi as elicitors of insect immune responses. Arch Insect Biochem Physiol 44:49–68

Godonou I, James B, Atcha-Ahowe C, Vodouhe S, Kooyman C, Ahanchede A, Korie S (2009) Potential of *Beauveria bassiana* and *Metarhizium anisopliae* isolates from Benin to control *Plutella xylostella* L. (Lepidoptera: Plutellidae). Crop Prot 28:220–224

Golshan H, Saber M, Majidi-Shilsar F, Bagheri M, Mahdavi V (2013) Effects of common pesticides used in rice fields on the conidial germination of several isolates of entomopathogenic fungus, *Beauveria bassiana* (Balsamo) Vuillemin. J Entomol Res Soc 15:17

Gopalkrishnan C, Mohan S (2000) Effect of certain insecticides and fungicides on the conidial germination of *Nomuraea rileyi* (Farlow) Samson. Entomon 25:217–223

Gothama AAA, Sikorowski PP, Lawrence GW (1995) Interactive effects of *Steinernema carpocapsae* and *Spodoptera exigua* nuclear polyhedrosis virus on *Spodoptera exigua* larvae. J Invertebr Pathol 66:270–276

Gouge DH, Hague NGM (1993) Effects of *Steinernema feltiae* against sciarids infesting conifers in a propagation house. Ann Appl Biol 122:184–185

Grewal PS, Richardson PN (1993) Effects of application rates of *Steinernema feltiae* (Nematoda: Steinernematidae) on biological control of the mushroom fly, *Lycoriella auripila* (Diptera: Sciaridae). Biocontrol Sci Tech 3:29–40

Guo HL, Ye BL, Yue YY, Chen QT, Fu CS (1986) Three new species of *Metarhizium*. Acta Mycol Sinica 5:185–190

Gupta VP (2003) Natural occurrence of the entomopathogenic fungus *Nomuraea rileyi* in the soybean green semilooper, *Chrysodeixis acuta*, in India. Online. Plant Health Prog. doi:10.1094/PHP-2003-0113-01-HN

Gupta S, Wang Y, Jang HB (2005) *Manduca sexta* proPhenoloxidase activation requires proPhenoloxidase activating proteinase (PAP) and serine protease homologues (HPSS) simultaneously. Insect Biochem Mol Biol 35:241–248

Gurmeet KB, Sanehdeep K (2013) Phenoloxidase activity in haemolymph of *Spodoptera litura* (Fabricius) mediating immune responses challenge with entomopathogenic fungus, *Beauveria bassiana* (Balsamo) Vuillmin. J Entomol Zool Stud 1:118–123

Hajek AE, Delalibera IJ (2010) Fungal pathogens as classical biological control agents against arthropods. BioControl 55:147–158

Hajek AE, St Leger RJ (1994) Interactions between fungal pathogens and insect hosts. Annu Rev Entomol 39:293–322

Hajek AE, Humber RA, Elkinton JS (1995) The mysterious origin of *Entomophaga maimaiga* in North America. Am Entomol 41:31–42

Hall RA (1982) Deuteromycetes: virulence and bioassay design. In: Invertebrate pathology and microbial control. Proceedings III international colloquium invertebrate pathology, XV Annual Medium Soc. of Invertebrate Pathology, Brighton, pp 191–196

Hatting JL (2012) Comparison of three entomopathogenic fungi against the bollworm, *Helicoverpa armigera* (Hubner) (Lepidoptera: Noctuidae), employing topical *vs* per os inoculation techniques. Afr Entomol 20:91–100

Hemasree E (2013) A critical review on the natural occurrence of entomopathogenic fungi in agricultural ecosystem. Int J Appl Biol Pharma Technol 4:372–375

Hodge KT (2003) Clavicipitaceous anamorphs. In: White JF, Bacon CW, Hywel-Jones NL, Spatafora JW (eds) Clavicipitalean fungi: evolutionary biology, chemistry, biocontrol and cultural impacts. Marcel Dekker, New York, pp 75–123

Hu G, St. Leger RJ (2002) Field studies using a recombinant mycoinsecticide (*Metarhizium anisopliae*) reveal that it is rhizosphere competent. Appl Environ Microbiol 68:6383–6387

Humber RA (1992) Collection of entomopathogenic fungi: catalog of strains. USDA-ARS Publication 110:1–177

Humber RA (1997) Fungi - Identification. In: Lacey L (ed) Manual of techniques in insect pathology. Academic, London, pp 153–185

Hung SY, Boucias DG, Vey AJ (1993) Effect of *Beauveria bassiana* and *Candida albicans* on the cellular defense response of *Spodoptera exigua*. J Invertebr Pathol 61:179–187

Hussain A, Tian MY, Ahmed S, Shahid M (2012) Current status of entomopathogenic fungi as mycoinsecticides and their inexpensive development in liquid cultures, zoology. In: Garcia M-D (ed) In Tech, pp 103–122. Available from: www.intechopen.com/books/zoology/current-status-of-entomopathogenic-fungi-as-mycoinecticides-andtheir-inexpensive-development-in-liq

Hussein HM, Zemek R, Habustova SO, Prenerova E, Adel MM (2013) Laboratory evaluation of a new strain CCM 8367 of *Isaria fumosorosea* (syn. *Paecilomyces fumosoroseus*) on *Spodoptera littoralis* (Boisd.). Arch Phytopath Plant Prot 46:1307–1319

Ignacimuthu IC (2008) Ecofriendly insect pest management. National symposium on 'ecofriendly insect pest management, Entomology Research Institute, Loyola College, Chennai. Curr Sci 94:10

Ignoffo CM, Garcia C, Hostetter DL, Pinnel RE (1977) Laboratory studies of the entomopathogenic fungus *Nomuraea rileyi*: soil born contamination of soybean seedlings and dispersal of diseased larvae of *Trichoplusia ni*. J Invertebr Pathol 29:147–152

Inceoglu AB, Kamita SG, Hammock BD (2006) Genetically modified baculoviruses: a historical overview and future outlook. Adv Virus Res 68:323–360

Ingle YV, Lande SK, Burgoni GK, Autkar SS (2004) Natural epizootic of *Nomuraea rileyi* on lepidopterous pests of soybean and green gram. J Appl Zool Res 15:160–162

Jackson TA, Alves SB, Pereira RM (2000) Success in biological control of soil dwelling insects by pathogens and Nematodes. In: Gurr G, Wratten S (eds) Biological control: measures of success. Kluwer Academic Dordrecht, The Netherlands, pp 271–296

Jandricic SE, Filotas M, Sanderson JP, Wraight SP (2014) Pathogenicity of conidia-based preparations of entomopathogenic fungi against the greenhouse pest aphids *Myzus persicae*, *Aphis gossypii*, and *Aulacorthum solani* (Hemiptera: Aphididae). J Invertebr Pathol 118:34–46

Jayaraj S (1978) All India symposium on insect pests management-present, past and future: an outlook (Abst.), Udaipur, pp 49–52

Joseph I, Edwin CD, Ranjit SAJA (2010) Studies on the influence *of Beauveria bassania* on survival and gut flora of groundnut caterpillar, *Spodoptera litura* Fab. J Biopestic 3:553–555

Jotwani MG, Ray BK, Pradhan S (1961) Bioassay of the comparative toxicity of some insecticides to the larvae of *Prodenia litura* Fabricius (Noctuidae: Lepidoptera). Ind J Entomol 23:50–53

Justin C, Leo G, Prem JJ, Jayasekhar M (2003) Comparative efficacy of *Bacillus thuringiensis* Berliner formulations with insecticides against *Plutella xylostella* (L.) and their effect on *Cotesia plutellae* Kurdj.on cauliflower. Agric Sci Digest 23:251–254

Kandibane M, Kumar K, Adiroubane D (2010) Effect of *Bacillus thuringiensis* Berliner formulation against the rice leaf folder *Cnaphalocrocis medinalis* Guenee (Pyralidae: Lepidoptera). J Biopestic 3:445–447

Karthikeyan A, Selvanarayanan V (2011) In vitro efficacy of *Beauveria bassiana* (Bals.)Vuill. and *Verticillium lecanii* (Zimm.) viegas against selected insect pests of cotton. Recent Res Sci Technol 3:142–143

Kaur S, Kaur HP, Kaur K, Kaur A (2011) Effect of different concentrations of *Beauveria bassiana* on development and reproductive potential of *Spodoptera litura* (Fabricius). J Biopestic 4: 161–168

Kavallieratos NG, Athanassiou CG, Aountala MM, Kontodimas D (2014) Evaluation of the entomopathogenic fungi *Beauveria bassiana*, *Metarhizium anisopliae*, and *Isaria fumosorosea* for control of *Sitophilus oryzae*. J Food Prot 77:87–93

Kaya HK (1985) Susceptibility of early larval stages of *Pseudaletia unipuncta* and *Spodoptera exigua* (Lepidoptera: Noctuidae) to the entomogenous nematode *Steinernema feltiae* (Rhabditida: Steinernematidae). J Invertebr Pathol 46:58–62

Kaya HK (2006) Status of entomopathogenic nematodes and their symbiotic bacteria from selected countries or regions of the world. Biol control 38:134–155

Kaya HK, Gaugler R (1993) Entomopathogenic nematodes. Annu Rev Entomol 38:181–206

Keller S, Zimmerman G (1989) Mycopathogens of soil insects. In: Wilding N, Collins NM, Hammond PM and Webber JF. (Eds.) Insect-fungus interactions. Academic Press, New York, pp 239–270

Khachatourians GG (1996) Biochemistry and molecular biology of entomopathogenic fungi. In: Howard DH, Miller JD (eds) Human and animal relationships, Mycota VI. Springer, Heidelberg, pp 331–363

Khachatourians GG, Sohail SQ (2008) Entomopathogenic fungi. In: Brakhage AA, Zipfel PF (eds) Biochemistry and molecular biology, human and animal relationships, 2nd edn. The Mycota VI. Springer, Berlin

Khan MA, Mumtaz R, Khan MA (2010) Management of *Spilarctia obliqua* through *Bt* and chlorpyrifos combinations. Ann Plant Prot Sci 18:499–500

Kilpatrick RA (1961) Fungi associated with larvae of *Sitona* spp. Phytopathology 51:640–641

Kleespies RG, Huger AM, Zimmermann G (2008) Diseases of insects and other arthropods: results of diagnostic research over 55 years. Biocontrol Sci Technol 18:439–484

Klingen I, Haukeland S (2006) The soil as a reservoir for natural enemies of pest insects and mites with emphasis on fungi and nematodes. In: Eilenberg J, Hokkanen, HMT (eds) An ecological and societal approach to biological control. Series: Progress in biological control, vol 2. Springer, Heidelberg, pp 145–211

Knowles A (2005) New developments in crop protection product formulation. Agrow reports, T & F Informa. UK Ltd. www.agrowreports.com

Korrat EEE, Abdelmonem AE, Helalia AA, Khalifa HMS (2012) Toxicological study of some conventional and nonconventional insecticides and their mixtures against cotton leaf worm, *Spodoptera littoralis* (Boisd.) (Lepidoptera: Noectudae). Ann Agric Sci 57:145–152

Kranthi KR, Jadhav DR, Wanjari RR, Ali SS, Russell D (2001) Carbamate and organophosphate resistance in cotton pests in India, 1995 to 1999. Bull Entomol Res 91:37–46

Krasnoff SB, Watson DW, Gibson DM, Kwan EC (1995) Behavioral effects of the entomopathogenic fungus, *Entomophthora muscae* on its host *Musca domestica*: postural changes in dying hosts and gated pattern of mortality. J Insect Physiol 41:895–903

Krutmuang P, Prakongsuk S, Visitpanich J (2008) Selection of entomopathogenic fungi for *Spodoptera litura* control. "Competition for resources in a changing world: New drive for rural development" *Tropentag*, 7–9 October 2008, Hohenheim

Kubicek CP, Druzhinina IS (2007) Environmental and microbial relationships, 2nd edn. The Mycota IV. Springer, Berlin, pp 159–187

Kulkarni NS, Lingappa S (2002) Bioefficacy of entomopathogenic fungus, *Nomuraea rileyi* (Farlow) Samson on *Spodoptera litura* and *Cydia ptychora* in Soybean and on *S. litura* in Potato. Karnataka J Agric Sci 15:47–52

Kumar V, Singh NP (2009) *Spodoptera litura* nuclear polyhedrosis virus (NPV-S) as a component in integrated pest management (IPM) of *Spodoptera litura* (Fab.) on cabbage. J Biopestic 2: 84–86

Kuno G, Ferrer MAC (1973) Pathogenicity of two *Fusarium* fungi to an armoured scale insect *Selenaspidus articulatus*. J Invertebr Pathol 22:473–474

Kurup AR, Joshi BG (1959) Investigation on the control of tobacco caterpillars (*Prodenia litura* F) in nursery and field. Indian J Entomol 21:10–14

Kuruvilla S, Jacob A (1979) Comparative susceptibility of nymphs and adults of *Nilaparvata lugens* to *Fusarium oxysporum* and its use in microbial control. Agric Res J Kerala 17:287–288

Lacey LA, Kaya HK (eds) (2000) Field manual of techniques in invertebrate pathology: application and evaluation of pathogens for control of insects and other invertebrate pests. Kluwer Academic, Dordrecht. ISBN 9781402059322

Lefroy HM (1908) The tobacco Catterpillar (Prudentia littoralis). Mem Dept Agric India Ent Ser 2:74–94

Lefroy HM (1909) Indian insect life: a manual of the insects of the plains (Tropical India). Thacher & Co., Calcutta, p 786

St. Leger RJ (2008) Studies on adaptations of *Metarhizium anisopliae* to life in the soil. J Invertebr Pathol 98:271–276

St. Leger RJ, Wang C (2009) Entomopathogenic fungi and the genomic era. In: Stock SP, Vandenberg J, Glazer I, Boemare N (eds) Insect pathogens: molecular approaches and techniques. CABI, Wallingford, pp 366–400

Lereclus D, Delecluse A, Lecadet MM (1993) Diversity of *Bacillus thuringiensis* toxins and genes. In: Entwistle PF, Cory JS, Bailey MJ, Higgs S (eds) *Bacillusthuringiensis,* an environmental biopesticide: theory and practice. John Wiley & Sons, Chichester, pp 37–69

Leslie JF, Summerell BA (2006) *Fusarium verticilliodes* (Saccardo) Nirenberg. In: Leslie JF, Summerell BA (eds) The Fusarium Laboratory manual. Blackwell, Iowa, pp 274–279

Lezama GR, Hamm JJ, Molina ORJ, Lopez EM, Pescad RA, Eloise L (2001) Occurrence of entomopathogens of *Spodoptera frugiperda* (Lepidoptera: Noctuidae) in the Mexican States of Michoacan, Colima, Jalisco and Tamaulipas. Fla Entomol 84:23–30

Li W, Sheng C (2007) Occurrence and distribution of entomo-phthoralean fungi infecting aphids in mainland China. Biocontrol Sci Technol 17:433–439

Liburd OE, Funderburk JE, Olson SM (2000) Effect of biological and chemical insecticides on *Spodoptera* species (Lep: Noctuidae) and marketable yields of tomatoes. J Appl Entomol 124: 19–25

Li-chang T, Roger F (2004) Potential application of the entomopathogenic fungus, shape *Nomuraea rileyi*, for control of the corn earworm, shape *Helicoverpa armigera*. J Entomol Exp Appl 88:25–30

Lin HP, Yang XJ, Gao YB, Li SG (2007) Pathogenicity of several fungal species on *Spodoptera litura*. Chin J Appl Ecol 18:937–940

Liu Y, Shen D, Zhou F, Wang G, An C (2014) Identification of immunity-related genes in *Ostrinia furnacalis* against entomopathogenic fungi by RNA-Seq analysis. PLoS ONE 9(1): e86436;1–24. doi:10.1371/journal.pone.0086436

Malarvannan S, Murali PD, Shanthakumar SP, Prabavathy VR, Nair S (2010) Laboratory evaluation of the entomopathogenic fungi, *Beauveria bassiana* against the Tobacco caterpillar, *Spodoptera litura* Fabricius (Noctuidae: Lepidoptera). J Biopest 3:126–131

Malo AR (1993) Estudio sobre la compatibilidad del hongo *Beauveria bassiana* (Bals.) Vuill.con formulaciones comerciales de funguicidase insecticidas. Rev Colomb Entomol 19:151–158

Manjula K, Nagalingam B, Rao PA (2003) Occurrence of *Nomuraea rileyi* on *Helicoverpa armigera* and *Spodoptera litura* in Guntur district of Andhra Pradesh. Ann Plant Prot Sci 11: 224–227

Maoye L, Shiguang L, Amei X, Huafeng L, Dexin C, Hui W (2014) Selection of *Beauveria* isolates pathogenic to adults of *Nilaparvata lugens*. J Insect Sci 14:1–12

Mariapackiam S, Ignacimuthu S (2008) Larvicidal & Histopathological effects of oil formulation on *Spodoptera litura*. In: Ignacimuthu S, Jeyarabbj S (eds) Recent trends in insect pest management. Elite Publishing House Pvt. Ltd., New Delhi

Matsuura H, Naito A (1997) Studies on the cold-hardiness and overwintering of *Spodoptera litura* F. (Lepidoptera: Noctuidae): VI. Possible overwintering areas predicted from meteorological data in Japan. Appl Entomol Zool 32:167–177

May RM (2000) The dimension of life on earth. In: Raven PH, Williams T (eds) Nature and human society: the quest for a sustainable world. National Academy Press, Washington, DC, pp 30–45

McClintock JT, Van-Beek NAM, Kough JL, Mendelsohn ML, Hutton PO (2000) Regulatory aspects of biological control agents and products derived by biotechnology. In: Rechcigl JE, Rechcigl NA (eds) biological and biotechnological control of insect pests, vol 1. Lewis Publishers Limited, Boca Raton (EUA), pp 305–357

Meshrif WS, Rohlfs M, Hegazi MA, Barakat EM, Ai S, Shehata MG (2011) Interactions of *Spodoptera littor*alis haemocytes following injection with the entomopathogenic fungi: *Beauveria bassiana* and *Nomuraea rileyi*. J Egyptian Soc Parasitol 41:699–714

Meyling NV, Eilenberg J (2007) Ecology of the entomopathogenic fungi *Beauveria bassiana* and *Metarhizium anisopliae* in temperate agro ecosystems: potential for conservation biological control. Biol Control 43:145–155

Milner RJ (1997) Prospects for biopesticides for aphid control. Entomophaga 42:227–239

Ming X, Yun-Hong P, Hong-Tao W, Qing-Liang L, Tong-Xian L (2010) Effects of four host plants on biology and food utilization of the cutworm, *Spodoptera litura*. J Insect Sci 10:1–22

Miranpuri GS, Khachatourians GG (1995) Application of *Beauveria bassiana* and *Verticillium lecanii* against Saskatoon berry leaf aphid *Acyrthosiphon macrosiphum*. J Insect Sci 8:93–95

Mirhaghparast SK, Zibaee A, Hajizadeh J (2013) Effects of *Beauveria bassiana* and *Metarhizium anisopliae* on cellular immunity and intermediary metabolism of *Spodoptera littoralis* Boisduval (Lepidoptera: Noctuidae). Invertebr Surviv J 10:110–119

Miyahara Y, Wakikado T, Tanaka A (1971) Seasonal changes in the number and size of the egg-masses of *Prodenia litura*. Jpn J Appl Entomol Zool 15:139–143

Moore D, Higgins PM, Lomer CJ (1996) The effects of simulated and natural sunlight on the viability of conidia of *Metarhizium flavoviride* Gams and Rozsypal and interactions with temperature. Biocontrol Sci Tech 7:87–94

Moraga QE, Carrasco DJA, Santiago AC (2006) Insecticidal and antifeedant activities of proteins secreted by entomopathogenic fungi against *Spodoptera littoralis* (Lep., Noctuidae). J Appl Entomol 130:442–452

Mudroncekova S, Marian M, Marek N, Ivan S (2013) Entomopathogenic fungus species *Beauveria bassiana* (bals.) and *Metarhizium anisopliae* (metsch.) used as mycoinsecticide effective in biological control of *Ips typographus* (L.). J Microbiol Biotechnol Food Sci 2: 2469–2472

Murlibaskaran MS, VenugopaL MS, Mahadevan NR (1997) Optical brighteners as UV protectants and their influence on the virulence of nuclear polyhedrosis virus of *Spodoptera litura* (Fabricius) (Lepidoptera : Noctuidae). J Biol Control 11:17–22

Murua GM, Molina OJ, Fidalgo P (2009) Natural distribution of parasitoids of larvae of the fall armyworm, *Spodoptera frugiperda,* in Argentina. J Insect Sci 9:1–17

Muthukrishnan N, Ganapathy N, Nalini R, Rajendran R (2005) Pest management in horticultural crops. New Madura Publishers, Madurai, p 325

Nandihalli BS, Patil BV, Somasekhar HP (1989) Influence of weather parameters on the population dynamics of *Spodoptera litura* (Fb.) in pheromone and light traps. Karnataka J Agric Sci 2:62–67

Naser M, Bijan H, Rahim E, Alireza A, Azidah BAA, Rouhollah R (2013) Effect of entomopathogenic fungi *Beauveria bassiana* (bals.) and *Lecanicillium muscarium* (petch) on *Trialeurodes vaporariorum* westwood. Indian J Entomol 75:95–98

Neelapu NR, Reineke A, Chanchala UM, Koduru UD (2009) Molecular phylogeny of asexual entomopathogenic fungi with special reference to *Beauveria bassiana* and *Nomuraea rileyi*. Rev Iberoam Micol 26:129–145

Nethravathi CJ, Hugar PS, Krishnaraj PU, Vastrad AS, Awaknavar JS (2010) Bioefficacy of native Sikkim *Bacillus thuringiensis* (Berliner) isolates against lepidopteran insects. J Biopestic 3:448–451

Neuman G, Shields EJ (2004) Survey for entomopathogenic nematodes and fungi in alfalfa snout beetle infested fields in Hungary and in New York state. Great Lakes Entomol 37:152–158

Nguya KM, Sunday E (2013) The use of entomopathogenic fungi in the control of tsetse flies. J Invertebr Pathol 112:583–588

Nguyen NC, Borgemeister HP, Zimmermann G (2007) Laboratory investigations on the potential of entomopathogenic fungi for biocontrol of *Helicoverpa armigera* (Lepidoptera: Noctuidae) larvae and pupae. Biocontrol Sci Technol 17:853–864

North JP, Cuthbertson AGS, Walters KFA (2006) The efficacy of two entomopathogenic biocontrol agents against adult *Thrips palmi* (Thysanoptera: Thripidae). J Invertebr Pathol 92: 89–92

Oerke EC, Dehne HW (2004) Safeguarding production-losses in major crops and the role of crop protection. Crop Prot 23:275–285

Ownley BH, Kimberly DG, Fernando EV (2010) Endophytic fungal entomopathogens with activity against plant pathogens: ecology and evolution. BioControl 55:113–128

Pandey SN (1970) Ph.D. thesis submitted to University of Udaipur, India

Pandey R, Hasan W (2009) Pathogenicity of entomopathogenic fungi, *Metarhizium anisopliae* against tobacco caterpillar, *Spodoptera litura* (Fabricius). Trends Biosci 2:29–30

Patil RK, Bhagat YS, Halappa B, Bhat RS (2013) Evaluation and characterization of entomopathogenic fungus, *Nomuraea rileyi* (Farlow) Samson for the control of *Spodoptera litura* (f.) and its compatibility with synthetic and botanical pesticides. Int J Recent Sci Res 4:2167–2172

Payne CC (1986) The control of insect pests by pathogens and insect-parasitic nematodes. In: Proc Agrobiotic Conf on Advance Biotechnology and Agriculture, Bologna

Pelizza SA, Stenglein SA, Cabello MN, Dinolfo MI, Lange CE (2011) First record of *Fusarium verticillioides* as an entomopathogenic fungus of grasshoppers. J Insect Sci 11:1–8

Pell JK, Eilenberg J, Hajek AE, Steinkraus DC (2001) Biology, ecology and pest management potential of Entomophthorales. In: Butt TM, Jackson CW, Magan N (eds) Fungi as biocontrol agents: progress, problems and potential. CABI International, Wallingford, pp 71–153

Petlamul W, Prasertsan P (2012) Evaluation of strains of *Metarhizium anisopliae* and *Beauveria bassiana* against *Spodoptera litura* on the basis of their virulence, germination rate, conidia production, radial growth and enzyme activity. Mycobiology 40:111–116

Powell KA, Jutsum AR (1993) Technical and commercial aspects of biocontrol products. Pestic Sci 37:315–321

Prabagaran SR, Rupesh KR, Nimal SJ, Sudha Rani S, Jayachandran S (2003) Advances in pest control: the role of *Bacillus thuringiensis*. Indian J Biotechnol 2:302–321

Purwar JP, Sachan GC (2005) Biotoxicity of *Beauveria bassiana* and *Metarhizium anisopliae* against *Spodoptera litura* and *Spilarctia oblique*. Ann Plant Prot Sci 13:360–364

Purwar JP, Sachan GC (2006) Synergistic effect of entomogenous fungi on some insecticides against Bihar hairy caterpillar *Spilarctia obliqua* (Lepidoptera: Arctiidae). Microbiol Res 161:38–42

Quesada-Moraga E, Carrasco-Diaz JA, Santiago-Alvarez C (2006) Insecticidal and antifeedant activities of proteins secreted by entomopathogenic fungi against *Spodoptera littoralis* (Lep., Noctuidae). J Appl Entomol 130:442–452

Rachappa V, Lingappa S, Patil RK (2007) Occurrence of entomopathogenic fungi in Northern Karnataka. J Ecobiol 20:85–91

Rajesh A, Bhupendra NT (2009) Pathogenicity of entomopathogenic fungi to eggs and larvae of *Spodoptera litura*, the common cutworm. Biocontrol Sci Technol 19:919–929

Rajesh A, Birendra P, Bhupendra NT (2009) Relative susceptibility of *Spodoptera litura* pupae to selected entomopathogenic fungi. BioControl 54:85–92

Ramakrishna N, Tiwari LD (1969) Polyhedrosis of *Prodenia litura* Fab. (Noctuidae: Lepidoptera). Indian J Entomol 31:191–192

Ramiro ERN, Ramiro ARE, Juan MSY, Jaime MO, Steven RS, Roberto CR, René PR, Francisco GH, John EF (2013) Occurrence of entomopathogenic fungi and parasitic nematodes on *Spodoptera frugiperda* (Lepidoptera: Noctuidae) larvae collected in central Chiapas, México. Fla Entomol 96:498–503

Rangel DN, Braga GL, Anderson AJ, Roberts DW (2005) Variability in Richard JS, Neal TD, Karl JK, Michael RK, 2010, Model reactions for insect cuticle sclerotization: participation of amino groups in the cross-linking of *Manduca sexta* cuticle protein MsCP36. Insect Biochem Mol Biol 40:252–258

Rath AC, Carr CJ, Graham BR (1995) Characterization of *Metarhizium anisopliae* strains carbohydrate utilization (AP150CH). J Invertebr Pathol 65:152–161

Ravensberg WJ (1994) Biological control of pests: current trends and future prospects. In: Proc. Brighton Crop Protection Conf., pests and diseases. British Crop Protection Council, Farnham, pp 591–600

Raymond JSL, Chengshu W (2010) Genetic engineering of fungal biocontrol agents to achieve greater efficacy against insect pests. Appl Microbiol Biotechnol 85:901–907

Rehner SA, Minnis D, Sung GH, Luangsa-ard JJ, DeVoto L, Humber RA (2011) Phylogeny and systematics of the anamorphic, entomopathogenic genus *Beauveria*. Mycologia 103:1055–1073

Rios-Velasco C, Cerna-Chavez E, Pena SS, Morales GG (2010) Natural epizootic of the entomopathogenic fungus *Nomuraea rileyi* (Farlow) Samson infecting *Spodoptera frugiperda* (Lepidoptera: Noctuidae) in Coahuila Mexico. J Res Lepidoptera 43:7–8

Roberts DW (1981) Toxins of entomopathogenic fungi. In: Burges HD (ed) Microbial control of pests and plant diseases 1970-1980. Academic, London, pp 441–464

Roberts DW, Humber RA (1981) Entomogenous fungi. In: Cole GT, Kendrick B (eds) The biology of conidial fungi, vol 2. Academic, New York, pp 201–236

Roberts DW, St. Leger RJ (2004) *Metarhizium* spp., cosmopolitan insect-pathogenic fungi: mycological aspects. Adv Appl Microbiol 54:1–7

Rosell G, Quero C, Coll J, Guerrero A (2008) Biorational insecticides in pest management. J Pestic Sci 33:103–121

Roy HE, Steinkraus DC, Eilenberg J, Hajek AE, Pell JK (2006) Bizarre interactions and endgames: entomopathogenic fungi and their arthropod hosts. Annu Rev Entomol 51:331–357

Saad ASA, Madhkour A, El-Bahrwawi A (1975) Notes on the effect of insecticides on different strains of *Spodoptera littoralis* Boisd. in Egypt. Indian J Agric Sci 45:231–232

Sadek MM (2003) Antifeedant and toxic activity of *Adhatoda vasica* leaf extract against *Spodoptera littoralis* (Lepidoptera: Noctuidae). J Appl Entomol 27:396–404

Saeedeh J, Ahmad SS, Rozi M, Lau WH (2013) Suitability of *Centella asiatica* (Pegaga) as a food source for rearing *Spodoptera litura* (F.) (Lepidoptera: Noctuidae) under laboratory conditions. J Plant Prot Res 53:184–189

Sahayaraj K, Borgio JF (2012) Screening of some mycoinsecticides for the managing hairy caterpillar, *Pericallia ricini* Fab. (Lepidoptera: Arctiidae) in castor. J Entomol 9:89–97

Sahayaraj K, Paulraj MG (1998) Screening the relative toxicity of some plant extracts to *Spodoptera litura* Fab. (Insecta: Lepidoptera: Noctuidae) of groundnut. Fresenius Environ Bull 7:557–560

Sahayaraj K, Namasivayam SKR, Rathi JM (2011) Compatibility of entomopathogenic fungi with extracts of plants and commercial botanicals. Afr J Biotechnol 10:933–938

Salama HS, Shoukry A (1972) Flight range of the moth of the cotton leaf worm *Spodoptera littoralis*. Zeitung Angew Entomol 71:181–184

Saleem MA, Ahmad M, Aslam M, Sayyed AH (2008) Resistance to selected organochlorine, organophosphate, carbamate and pyrethroid, in *Spodoptera litura* (Lepidoptera: Noctuidae) from Pakistan. J Econ Entomol 101:1667–1675

Samson RA, Evans HC, Latg'e JP (1988) Atlas of entomopathogenic fungi. Springer, Berlin, p 187

Sanchez-Pena SR (2000) Entomopathogens from two Chihuahuan desert localities in Mexico. BioControl 45:63–78

Sarup P (1970) Effect of formulation on the toxicity of insecticidal dusts to the larva of *Prodenia litura* Fabricius-1.Choice of diluents. Indian J Entomol 32:356–375

Sarup P, Singh DS (1974) Intra- and inter-specific variations in the effectiveness of diluents in insecticidal dust formulations. Indian J Entomol 34:368–369

Satti AA, Gorashi NE (2013) Isolation and characterization of new entomopathogenic fungi from the Sudan. Int J Sci Innov Disc 3:326–329

Serigo RSP, Jorge SJL, Raul F (2010) Occurrence of entomopathogenic fungi from agricultural ecosystems in Saltillo, Mexico, and their virulence towards thrips and white flies. J Insect Sci 11:1

Sertkaya E, Bayram A, Kornosor S (2004) Egg and larval parasitoids of the beet armyworm *Spodoptera exigua* on maize in Turkey. Phytoparasitica 32:305–312

Shankarganesh K, Walia S, Dhingra S, Subrahmanyam B, Babu SR (2012) Effect of dihydrodillapiole on pyrethroid resistance associated esterase inhibition in an Indian population of *Spodoptera litura* (Fabricius). Pestic Biochem Physiol 102:86–90

Sharma RK, Bisht RS (2008) Antifeedant activity of indigenous plant extracts against *Spodoptera litura* Fabricius. J Insect Sci 21:56–60

Shaw KE, Davidson G, Clark SJ, Ball BV, Pell JK, Chandler D, Sunderland KD (2002) Laboratory bioassay to assess the pathogenicity of microscopic fungi to *Varroa destructor*, an ectoparasitic mite of the honey bee, *Apis mellifera*. Biol Control 24:266–276

Shereen ME, Nabawia ME, Fayez MS, Tayseer AR (2012) Physiological and biochemical effect of entomopathogenic fungus *Metarhizium anisopliae* on the 5th instar of *Schistcereca gregaria* (Orthoptera: Acrididae). J Res Environ Sci Toxicol 1:7–18

Shieh TR (1989) Industrial production of viral pesticides. Adv Virus Res 36:315–343

Sierotzki H, Camastral F, Shah PA, Aebi M, Tuor U (2000) Biological characteristics of selected *Erynia neoaphidis* isolates. Mycol Res 104:213–219

Silva CAD (2000) Microorganismos entomopatogênicos associados a insetos e ácaros do algodoeiro. Documentos. Campina Grande: EMBRAPA-CNPA 77, p 45

Sookar P, Bhagwant S, Awuor OE (2008) Isolation of entomopathogenic fungi from the soil and their pathogenicity to two fruit fly species. J Appl Entomol 32:778–788

Sreedhar V, Devaprasad V (1995) Mycosis of *Nomuraea rileyi* in field populations of *Spodoptera litura* in relation to four host plants. Indian J Entomol 58:192–195

Srisukchayakul P, Wiwat C, Pantuwatana S (2005) Studies on the pathogenesis of the local isolates of *Nomuraea rileyi* against *Spodoptera litura*. Sci Asia 31:273–276

Steinhaus EA (1949) Principles of insect pathology. McGraw-Hill, New York, p 757

Steinhaus EA (1964) Microbial diseases of insects. In: DeBach P (ed) Biological control of insect pests and weeds. Chapman and Hall, London, pp 515–547

Strasser H, Vey A, Butt TM (2000) Are there any risks in using entomopathogenic fungi for pest control, with particular reference to the bioactive metabolites of *Metarhizium*, *Tolypocladium* and *Beauveria* species? Biocontrol Sci Technol 10:717–735

Sun BD, Liu XZ (2008) Occurrence and diversity of insect-associated fungi in natural soils in China. Appl Soil Ecol 39:100–108

Tamez-Guerra P, Castro-Franco R, Medrano-Roldan H, Mcguire MR, Galan-Wong LJ, Luna-Olvera HA (1998) Laboratory and field comparisons of strains of *Bacillus thuringiensis* for activity against noctuid larvae using granular formulations (Lepidoptera). J Econ Entomol 91: 86–93

Tanada Y, Kaya HK (1993) Insect Pathology. Academic, San Diego, 666

Teetor-Barsch GH, Roberts WD (1983) *Fusarium* species pathogens of insects review. Mycopathologia 84:3–16

Thackar JRM (2002) An introduction to arthropod pest control. Cambridge University Press, Cambridge, p 144

Thomas DB (1992) Taxonomic synopsis of the *Asopinae pentatomidae* (Heteroptera) of the western hemisphere, vol 16. The Thomas Say Foundation, Entomological Society of America, Lanham

Thomas MB, Read AF (2007) Can fungal biopesticides control malaria? Nature Rev Microbiol 5: 377–383

Tigano MS, Faria MR, Lecuona RE, Sartori MR, Arima EY, Diaz BM, De-Faria MR (1995) Analysis of pathogenicity and germination of the fungus *Nomuraea rileyi* isolated in Federal district. Anais Soc Entomológ Brasil 24:53–60

Tingle FC, Mitchell ER (1977) Seasonal populations of armyworms and loopers at Hastings, Florida. Fla Entomol 60:115–122

Tong H, Su Q, Zhou X, Bai L (2013) Field resistance of *Spodoptera litura* (Lepidoptera: Noctuidae) to organophosphates, pyrethroids, carbamates and four newer chemistry insecticides in Hunan, China. J Pestic Sci 86:599–609

Torasco E, Poliseno M (2005) Preliminary survey on the occurrence of entomopathogenic nematodes and fungi in Albanian soils. Bull OILB/SROP 28:165–168

Tucker MR (1983) Light-trap catches of African armyworm moths, *Spodoptera exempta* (Walker) (Lepidoptera: Noctuidae), in relation to rain and wind. Bull Entomol Res 73:315–319

Udayababu P, Sunil Z, Goud CR (2012) Evaluation of entomopathogenic fungi for the management of tobacco caterpillar *Spodoptera litura* (Fabricius). Indian J Plant Prot 40:214–220

Ujian AA, Shahzad S (2007) Pathogenicity of *Metarhizium* anisopliae Var. *Acridum* strains on Pink Hibiscus Mealy bug (*Maconellicoccus hirsutus*) affecting cotton crop. Pak J Bot 39: 967–973

Uribe D, Khachatourians GG (2008) Identification and characterization of an alternative oxidase in the entomopathogenic fungus *Metarhizium anisopliae*. Can J Microbiol 54:1–9

Vandenberg JD, Shelton AM, Wilsey WT, Ramos M (1998) Assessment of *Beauveria bassiana* sprays for control of diamond back moth (Lepidoptera: Plutellidae) on crucifers. J Econ Entomol 91:624–630

Vanninen I, Hokkanen H (1988) Effects of pesticides on four species of entomopathogenic fungi. Annales Agriculturae Fennici 27:345–353

Veena KK, Rabindra RJ, Srinivasa NCD, Subha MR (2006) First report of *Beauveria bassiana* on *Amsacta albistriga* Walker from Karnataka. Indian J Biol Control 20:95–96

Vey A, Hoagland R, Butt TM (2001) Toxic metabolites of fungal biocontrol agents. In: Butt TM, Jackson CW, Magan N (eds) Fungi as biocontrol agents. CAB International, Wallingford, pp 311–345

Vimaladevi PS (1994) Conidia production of the entomopathogenic fungus *Nomuraea rileyi* and its evaluation for control of *Spodoptera litura* (Fab) on *Ricinus communis*. J Invertebr Pathol 63:145–150

Vimaladevi PS, Prasad YG (1997) The entomofungal pathogen *Nomuraea rileyi*. Info Bull Direc Oilseeds Res, Hyderabad, p 3

Vimaladevi PS, Prasad YG (2001) *Nomuraea rileyi*: A potential mycoinsecticide. In: Upadyay RK, Mukherji KG, Chamola BP (eds) Biocontrol potential and its exploitation in sustainable agriculture insect Pests, vol 2. Kluwer Academic/Plenum, New York, pp 23–38

Vimaladevi PS, Prasad RD (2008) Isolation, maintenance and preservation techniques. In: Hands-on training on microbial biocontrol agents of major insect pests and disease of crops, Hyderabad, pp 14–21

Vimaladevi PS, Prasad YG, Chowdary DA, Mallikarjuna RL, Balakrishnan K (2003) Identification of virulent isolates of the entomopathogenic fungus *Nomuraea rileyi* (F.) Samson for the management of *Helicoverpa armigera* and *Spodoptera litura*. Mycopathologia 156:365–373

Vimaladevi PS, Prasad YG, Rajeswari B, Vijaya BL (1996) Epizootics of the entomofungal pathogen, *Nomuraea rileyi* on lepidopterous pests of oil seed crops. J Oil Seeds Res 13:144–148

Vincent C, Goettel MS, Lazarovits G (2007) Biological control, a global perspective. CABI, Oxfordshire

Vlak JH, Groner A (1980) Identification of two nuclear polyhedrosis viruses from the cabbage moth, *Mamestra brassicae* (Lepidoptera: Noctuidae). J Invertebr Pathol 35:269–278

Vlak JH, Beider Eden, Peters D, Vrie V (1982) Bekampfung eines eingeschleppten Schadlings, *Spodoptera exiqua*, in Gewachshausern mit dem autochtonen virus. Mededelingen. Faculteit. Landbouwwetenschappen Rijksuniversiteit Gent 47:1005–1016

Wang CS, St. Leger RJ (2007a) A scorpion neurotoxin increases the potency of a fungal insecticide. Nature Biotechnol 25:1455–1456

Wang CS, St. Leger RJ (2007b) The *Metarhizium anisopliae* perilipin homolog MPL1 regulates lipid metabolism, appressorial turgor pressure, and virulence. J Biol Chem 282:21110–21115

Wang CS, St. Leger RJ (2007c) The MAD1 adhesin of *Metarhizium anisopliae* links adhesion with blastospore production and virulence to insects, and the MAD2 adhesin enables attachment to plants. Eukaryot Cell 6:808–816

Wang CS, Skrobek A, Butt TM (2004) Investigations on the destruxin production of the entomopathogenic fungus *Metarhizium anisopliae*. J Invertebr Pathol 85:168–174

Wang S, Miao X, Zhao W, Huang B, Fan M, Li Z, Huang Y (2005) Genetic diversity and population structure among strains of the entomopathogenic fungus, *Beauveria bassiana*, as revealed by inter-simple sequence repeats (ISSR) Mycol Res 109:1364–1372

Wang L, Huang J, You M, Guan X, Liu B (2007) Toxicity and feeding deterrence of crude toxin extracts of *Lecanicillium* (*Verticillium*) *lecanii* (Hyphomycetes) against sweet potato whitefly, *Bemisia tabaci* (Homoptera: Aleyrodidae). Pest Manage Sci 63:381–387

Wheeler DA, Isman MB (2001) Antifeedant and toxic activity of *Trichilia americana* extract against the larvae of *Spodoptera litura*. Entomol Exp Appl 98:9–16

Willcock FC (1905) Yaer book Khedivial. Agricultural Society, Egypt

Wilson JW (1934) The asparagus caterpillar: its life history and control. Fla Agric Exp St Bull 271: 1–26

Wraight SP, Carruthers RI, Bradley CA, Jaronski ST, Lacey LA, Wood P, Galaini-Wraight S (1998) Pathogenicity of the entomopathogenic fungi *Paecilomyces* spp. and *Beauveria bassiana* against the silver leaf whitefly, *Bemisia argentifolii*. J Invertebr Pathol 71:217–226

Wraight SP, Carruthers RI, Jaronski ST, Bradley CA, Garza CJ, Galaini-Wraight S (2000) Evaluation of the entomopathogenic fungi *Beauveria bassiana* and *Paecilomyces fumosoroseus* for microbial control of the silver leaf whitefly, *Bemisia argentifolii*. Biol Control 17:203–217

Wraight SP, Inglis GD, Goettel MS (2007) Fungi. In: Lacey LA, Kaya HK (eds) Field manual of techniques in invertebrate pathology, 2nd edn. Springer, Dordrecht, pp 223–248. ISBN 978-1-4020-5931-5

Wraight SP, Ramos ME, Avery PB, Jaronski ST, Vandenburg JD (2010) Comparative virulence of *Beauveria bassiana* isolates against lepidopteran pests of vegetable crops. J Invertebr Pathol 103:186–199

Yang S, Yang S, Sun W, Jianping LV, Kuang R (2009) Use of sex Pheromone for control of *Spodoptera litura* (Lepidoptera: Noctuidae). J Entomol Res Soc 11:27–36

Yathom S, Rivnay E (1960) Field trials against noctuidaes in cotton (In Hebraw). Rep Nat Inst Agri Minist Serial No. 318, p 22

Yi Q, Sheng HY, Ying C, Zhi CS, Ming GF (2010) Integration of Insecticidal Protein Vip3Aa1 into *Beauveria bassiana* enhances fungal virulence to *Spodoptera litura* larvae by cuticle and Per Os Infection. Appl Environ Microbiol 76:4611–4618

Zalom FG, Wilson LT, Hoffmann MP (1986) Impact of feeding by tomato fruitworm, *Heliothis zea* (Boddie) (Lepidoptera: Noctuidae), and beet armyworm, *Spodoptera exigua* (Hübner) (Lepidoptera: Noctuidae), on processing tomato fruit quality. J Econ Entomol 79:822–826

Zanuncio JC, Batalha VC, Guedes RNC, Picanco M (1998) Insecticide selectivity to *Supputius cincticeps* (Stal) (Het.: Pentatomidae) and its prey *Spodoptera frugiperda* (J.E. Smith) (Lep.: Noctuidae). J Appl Entomol 122:457–460

Zanuncio JC, da Silva CAD, de Lima ER, Pereira FF, Ramalho FDS, Serrao JE (2008) Predation rate of *Spodoptera frugiperda*(Lepidoptera: Noctuidae) larvae with and without defense by *Podisus nigrispinus* (Heteroptera: Pentatomidae). Braz Arch Biol Technol 51(1), Curitiba. http://dx.doi.org/10.1590/S1516-89132008000100015

Zare R, Gams W (2001) A revision of Verticillium section. Prostata IV. The genera *Lecanicillium* and *Simplicillium* gen.nov. Nova Hedwigia 73:1–50

Zemek R, Hussein HM, Prenerova E (2012) Laboratory evaluation of *Isaria fumosorosea* against *Spodoptera littoralis*. Commu Agric Appl Biol Sci 77:685–689

Zhang S (2001) A species of entomogenous fungus *Fusarium lateritium* isolated from citrus aphid. Scientia Silvae Sinicae 37:66–70

Zimmermann G (1986) The 'Galleria bait method' for detection of entomopathogenic fungi in soil. J Appl Entomol 102:213–215

Zimmermann G (2007a) Review on Safety of the Entomopathogenic fungi *Beauveria bassiana* and *Beauveria brongniartii*. Biocontrol Sci Technol 17:553–596

Zimmermann G (2007b) Review on safety of the entomopathogenic fungus *Metarhizium anisopliae*. Biocontrol Technol 17:879–920

Zimmermann G (2008) The entomopathogenic fungi *Isaria farinosa* (formerly *Paecilomyces farinosus*) and the *Isaria fumosorosea* species complex (formerly *Paecilomyces fumosoroseus*): biology, ecology and use in biological control. Biocontrol Sci Technol 18: 865–901

Chapter 7
Comparative Account of Generalist and Specialist Species of the Entomopathogenic Fungus, *Metarhizium*

K. Sowjanya Sree and Hemesh Joshi

7.1 Introduction

Entomopathogenic fungi are one of the most extensively studied microorganisms with respect to biological control of insect pests. They are one amongst the many natural enemies of insect species worldwide (Hajek 2004; Kleespies et al. 2008). There are around 750 entomopathogenic fungal species known on this planet, amongst which species belonging to the genera *Metarhizium* (Pandey and Hasan 2009), *Beauveria* (Garcia et al. 2011), *Nomuraea* (Ingle et al. 2004), *Lecanicillium* (Anand et al. 2009) and *Isaria* (Zimmermann 2008) have been well investigated. These are also considered as model organisms for host–pathogen interaction studies (Thomas and Read 2007). Their huge biodiversity allows nature as well as researchers to choose a high virulent strain capable of killing the host in an effective manner for use as biocontrol agent (Thomas and Read 2007). Much of the research on these entomopathogenic fungi at present is focused on the factors determining host specificity. In the present chapter, we will discuss on a few such recent findings on the entomopathogenic fungal species belonging to the genus *Metarhizium*.

Metarhizium also known as the green muscardine fungus causes green muscardine disease in their target insect hosts. A number of well-characterised species are grouped under this genus; a few include *Metarhizium anisopliae*, *M. acridum*, *M. flavoviride*, *M. album* and so on (Driver et al. 2000). Elie Metchnikoff in 1879 discovered *M. anisopliae* and named it as *Oospora destructor* and also conducted the very first experiments on testing this fungus against a coleopteran grain beetle, *Anisoplia austriaca*. After realising the importance of biocontrol agents, researchers started investigating this fungus from both application point of view and for basic research on its infection process in the host and its host

K.S. Sree (✉) • H. Joshi
Amity Institute of Microbial Technology, Amity University Uttar Pradesh, Noida 201303, Uttar Pradesh, India
e-mail: ksowsree@gmail.com

© Springer International Publishing Switzerland 2015
K.S. Sree, A. Varma (eds.), *Biocontrol of Lepidopteran Pests*, Soil Biology 43,
DOI 10.1007/978-3-319-14499-3_7

specificity (Wang et al. 2005; Anand and Tiwary 2009; Schrank and Vainstein 2010; Contreras et al. 2014). *Metarhizium* species have a worldwide distribution from the tropics to the arctic regions and infect the insects and inhabit the soils in varied climatic conditions like forests, coasts, swamps and deserts (Zimmerman 2007). Its life stages alternate between saprophytic phase in the soil and pathogenic phase in the insect host. More recent studies showed that *M. anisopliae* colonises plant roots, acting as both biopesticide and biofertiliser (St. Leger 2008).

7.2 Host Specificity of *Metarhizium* Species: Generalist vs Specialist

The species of this genus exhibit a wide array of host specificity from being a broad host range pathogen (e.g. *M. anisopliae*) to a very narrow host range pathogen (e.g. *M. acridum*) (Driver et al. 2000). *M. anisopliae* is known to be pathogenic to more than 100 insect pests, whereas its close relative *M. acridum* is known to infect only locusts and grasshoppers (Zimmermann 1993; Driver et al. 2000; Peng et al. 2008). *M. acridum* was earlier considered as a variety of *M. anisopliae* until the investigations of Bischoff et al. (2009) came forward. They used a multigene phylogenetic approach using nuclear-encoded gene regions of EF-1α, RPB1, RPB2 and β-tubulin together with the morphological markers. The phylogenetic evidence from this study suggested the recognition of *M. acridum* at a species rank instead of as a variety of *M. anisopliae*.

Recently, Gao et al. (2011) carried out very interesting investigation on these fungal relatives. They made a comparative analysis of the genome and transcriptome of *M. anisopliae* and *M. acridum*. Being very closely related, a number of similarities were observed. Genomes of both the species were closer to fungal pathogens and endophytes of plants than that of animals. This suggests that *Metarhizium* might have evolved from plant fungal pathogens or endophytes. The probable transition of fungi from plant to insect habitat might involve adaptation to feed on insects. This was supported by the presence of huge number of genes for proteases, lipases and chitinases which facilitate the digestion of insect cuticle and body. The number of genes encoding secreted proteins was noticeably high in both the species when compared to other plant fungal pathogens. These fungi produce a whole set of proteases belonging to different types like subtilisins, chymotrypsins, trypsins, metalloproteases, aspartyl proteases, cysteine proteases and exopeptidases; however, the chymotrypsins are specific to *M. anisopliae*. This might have occurred via horizontal gene transfer (Screen and St. Leger 2000).

Although an array of proteases is produced by both the species of *Metarhizium*, the levels are much too low in *M. acridum* when compared to *M. anisopliae*. The high amounts of varied secreted proteases might allow *M. anisopliae* to survive in different nutritional environments or, in other terms, in different hosts (Gao et al. 2011). In addition, *M. anisopliae* encodes for dehydrogenases unlike

M. acridum and produces far more cytochrome P450s when compared to the latter which might aid *M. anisopliae* to fight against a multitude of insect hosts.

Comparative transcriptome analysis revealed that *M. anisopliae* transcribed the same G-protein-coupled receptors on cuticles of different hosts like locust and cockroach, but *M. acridum* transcribed a G-protein-coupled receptor unique to its host, i.e. a distinct receptor on cuticle of locust and another one on that of cockroach (Gao et al. 2011). The comparative genome analysis also suggested that *M. anisopliae* possesses greater potential to produce secondary metabolites than *M. acridum*. More details will be discussed in the sections below.

7.2.1 Mechanical Kill of the Pest

Although there is a huge variation in the host range of the two species, *M. anisopliae* and *M. acridum*, they both follow the same basic process of infecting their host. When the spores come in contact with a susceptible host cuticle, they adhere to the cuticle owing to the hydrophobic interactions between the hydrophobins on the spore surface and the lipid layer on the host cuticle (Fang et al. 2007). Germination of these conidia starts under suitable temperature and humidity giving rise to a germ tube. They penetrate the host cuticle facilitated by appressorium and penetration peg formation. Together with the help of turgor pressure and the action of the cuticle-degrading enzymes as described above (da Silva et al. 2005; Gao et al. 2011), the pathogen makes its way into the haemolymph of the host where its hyphae proliferate absorbing nutrients from the host body. The functions of several pathogenicity-related genes during this process have been revealed recently (Wang and St. Leger 2006, 2007a, b; Duan et al. 2009; Fang et al. 2010). Mycelium is septate and the hyphal network spreads throughout the host body (Bechara et al. 2011), eventually killing the host through mechanical injury. Later, the hyphae pierce out of the cadaver and sporulate resulting in mycosis. The infection process is described in more details in Chap. 6 of this volume.

7.2.2 Toxic Action of Secondary Metabolites

Destruxin (dtx) is one of the toxic secondary metabolites extracellularly secreted by the fungus *M. anisopliae*. Chemically, it is cyclodepsipeptidic in nature comprising of five amino acids and an α-hydroxy acid. The name destruxin was derived from "destructor" from the species *Oospora destructor* (Metsch.) Delac., the entomopathogenic fungus from which these metabolites were first isolated (Kodaira 1961). Later on *O. destructor* was renamed as *M. anisopliae* (Metsch.) Sorokin, but as customary the compound's trivial name was retained (Suzuki et al. 1970). Individual dtxs differ on the hydroxy acid, *N*-methylation and R group of the amino acid residues. Dtxs A, B and E have the same amino acid sequence (proline-isoleucine-methyl valine-methyl alanine-beta alanine) but differ in the R group of the hydroxy acid residue.

The EST analysis of *M. anisopliae* genome showed the presence of the peptide synthases, reductases and other enzymes involved in the production of dtxs, enniatins, trichothecenes and cytochalasins (Freimoser et al. 2003). Of all these, dtxs have been widely exploited for their insecticidal activity. The great majority of dtxs and its analogues were isolated from cultures of *M. anisopliae* (Pedras et al. 2002), although this mycotoxin is produced by other fungi as well (Liu and Tzeng 2012). Out of all the variants of this peptide produced in vitro, the most predominant ones are dtxs A, B and E. Dtxs A and E have been proven to have active insecticidal properties against different orders of insects (Dumas et al. 1994; Pedras et al. 2002; Padmaja and Sree 2008). These mycotoxins which are extracellularly secreted play an important role in pathogenesis (Kershaw et al. 1999).

Recently, the non-ribosomal peptide synthetase gene cluster coding for different dtxs has been successfully revealed (Wang et al. 2012). Different strains of *M. anisopliae* produce different amounts of dtxs in varying combinations (Sree et al. 2008). However, its close relative and a specialist entomopathogen, *M. acridum*, does not produce dtx (Wang et al. 2012). This is one of the major differences between these two species. Dtxs suppress the host innate immune response (Pal et al. 2007) and are also reported to induce oxidative stress in the dtx-treated larvae (Sree and Padmaja 2008; Sree et al. 2010). The other cellular effects of dtx on insects are briefed in Chap. 9 of this volume. Thus, the fungus *M. anisopliae* kills its host by mechanical injury which is accompanied by the toxic effects of dtx, whereas *M. acridum* kills its host only by mechanical injury. This is also supported by the recent comparative genomic analysis of the two species (Gao et al. 2011).

7.3 Manipulation of Host Range

In a recent study, Wang et al. (2011) demonstrated that expression of the gene *Mest1* coding for an esterase in the specialist *M. acridum*, transformed with this gene, increased the host range of the fungus. In *M. robertsii*, under natural conditions, MEST1, an esterase, is localised in the conidial lipid droplets. *Mest1* expression mobilises the stored lipids, subjecting them to hydrolysis and thereby supporting the germination and infection processes like appressorium formation. It was found that *Mest1* gene was upregulated in *M. robertsii* grown on *Manduca sexta* cuticle-containing medium and its expression correlated with the virulence of the fungus. This gene is reported to be absent in *M. acridum*. In the locust-specific *M. acridum* when transformed with *Mest1* gene from *M. robertsii*, it was found that the transformant's host range was broadened, and these transformants could now infect and colonise lepidopteran caterpillars (Wang et al. 2011).

7.4 Conclusions and Future Prospects

Two of the species of the entomopathogenic fungus *Metarhizium*, *M. anisopliae* and *M. acridum*, differ considerably in their host range. Advancements in the fields of genomics and transcriptomics led to an insight into the host specificity factors involved in pathogenicity of *Metarhizium*. Nevertheless an in-depth understanding of the factors inhibiting *M. acridum* from infecting insects other than locusts and grasshoppers will be highly appreciated and will facilitate the development and improvement of broad- and narrow-range bio-insecticides.

Acknowledgements KSS is grateful to SERB, Govt. of India, for financial assistance through the Fast Track Young Scientist scheme.

References

Anand R, Tiwary BN (2009) Pathogenicity of entomopathogenic fungi to eggs and larvae of *Spodoptera litura*, the common cutworm. Biocontrol Sci Technol 19:919–929

Anand R, Prasad B, Tiwary BN (2009) Relative susceptibility of *Spodoptera litura* pupae to selected entomopathogenic fungi. Biocontrol 54:85–92

Bechara IJ, Destefano RHR, Bresil C, Messias CL (2011) Histopathological events and detection of *Metarhizium anisopliae* using specific primers in infected immature stages of the fruit fly *Anastrepha fraterculus* (Wiedemann, 1830) (Diptera: Tephritidae). Braz J Biol 71:91–98

Bischoff JF, Rehner SA, Humber RA (2009) A multilocus phylogeny of the *Metarhizium anisopliae* lineage. Mycologia 101:512–530

Contreras J, Mendoza JE, Martínez-Aguirre MR, García-Vidal L, Izquierdo J, Bielza P (2014) Efficacy of enthomopathogenic fungus *Metarhizium anisopliae* against *Tuta absoluta* (Lepidoptera: Gelechiidae). J Econ Entomol 107:121–124

da Silva MV, Santi L, Staats CC, da Costa AM, Colodel EM, Driemeier D, Vainstein MH, Schrank A (2005) Cuticle-induced endo/exoacting chitinase CHIT30 from *Metarhizium anisopliae* is encoded by an ortholog the chi3 gene. Res Microbiol 156:382–392

Driver F, Milner JR, Trueman HWJ (2000) A taxonomic revision of *Metarhizium* based on a phylogenetic analysis of rDNA sequence data. Mycol Res 104:134–150

Duan ZB, Shang YF, Gao Q, Zheng P, Wang CS (2009) A phosphoketolase Mpk1 of bacterial origin is adaptively required for full virulence in the insect-pathogenic fungus *Metarhizium anisopliae*. Environ Microbiol 11:2351–2360

Dumas C, Rober P, Pais M, Vey A, Quiot JM (1994) Insecticidal and cytotoxic effects of natural and hemisynthetic destruxins. Comp Biochem Physiol C 108:195–203

Fang W, Pei Y, Bidochka MJ (2007) A regulator of a G protein signalling (RGS) gene, cag8, from the insect-pathogenic fungus *Metarhizium anisopliae* is involved in conidiation, virulence and hydrophobin synthesis. Microbiology 153:1017–1025

Fang W, Fernandes EK, Roberts DW, Bidochka MJ, St. Leger RJ (2010) A laccase exclusively expressed by *Metarhizium anisopliae* during isotropic growth is involved in pigmentation, tolerance to abiotic stresses and virulence. Fungal Genet Biol 47:602–607

Freimoser FM, Screen S, Bagga S, Hu G, St Leger RJ (2003) Expressed sequence tag (EST) analysis of two subspecies of *Metarhizium anisopliae* reveals a plethora of secreted proteins with potential activity in insect hosts. Microbiology 149:239–247

Gao Q, Jin K, Ying SH, Zhang Y, Xiao G et al (2011) Genome sequencing and comparative transcriptomics of the model entomopathogenic fungi *Metarhizium anisopliae* and *M. acridum*. PLoS Genet 7:e1001264

Garcia GC, Berenice GMM, Nestor BM (2011) Pathogenicity of isolates of entomopathogenic fungi against *Spodoptera frugiperda* (Lepidoptera: Noctuidae) and *Epilachna varivestis* (Coleoptera: Coccinellidae). Rev Colomb Entomol 37:217–222

Hajek A (2004) Natural enemies: an introduction to biological control. Cambridge University Press, Cambridge

Ingle YV, Lande SK, Burgoni GK, Autkar SS (2004) Natural epizootic of *Nomuraea rileyi* on lepidopterous pests of soybean and green gram. J Appl Zool Res 15:160–162

Kershaw MJ, Moorhouse ER, Bateman RP, Reynolds SE, Charnley AK (1999) The role of destruxins in the pathogenicity of *Metarhizium anisopliae* for three species of insect. J Invertebr Pathol 74:213–223

Kleespies RG, Huger AM, Zimmermann G (2008) Diseases of insects and other arthropods: results of diagnostic research over 55 years. Biocontrol Sci Technol 18:439–484

Kodaira Y (1961) Biochemical studies on the muscardine fungi in the silkworms, *Bombyx mori* L. J Fac Text Sci Technol Sinshu Univ 5:1–68

Liu BL, Tzeng YM (2012) Development and applications of destruxins: a review. Biotechnol Adv 30:1242–1254

Padmaja V, Sree KS (2008) Role of mycotoxin from the entomopathogenic fungus, *Metarhizium anisopliae* for insect pest management- current status. In: Ignacimuthu sj S, Jayaraj S (eds) Recent trends in insect pest management. Elite Publishers, New Delhi, pp 146–154

Pal S, St. Leger RJ, Wu LP (2007) Fungal peptide destruxin a plays a specific role in suppressing the innate immune response in *Drosophila melanogaster*. J Biol Chem 282:8969–8977

Pandey R, Hasan W (2009) Pathogenicity of entomopathogenic fungi, *Metarhizium anisopliae* against tobacco caterpillar, *Spodoptera litura* (Fabricius). Trends Biosci 2:29–30

Pedras MSC, Irina ZL, Ward DE (2002) The destruxins: synthesis, biosynthesis, biotransformation and biological activity. Phytochemistry 59:579–596

Peng GX, Wang ZK, Yin YP, Zeng DY, Xia YX (2008) Field trials of *Metarhizium anisopliae* var. acridum (Ascomycota: Hypocreales) against oriental migratory locusts, *Locusta migratoria manilensis* (Meyen) in Northern China. Crop Prot 27:1244–1250

Schrank A, Vainstein MH (2010) *Metarhizium anisopliae* enzymes and toxins. Toxicon 56:1267–1274

Screen SE, St. Leger RJ (2000) Cloning, expression, and substrate specificity of a fungal chymotrypsin. Evidence for lateral gene transfer from an actinomycete bacterium. J Biol Chem 275:6689–6694

Sree KS, Padmaja V (2008) Oxidative stress induced by destruxin from *Metarhizium anisopliae* (Metch.) involves changes in glutathione and ascorbate metabolism and instigates ultrastructural changes in the salivary glands of *Spodoptera litura* (Fab.) larvae. Toxicon 51:1140–1150

Sree KS, Padmaja V, Murthy LNY (2008) Insecticidal activity of destruxin, a mycotoxin from *Metarhizium anisopliae* (Hypocreales), against *Spodoptera litura* (Lepidoptera: Noctuidae) larval stages. Pest Manag Sci 64:119–125

Sree KS, Sachdev B, Padmaja V, Bhatnagar RK (2010) Electron spin resonance spectroscopic studies of free radical generation and tissue specific catalase gene expression in *Spodoptera litura* (Fab.) larvae treated with the mycotoxin, destruxin. Pestic Biochem Physiol 97:168–176

St. Leger RJ (2008) Studies on adaptations of *Metarhizium anisopliae* to life in the soil. J Invertebr Pathol 98:271–276

Suzuki A, Taguchi H, Tamura S (1970) Isolation and structure elucidation of three new insecticidal cyclodepsipeptides, destruxin C, D and desmethyl-destruxin B, produced by *Metarhizium anisopliae*. Agric Biol Chem 34:813–816

Thomas MB, Read AF (2007) Can fungal biopesticides control malaria? Nat Rev Microbiol 5:377–383

Wang C, St. Leger RJ (2006) A collagenous protective coat enables *Metarhizium anisopliae* to evade insect immune responses. Proc Natl Acad Sci USA 103:6647–6652

Wang C, St. Leger RJ (2007a) The MAD1 adhesin of *Metarhizium anisopliae* links adhesion with blastospore production and virulence to insects, and the MAD2 adhesin enables attachment to plants. Eukaryot Cell 6:808–816

Wang C, St. Leger RJ (2007b) The *Metarhizium anisopliae* perilipin homolog MPL1 regulates lipid metabolism, appressorial turgor pressure, and virulence. J Biol Chem 282:21110–21115

Wang C, Butt TM, St. Leger RJ (2005) Colony sectorization of *Metarhizium anisopliae* is a sign of ageing. Microbiology 151:3223–3236

Wang S, Fang W, Wang CS, St Leger RJ (2011) Insertion of an esterase gene into a specific locust pathogen (*Metarhizium acridum*) enables it to infect caterpillars. PLoS Pathog 7:e1002097

Wang B, Kang Q, Lu Y, Bai L, Wang C (2012) Unveiling the biosynthetic puzzle of destruxins in *Metarhizium* species. Proc Natl Acad Sci USA 109:1287–1292

Zimmerman G (2007) Review on safety of the entomopathogenic fungus *Metarhizium anisopliae*. Biocontrol Sci Technol 17:879–920

Zimmermann G (1993) The entomopathogenic fungus *Metarhizium anisopliae* and its potential as a biocontrol agent. Pestic Sci 37:375–379

Zimmermann G (2008) The entomopathogenic fungi *Isaria farinosa* (formerly *Paecilomyces farinosus*) and the *Isaria fumosorosea* species complex (formerly *Paecilomyces fumosoroseus*): biology, ecology and use in biological control. Biocontrol Sci Technol 18:865–901

Chapter 8
Non-ribosomal Peptides from Entomogenous Fungi

Qiongbo Hu and Tingyan Dong

8.1 Introduction

Entomogenous fungi, the pathogenic fungi of insects, are ubiquitous in natural environment. They play a very important role in controlling the natural population of insect pests. Many of their species (e.g., *Beauveria bassiana*, *Metarhizium anisopliae*) have been developed as myco-insecticides worldwide. Entomogenous fungi are able to produce mycotoxins such as non-ribosomal peptides (NRPs) which are toxic to tissues of host insects. Chemically, NRPs are a kind of secondary compounds biosynthesized by non-ribosomal peptide synthetase (NRPS) existing widely in fungi. NRPS gene of fungi is an open reading frame encoding a peptide chain composed of several modules, which activate amino acids and combined with a specific peptide product. Each module has a number of domains, and a specific reaction is catalyzed by one domain. The main domains include: adenylation domains (A domains), thiotion domains (T domains), condensation domains (C domains), epimerization domains (E domains), and methylation domains (M domains). The enzyme has been documented in many literature reviews (Boettger and Hertweck 2013; Hur et al. 2012).

Entomogenous fungi produce various kinds of NRPs, and each NRP consists of many analogues. According to the different molecular structures, entomogenous fungal NRPs could be divided into chain peptides (e.g., cicadapeptin and efrapeptin) and cyclic peptides including a subdivision of cyclopeptides and cyclodepsipeptides. Cyclopeptides are cyclic structures built by amino acid residues through peptide bonding (e.g., cyclosporin), while cyclodepsipeptides are lactone compounds consisting of amino acids and hydroxyl acids which are connected by peptide bonds. Most of the NRPs such as destruxin, beauvericin,

Q. Hu (✉) • T. Dong
College of Agriculture, South China Agricultural University, Guangzhou 510642, China
e-mail: hqbscau@126.com

© Springer International Publishing Switzerland 2015   169
K.S. Sree, A. Varma (eds.), *Biocontrol of Lepidopteran Pests*, Soil Biology 43,
DOI 10.1007/978-3-319-14499-3_8

enniatin, bassianolide, beauverolide, and serinocyclin belong to the group, cyclodepsipeptides.

8.2 Chain Peptides

8.2.1 Cicadapeptins

Cicadapeptins were isolated from the entomogenous fungi *Cordyceps heteropoda* (ARSEF #1880) and *Isaria sinclairii* (Krasnoff et al. 2005; Nagaoka et al. 2006). It was structurally elucidated that cicadapeptins I and II are acylated at the N-terminus by *n*-decanoic acid and amidated at the C-terminus by 1,2-diamino-4-methylpentane. The amino acid sequence of cicadapeptin I is N-terminus-Hyp-Hyp-Val-Aib-Gln-Aib-Leu-C-terminus, while in cicadapeptin II the Leu residue is replaced by Ile substitutes (Krasnoff et al. 2005) (Fig. 8.1 and Table 8.1). Cicadapeptins inhibited the acetylcholine (ACh)-evoked secretion of catecholamines in a concentration-dependent manner (Nagaoka et al. 2006). Cicadapeptins I and II showed antibacterial activity against *Bacillus cereus*, *B. subtilis*, and *Escherichia coli* (Rivas et al. 2008).

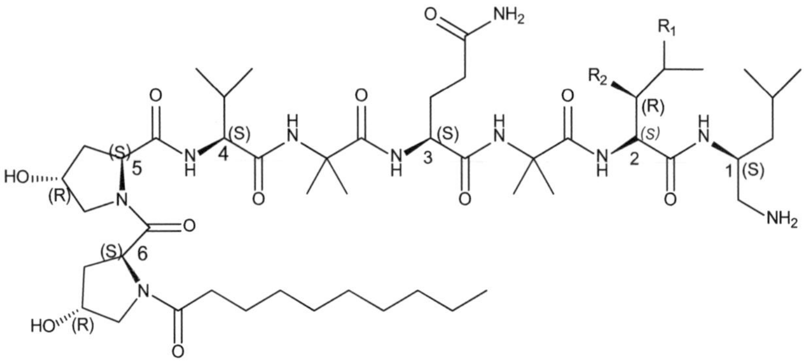

Fig. 8.1 Structure of cicadapeptins

Table 8.1 Analogues of cicadapeptins

Analogues name	CAS registry no.	Molecular formula	Geometry configuration	R1	R2
Cicadapeptin I	845626-76-8	$C_{50}H_{90}N_{10}O_{11}$	1S, 2S, 3S, 4S, 5S, 6S	Me	H
Cicadapeptin II	845626-81-5	$C_{50}H_{90}N_{10}O_{11}$	1S, 2S, 3S, 4S, 5S, 6S	H	Me

Note: Me, –CH$_3$

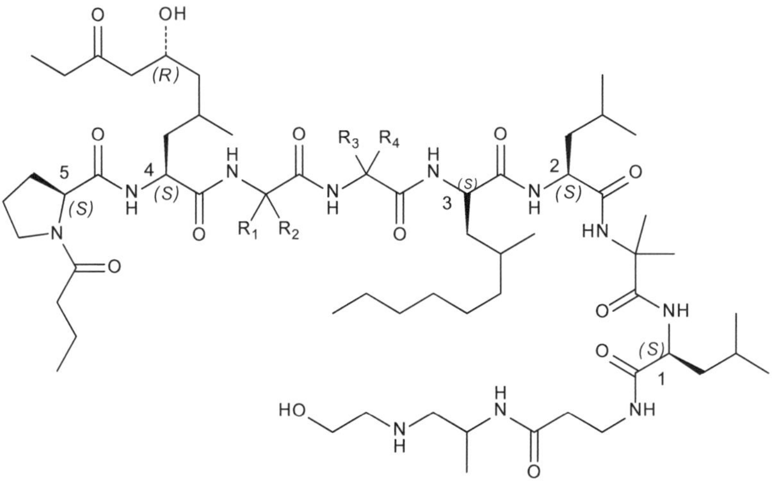

Fig. 8.2 Basic structure of culicinins

8.2.2 Culicinins

Culicinins were isolated from *Culicinomyces clavisporus*, a fungal pathogen of a wide range of mosquito larvae (He et al. 2006). The chemistry, structure, and synthesis of culicinins were studied by Zhang et al. (2008, 2009, 2011) (Fig. 8.2 and Table 8.2). The major analogue, anticancer agent culicinin D, was tested in oncology assays and exhibited selective inhibitory activity against PTEN-negative MDA468 breast tumor cells versus PTEN-positive MDA468 cells (He et al. 2006).

8.2.3 Efrapeptins and Neoefrapeptins

Efrapeptin was firstly discovered in the 1970s and was originally named A23871 (Lardy et al. 1975). They were isolated from *Tolypocladium* and other fungal species (Jackson et al. 1979). There are ten efrapeptin analogues known till date (Fig. 8.3 and Table 8.3). Structurally, efrapeptins are rich in Cα-dialkyl amino acids such as α-aminoisobutyric acid (Aib) or isovaline (Iva) and contain one β-alanine and several pipecolic acid residues (Krasnoff and Gupta 1991; Hayakawa et al. 2008; Boot et al. 2007). The C-terminus bears an unusual heterocyclic cationic cap. All efrapeptins were shown to adopt helical conformations in solvent (Weigelt et al. 2012).

Efrapeptins have insecticidal activity against *Tetranychus telarius*, *Musca domestica*, *Leptinotarsa decemlineata*, *Tetranychus urticae*, *Helicoverpa assulta*, etc. (Krasnoff et al. 1991). Efrapeptins have anti-immune action suppressing

Table 8.2 Analogues of culicinins

Analogues name	CAS registry no.	Molecular formula	Geometry configuration	R1	R2	R3	R4
Culicinin A	889128-37-4	$C_{61}H_{111}N_{11}O_{13}$	1S, 2S, 3S, 4S, 5S	H	Me	H	Me
Culicinin B	889128-38-5	$C_{62}H_{113}N_{11}O_{13}$	1S, 2S, 3S, 4S, 5S	Me	Me	Me	H
Culicinin C	889128-39-6	$C_{62}H_{113}N_{11}O_{13}$	1S, 2S, 3S, 4S, 5S	H	Me	Me	Me
Culicinin D	889128-40-9	$C_{63}H_{115}N_{11}O_{13}$	1S, 2S, 3S, 4S, 5S	Me	Me	Me	Me

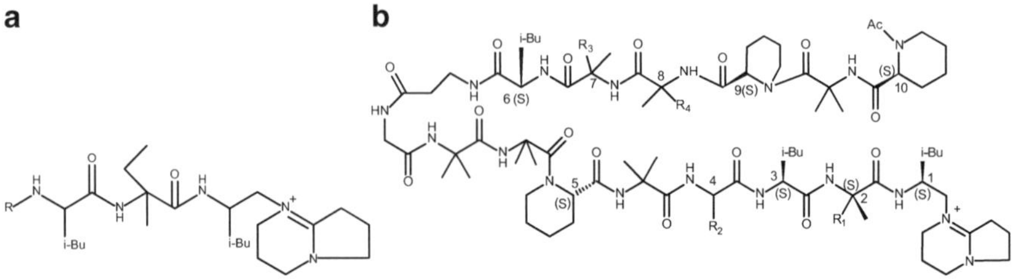

Fig. 8.3 Basic structure of efrapeptins; (**a**) efrapeptins A–B and (**b**) efrapeptins C–G

agglutination in *Galleria mellonella*, and this may be a result of its interference in the ligand-receptor interactions at the membrane of specific hemocytes (Bandani 2004). They also exhibit antifungal activities against *M. anisopliae* and *Tolypocladium niveum* (Krasnoff et al. 1991).

Efrapeptins are known as inhibitors of F1F0-ATPase, a mitochondrial enzyme (Papathanassiu et al. 2006). The breast cancer cells treated with efrapeptins show a disruption of the Hsp90:F1F0-ATPase complex and inhibition of Hsp90 chaperone activity (Papathanassiu et al. 2011). Efrapeptins displayed potent cytotoxicity against murine cancer cell lines and also demonstrated antibacterial properties (Boot et al. 2007). It was found that efrapeptin F had preferential cytotoxicity to nutrient-deprived cells compared to nutrient-sufficient cells. Efrapeptin F acts as a mitochondrial complex V inhibitor (Momose et al. 2010). Efrapeptins F, G, and J inhibited, in a dose-dependent manner, the 2-deoxyglucose-induced luciferase expression in HT1080 human fibrosarcoma cells transfected with a luciferase reporter plasmid containing the GRP78 promoter. Efrapeptin J also inhibited the protein expression of GRP78 in HT1080 cells and MKN-74 human gastric cancer cells (Hayakawa et al. 2008).

Neoefrapeptins, another chain peptide structurally similar to efrapeptins, were recently isolated from *Geotrichum candidum* (Molleyres et al. 2004; Fredenhagen et al. 2006) and displayed insecticidal activities. The structures of neoefrapeptins A to N have been elucidated (Fredenhagen et al. 2006; De Zotti et al. 2012) (Fig. 8.4 and Table 8.4).

Table 8.3 Analogues of efrapeptins

Analogues name	CAS registry no.	Molecular formula	Geometry configuration	R	R1	R2	R3	R4
Efrapeptin A	138145-52-5	$C_{32}H_{58}N_7O_5$	–	–COCH$_2$NHCOC(CH$_3$)$_2$NHAc	–	–	–	–
Efrapeptin B	138145-53-6	$C_{26}H_{48}N_5O_3$	–	–Ac	–	–	–	–
Efrapeptin C	138145-54-7	$C_{80}H_{137}N_{18}O_{16}$	1S, 3S, 5S, 6S, 9S, 10S	–	Me	H	Me	Me
Efrapeptin D	71503-60-1	$C_{81}H_{139}N_{18}O_{16}$	1S, 2S, 3S, 5S, 6S, 9S, 10S	–	Et	H	Me	Me
Efrapeptin E	138168-07-7	$C_{82}H_{141}N_{18}O_{16}$	1S, 2S, 3S, 5S, 6S, 8S, 9S, 10S	–	Et	H	Me	Et
Efrapeptin F	131353-66-7	$C_{82}H_{141}N_{18}O_{16}$	1S, 2S, 3S, 4S, 5S, 6S, 9S, 10S	–	Et	Me	Me	Me
Efrapeptin G	138145-55-8	$C_{83}H_{143}N_{18}O_{16}$	1S, 2S, 3S, 4S, 5S, 6S, 8S, 9S, 10S	–	Et	Me	Me	Et
Efrapeptin H	138264-31-0	$C_{84}H_{145}N_{18}O_{16}$	1S, 2S, 3S, 4S, 5S, 6S, 7S, 8S, 9S, 10S	–	Et	Me	Et	Et
Efrapeptin J	1058669-41-2	$C_{81}H_{139}N_{18}O_{16}$	1S, 3S, 4S, 5S, 6S, 9S, 10S	–	Me	Me	Me	Me

Note: Ac, HOOC–

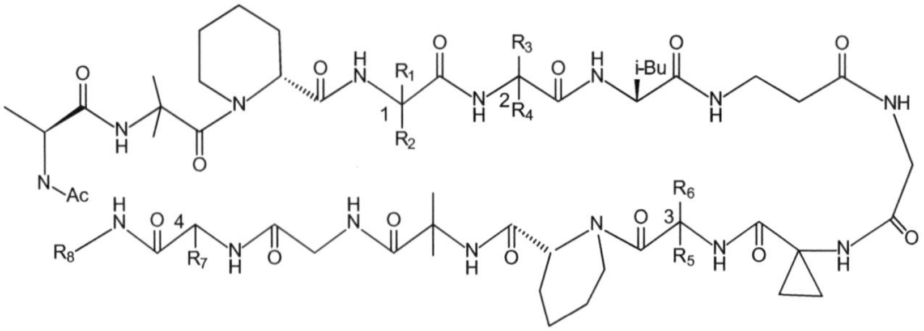

Fig. 8.4 Basic structure of neoefrapeptins

8.3 Cyclodepsipeptides

8.3.1 Bassianolides

Bassianolides were first isolated in the 1970s; they are reported to be produced by *B. bassiana* and *Verticillium lecanii* (Suzuki et al. 1977; Kanaoka et al. 1978) (Fig. 8.5 and Table 8.5). Bassianolide analogues are octadepsipeptidic derivatives with a 24-membered macrolactone ring that is formed as the cyclic tetrameric ester of the dipeptidol monomer, D-hydroxyisovalerate (D-Hiv)-N-methyl-L-leucine (N-Me-Leu) (Xu et al. 2009). Bassianolides could inhibit muscle contraction (Nakajyo et al. 1982) and kill silkworm when fed with an artificial diet containing 13 ppm of bassianolides (Suzuki et al. 1977). Bassianolides had no influence on intracellular Na^+ and K^+ contents and, therefore, might not be ionophoric (Nakajyo et al. 1983). Bassianolide analogues also have anthelmintic efficacy against the parasitic nematode *Ascaridia galli* in chicken (Ohyama et al. 2011). Bassianolide synthetases (348 kDa) from *Beauveria bassiana* ATCC 7159 was reconstituted in *Saccharomyces cerevisiae* BJ5464-NpgA, leading to the production of bassianolide (Yu et al. 2013b; Yu et al. 2013a).

8.3.2 Beauvericins

Beauvericins were first isolated in the 1960s (Hamill et al. 1969). They were produced by the entomogenous fungal genera, *Beauveria*, *Paecilomyces*, *Fusarium*, etc. (Bernardini et al. 1975; Gupta et al. 1991, 1995; Plattner and Nelson 1994). Chemically, beauvericins are a kind of cyclic hexadepsipeptide with alternating methyl-phenylalanyl and hydroxy-isovaleryl residues (Figs. 8.6 and 8.7, Tables 8.6 and 8.7).

Beauvericins G1–G3 and H1–H3 were biosynthesized and isolated from *B. bassiana* ATCC 7159 cultured with analogues of D-2-hydroxyisovalerate and L-phenylalanine. Beauvericins G1–G3 caused a parallel decline of cell migration

Table 8.4 Analogues of neoefrapeptins

Analogues name	CAS registry no.	Molecular formula	Geometry configuration	R1	R2	R3	R4	R5	R6	R7	R8
Neoefrapeptin A	695200-77-2	$C_{82}H_{139}N_{18}O_{16}$	1R, 4S	sBu	H	Me	Me	Me	Me	iBu	Rx
Neoefrapeptin B	695200-79-4	$C_{83}H_{141}N_{18}O_{16}$	1R, 2R, 4S	sBu	H	sBu	H	Me	Me	iBu	Rx
Neoefrapeptin C	695200-81-8	$C_{83}H_{141}N_{18}O_{16}$	1R, 3S, 4S	sBu	H	Me	Me	sBu	H	iBu	Rx
Neoefrapeptin D	695200-83-0	$C_{81}H_{137}N_{18}O_{16}$	4S	Me	Me	Me	Me	Me	Me	iBu	Rx
Neoefrapeptin E	695200-85-2	$C_{84}H_{143}N_{18}O_{16}$	1R, 2R, 3S, 4S	sBu	H	sBu	H	sBu	H	iBu	Rx
Neoefrapeptin F	695200-87-4	$C_{82}H_{139}N_{18}O_{16}$	–	sBu	H	Me	Me	Me	Me	iBu	Rx
Neoefrapeptin G	909093-58-9	$C_{58}H_{96}N_{14}O_{14}$	1R	sBu	H	Me	Me	Me	Me	H	H
Neoefrapeptin H	909093-59-0	$C_{59}H_{98}N_{14}O_{14}$	1R, 2R	sBu	H	sBu	H	Me	Me	H	H
Neoefrapeptin I	695200-89-6	$C_{83}H_{141}N_{18}O_{16}$	–	sBu	H	Me	H	Me	Me	iBu	Rx
Neoefrapeptin L	909093-57-8	$C_{83}H_{141}N_{18}O_{16}$	1R, 2R, 3S, 4S	sBu	H	sBu	H	sBu	H	iBu	Rx
Neoefrapeptin M	909093-56-7	$C_{83}H_{141}N_{18}O_{16}$	1R, 3S, 4S	sBu	H	Me,	Me	sBu	H	iBu	Rx
Neoefrapeptin N	909093-55-6	$C_{80}H_{135}N_{18}O_{16}$	–	Me	Me	Me	Me	Me	Me	iBu	Ry

Note: Rx, ; Ry, ; Et, $-CH_2CH_3$; iBu, $-CH_2CH(CH_3)_2$; sBu, $-CH(CH_3)CH_2CH_3$

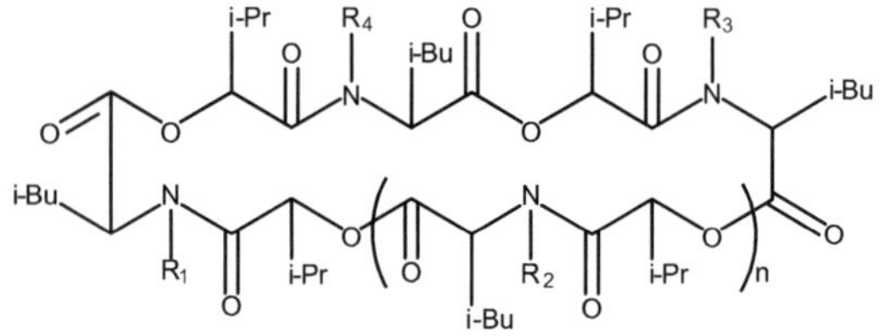

Fig. 8.5 Basic structure of bassianolides

Table 8.5 Analogues of bassianolides

Analogues name	CAS registry no.	Molecular formula	n	R1	R2	R3	R4
Bassianolide (9CI); NSC 321804	64763-82-2	$C_{48}H_{84}N_4O_{12}$	1	Me	Me	Me	Me
Bassianolide,2-L-leucine-(9CI)	76646-32-7	$C_{47}H_{82}N_4O_{12}$	1	H	Me	Me	Me
Bassianolide,2-L-leucine-4-L-leucine-(9CI)	76646-33-8	$C_{46}H_{80}N_4O_{12}$	1	H	Me	Me	H
Bassianolide,2-L-leucine-6-L-leucine-(9CI)	76657-98-2	$C_{46}H_{80}N_4O_{12}$	1	Me	H	Me	H
Bassianolide,2-L-leucine-4-L-leucine-6-L-leucine-(9CI)	76646-34-9	$C_{45}H_{78}N_4O_{12}$	1	H	H	Me	H
Cyclo(3-methyl-D-2-hydroxybutanoyl-L-leucyl-3-methyl-D-2-hydroxybutanoyl-L-leucyl-3-methyl-D-2-hydroxybutanoyl-L-leucyl-3-methyl-D-2-hydroxybutanoyl-L-leucyl)	76646-35-0	$C_{44}H_{76}N_4O_{12}$	1	H	H	H	H
Bassianolide,8a-endo-(D-2-hydroxy-3-methylbutanoic acid)-8b-endo-(N-methyl-L-leucine)-; 1,7,13,19,25-pentaoxa-4,10,16,22,28-pentaazacyclo-triacontane, cyclic peptide deriv.; decabassianolide	71326-79-9	$C_{60}H_{105}N_5O_{15}$	2	Me	Me	Me	Me

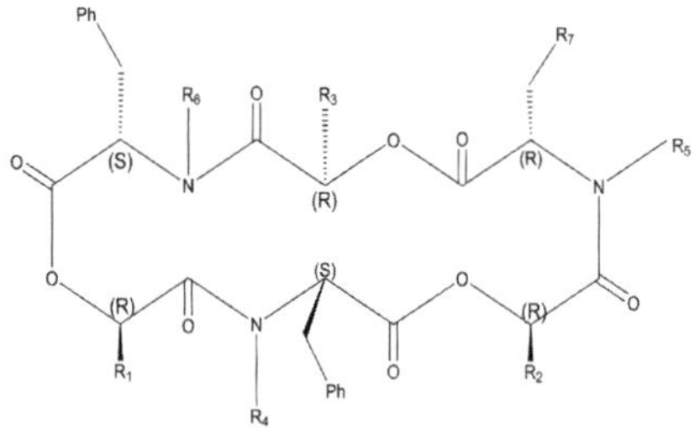

Fig. 8.6 Basic structure of beauvericins A–F

inhibitory activity and cytotoxicity of metastatic prostate cancer cell line PC-3M. Beauvericins H1–H3 increased cytotoxicity without affecting antihaptotactic activity (Xu et al. 2007).

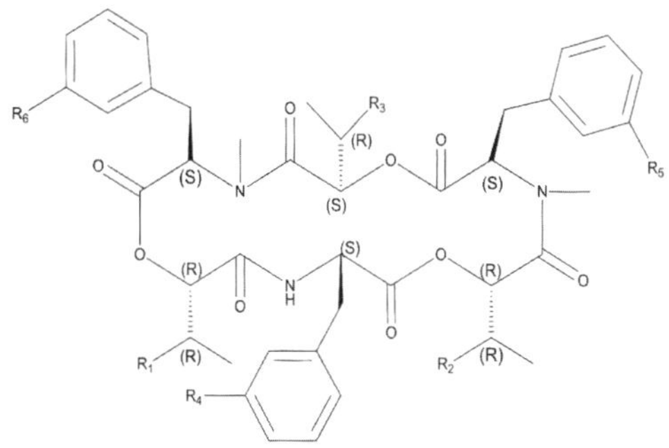

Fig. 8.7 Basic structure of beauvericins

Table 8.6 Analogues of beauvericins A–F

Analogues name	CAS registry no.	Molecular formula	R1	R2	R3	R4	R5	R6	R7
Beauvericin A	165467-50-5	$C_{46}H_{59}N_3O_9$	iPr	iPr	iBu	Me	Me	Me	Ph
Beauvericin B	165467-51-6	$C_{47}H_{61}N_3O_9$	iPr	iBu	iBu	Me	Me	Me	Ph
Beauvericin C	444585-79-9	$C_{48}H_{63}N_3O_9$	iBu	iBu	iBu	Me	Me	Me	Ph
Beauvericin D	728912-25-2	$C_{44}H_{55}N_3O_9$	iPr	iPr	iPr	H	Me	Me	Ph
Beauvericin E	728912-26-3	$C_{41}H_{57}N_3O_9$	iPr	iPr	iPr	Me	H	Me	iBu
Beauvericin F	728912-27-4	$C_{46}H_{59}N_3O_9$	iBu	iPr	iPr	iPr	Me	Me	Ph

Note: iPr, $-CH(CH_3)_2$; Ph, $-C_6H_5$

Table 8.7 Analogues of beauvericin and beauvericins G–H

Analogues name	CAS registry no.	Molecular formula	R1	R2	R3	R4	R5	R6
Beauvericin	26048-05-5	$C_{45}H_{57}N_3O_9$	Me	Me	Me	H	H	H
Beauvericin G1	CID 23643017	$C_{44}H_{55}N_3O_9$	H	Me	Me	H	H	H
Beauvericin G2	CID 23643107	$C_{43}H_{53}N_3O_9$	H	Me	Me	H	H	H
Beauvericin G3	CID 23643015	$C_{42}H_{51}N_3O_9$	H	H	H	H	H	H
Beauvericin H1	CID 23643108	$C_{45}H_{56}FN_3O_9$	Me	Me	Me	F	H	H
Beauvericin H2	CID 23643016	$C_{45}H_{55}F_2N_3O_9$	Me	Me	Me	F	F	H
Beauvericin H3	CID 23643109	$C_{45}H_{54}F_3N_3O_9$	Me	Me	Me	F	F	F

Note: CID, Identity of PubChem Compound

The insecticidal effects of beauvericins at a microgram level were reported in several insects such as *Calliphora erythrocephala*, *Aedes aegypti*, *Lygus* spp., *Spodoptera frugiperda*, and *Schizaphis graminum* (Wang and Xu 2012).

Besides being insecticidal, beauvericins have other bioactivities. Beauvericins strongly inhibited mycelial growth of both *Phytophthora sojae* and *Aphanomyces cochlioides* (Putri et al. 2013). They could significantly strengthen the fungicidal activity of ketoconazole (KTC); combinations of beauvericin (0.5 mg/kg) and KTC (0.5 mg/kg) prolonged the survival of the host infected with *Candida parapsilosis* and reduced fungal colony counts in animal organs including the kidneys, lungs, and brains. Such an effect was not achieved even with a high dose of 50 mg/kg

KTC. This supports a prospective strategy for antifungal therapy (Zhang et al. 2007). The cytotoxicity of beauvericins on human colon adenocarcinoma (Caco-2) cells was also reported (Prosperini et al. 2013).

Acetyl coenzyme A (acyl-CoA:cholesterol acyltransferase, ACAT) might be a target protein of beauvericins (Tomoda et al. 1992), and ionophores produced by beauvericins have been studied (Steinrauf 1985; Makrlik et al. 2013a, b, c, d).

Traces of beauvericins have also been detected in animal tissues (Jestoi et al. 2007); this might be because of the fact that various *Fusarium* species produce beauvericins which then enter the food chain (Moretti et al. 2007). Therefore, more and more attention is being paid to beauvericins as a food safety risk factor (Vaclavik et al. 2013; Juan et al. 2014; Luciano et al. 2014).

8.3.3   Beauverolides and Beauveriolides

Beauverolides were isolated from the genera *Beauveria* and *Paecilomyces* (Elsworth and Grove 1977, 1980; Grove 1980; Mochizuki et al. 1993b; Jegorov et al. 1994; Matsuda et al. 2004) (Fig. 8.8 and Table 8.8). They have a four-membered cyclopeptide molecular structure containing L-phenylalanine, L-alanine, D-leucine, and 3-hydroxy-4-methyl decylic acid with the molecular weight approximately 500 Da. Beauverolide L inhibited the adhesion, extension, and phagocytosis of plasmatocyte of *G. mellonella* (Vilcinskas et al. 1999). The target of its action might be acetyl coenzyme A (acyl-CoA:cholesterol acyltransferase, ACAT).

Beauveriolides, having a similar structure as beauverolides, were isolated from the culture broth of *Beauveria* spp. (Mochizuki et al. 1993a; Omura and Tomoda 2002; Matsuda et al. 2004) (Fig. 8.9 and Table 8.9). Beauveriolide I and beauveriolide III show promising antiatherogenic and anti-obesity activities (Namatame et al. 2004; Tomoda and Omura 2007). Study on its mechanism of action revealed that beauveriolides inhibited macrophage acyl-CoA:cholesterol acyltransferase (ACAT) to block the synthesis of cholesteryl ester (CE), leading to a reduction of lipid droplets in macrophages (Ohshiro et al. 2009; Tomoda et al. 2010). It was reported that beauveriolide I is a potent antiaging agent (Nakaya et al. 2012).

Beauveriolides I and III can potently decrease Abeta secretion from cells expressing human amyloid precursor protein; this offers a potential new scaffold for the development of compounds with proven bioavailability for the treatment of Alzheimer's disease (AD) (Witter et al. 2009).

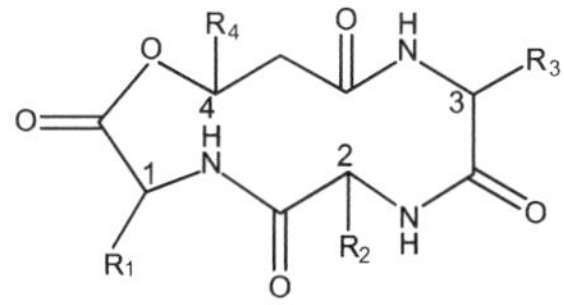

Fig. 8.8 Basic structure of
beauverolides

8.3.4 Conoideocrellides and Paecilodepsipeptides

Conoideocrellide A and its linear derivatives, conoideocrellides B–D (Fig. 8.10 and
Table 8.10), were isolated from the scale insect pathogenic fungi *Conoideocrella
tenuis* BCC 18627 and *Paecilomyces militaris* (Isaka et al. 2011; Zhang et al. 2012).
Conoideocrellide A is structurally very similar to paecilodepsipeptide A, which was
previously isolated from *T. luteorostrata* BCC 9617 and its anamorph
Paecilomyces cinnamomeus BCC 9616 (Isaka et al. 2007a, b).

Paecilodepsipeptide A, a cyclohexadepsipeptide possessing three D-amino acid
residues, together with its linear analogues paecilodepsipeptides B and C (Fig. 8.11
and Table 8.11), was isolated from the insect pathogenic fungus *P. cinnamomeus*
BCC 9616 (Isaka et al. 2007b). The products have remarkable antiproliferative
activity on human hepatocellular carcinoma cell line (SMMC-7721) and human
lymphoma cell line (Raji), which showed good antitumor activity (Yang
et al. 2013). Paecilodepsipeptide A showed activity against the malarial parasite
Plasmodium falciparum and breast cancer MCF-7 cell lines (Isaka et al. 2007b).

8.3.5 Cordycommunin

Cordycommunin was isolated from the insect pathogenic fungus *Ophiocordyceps
communis* BCC 16475 (Haritakun et al. 2010). The molecular formula of
cordycommunin was established as $C_{43}H_{69}N_7O_{11}$. It contains the amino acids
valine (Val), alanine (Ala), glutamine (Gln), threonine (Thr), and tyrosine (Tyr).
Cordycommunin showed growth inhibition of *Mycobacterium tuberculosis* H37Ra
with an MIC value of 15 μM. This compound also exhibited weak cytotoxicity to
KB cells with an IC_{50} of 45 μM, while it was inactive against BC, NCI-H187, and
Vero cell lines at a concentration of 88 μM (50 μg/mL) (Haritakun et al. 2010).
Figure 8.12 shows the structure of cordycommunin.

8.3.6 Destruxins

Destruxins, a series of cyclohexadepsipeptidic mycotoxins, were first isolated from
a culture medium of entomogenous fungus *Oospora destructor* (later renamed as
Metarhizium anisopliae) (Kodaira 1961, 1962). They were also isolated from

Table 8.8 Analogues of beauverolides

Analogues name	CAS registry no.	Molecular formula	Geometry configuration	R1	R2	R3	R4
Beauverolide A	75920-37-5	$C_{30}H_{47}N_3O_5$	–	iPr	iPr	–CH2-Ph	–CH(CH$_3$)(CH$_2$)$_5$CH$_3$
Beauverolide B	75947-02-3	$C_{31}H_{49}N_3O_5$	–	sBu	–CH$_2$-Ph	iPr	–CH(CH$_3$)(CH$_2$)$_5$CH$_3$
Beauverolide Ba	13594-27-9	$C_{31}H_{49}N_3O_5$	–	sBu	–CH$_2$-Ph	iPr	–CH(CH$_3$)(CH$_2$)$_5$CH$_3$
Beauverolide C	75899-64-8	$C_{35}H_{49}N_3O_5$	–	sBu	–CH$_2$-Ph	–CH$_2$-Ph	–CH(CH$_3$)(CH$_2$)$_5$CH$_3$
Beauverolide Ca	13594-29-1	$C_{35}H_{49}N_3O_5$	–	sBu	–CH$_2$-Ph	–CH$_2$-Ph	–CH(CH$_3$)(CH$_2$)$_5$CH$_3$
Beauverolide D	75899-63-7	$C_{28}H_{43}N_3O_5$	–	iPr	iPr	–CH$_2$-Ph	–CH(CH$_3$)(CH$_2$)$_3$CH$_3$
Beauverolide E	75947-01-2	$C_{29}H_{45}N_3O_5$	1R, 2S, 3S	A	–CH$_2$-Ph	iPr	–CH(CH$_3$)(CH$_2$)$_3$CH$_3$
Beauverolide Ea	75899-62-6	$C_{29}H_{45}N_3O_5$	1R, 2S, 3S	B	–CH$_2$-Ph	iPr	–CH(CH$_3$)(CH$_2$)$_3$CH$_3$
Beauverolide F	75947-00-1	$C_{33}H_{45}N_3O_5$	–	sBu	–CH$_2$-Ph	–CH$_2$-Ph	–CH(CH$_3$)(CH$_2$)$_3$CH$_3$
Beauverolide Fa (Beauveriolide IX)	75899-61-5	$C_{33}H_{45}N_3O_5$	–	sBu	–CH$_2$-Ph	–CH$_2$-Ph	–CH(CH$_3$)(CH$_2$)$_3$CH$_3$
Beauverolide H	62995-90-8	$C_{27}H_{41}N_3O_5$	–	iBu	Me	–CH$_2$-Ph	–(CH$_2$)$_5$CH$_3$
Beauverolide I	62995-91-9	$C_{29}H_{45}N_3O_5$	–	iBu	Me	–CH$_2$-Ph	–(CH$_2$)$_7$CH$_3$
Beauverolide Ja	76265-41-3	$C_{35}H_{46}N_4O_5$	1R, 2S, 3S	C	–CH$_2$-Ph	D	–CH(CH$_3$)(CH$_2$)$_3$CH$_3$
Beauverolide Ka	76265-42-4	$C_{37}H_{50}N_4O_5$	–	sBu	–CH$_2$-Ph	D	–CH(CH$_3$)(CH$_2$)$_5$CH$_3$

Beauverolide L	154491-56-2	$C_{29}H_{45}N_3O_5$	1R, 2S, 3S, 4S	iBu	Me	–CH$_2$–Ph	E
Cyclo[L-alanyl-D-leucyl-(3S,4S)-3-hydroxy-4-methyloctanoyl-L-tyrosyl] (9CI)	413579-49-4	$C_{27}H_{41}N_3O_6$	1R, 2S, 3S, 4S	iBu	Me	–CH$_2$–Ph–OH	G

Note: nBu, –CH(CH$_2$)$_2$CH$_3$

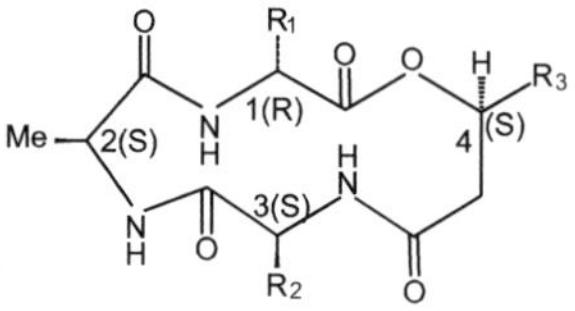

Fig. 8.9 Basic structure of
beauveriolides

Table 8.9 Analogues of beauveriolides

Analogues name	CAS registry no.	Molecular formula	Geometry configuration	R1	R2	R3
Beauveriolide I	154491-55-1	$C_{27}H_{41}N_3O_5$	1R, 2S, 3S, 4S, 5S	iBu	–CH$_2$–Ph	nBu
Beauveriolide III	221111-70-2	$C_{27}H_{41}N_3O_5$	1R, 2S, 3S, 4S, 5S	B	Me	Me
Beauveriolide IV	460352-01-6	$C_{22}H_{39}N_3O_5$	1R, 2S, 3S	iPr	iPr	nBu
Beauveriolide V	460352 02 7	$C_{23}H_{41}N_3O_5$	1R, 2S, 3S	B	iPr	nBu
Beauveriolide VI	413579-48-3	$C_{23}H_{41}N_3O_5$	1R, 2S, 3S, 4S, 5S	iBu	iPr	nBu
Beauveriolide VII	460352-03-8	$C_{26}H_{39}N_3O_5$	1R, 2S, 3S	iPr	–CH$_2$–Ph	nBu
Beauveriolide VII	460352-04-9	$C_{25}H_{45}N_3O_5$	1R, 2S, 3S	B	iPr	–(CH$_2$)$_5$– Me

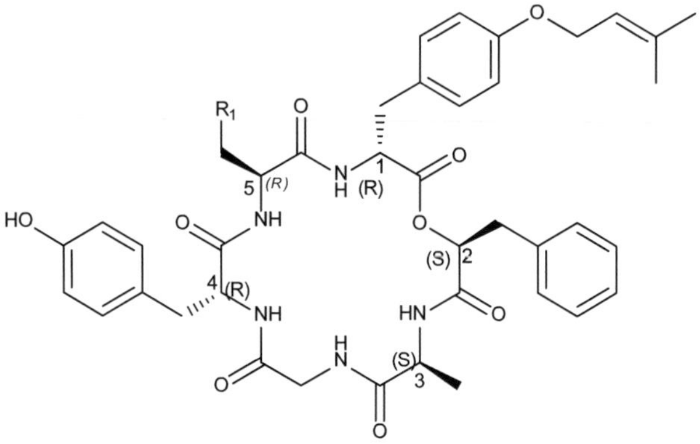

Fig. 8.10 Conoideocrellide A and paecilodepsipeptide A

Aschersonia and other fungal species (Pedras et al. 2002). The general formula of
destruxin is cyclo(-D-HA-L-Pro-L-Ile-L-MeVal-L-MeAla-b-Ala-), where HA represents a D-α-hydroxyl acid group (Fig. 8.13 and Table 8.12). Among 39 destruxin
analogues, destruxins A, B, and E (DA, DB, and DE, respectively) show substantial
bioactivity (Liu and Tzeng 2012). However, the linear molecule resulting from the

Table 8.10 Analogues of conoideocrellides

Analogues name	CAS registry no.	Molecular formula	Geometry configuration	R1
Conoideocrellide A	1296607-05-0	$C_{40}H_{47}N_5O_{10}$	1R, 2S, 3S, 4R, 5R	OH
Paecilodepsipeptide A	931423-99-3	$C_{40}H_{47}N_5O_9$	1R, 2S, 3S, 4R, 5R	H

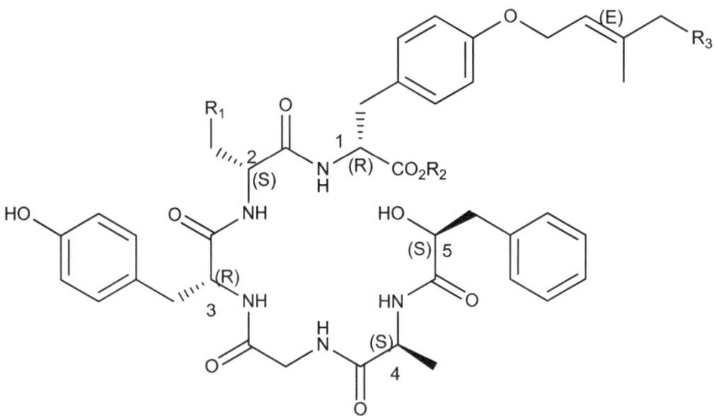

Fig. 8.11 Conoideocrellides B–D and paecilodepsipeptides B–C

Table 8.11 Analogues of paecilodepsipeptides

Analogues name	CAS registry no.	Molecular formula	Geometry configuration	R1	R2	R3
Conoideocrellide B	1296607-06-1	$C_{40}H_{49}N_5O_{10}$	1R, 2S, 3S, 4R, 5S	OH	H	H
Conoideocrellide C	1296607-07-2	$C_{41}H_{51}N_5O_{10}$	1R, 2S, 3S, 4R, 5S	OH	Me	H
Conoideocrellide D	1296607-08-3	$C_{41}H_{51}N_5O_{12}$	1R, 2S, 3S, 4R, 5S	OH	Me	OH
Paecilodepsipeptide B	931423-29-9	$C_{40}H_{49}N_5O_{10}$	1R, 2S, 3S, 4R, 5S	H	H	H
Paecilodepsipeptide C	931423-31-3	$C_{41}H_{51}N_5O_{10}$	1R, 2S, 3S, 4R, 5S	H	Me	H

opening of the DA cycle is not toxic, and DE would degrade to less toxic DE-diol upon enzymatic action (Jegorov et al. 1992; Dumas et al. 1994).

Destruxins have insecticidal activity against many pests with various modes of action. Destruxins via hemocoel injection exhibited promising insecticidal activity in early studies. Further researches discovered that destruxins showed contact action (Hu et al. 2007b), gut toxicity (Brousseau et al. 1996), antifeedant effect (Amiri-Besheli et al. 2000, Hu et al. 2007a), and ovicidal and oviposition-deterrent activity (Pedras et al. 2002). The mixture of destruxins and Bt, *Paecilomyces*, and

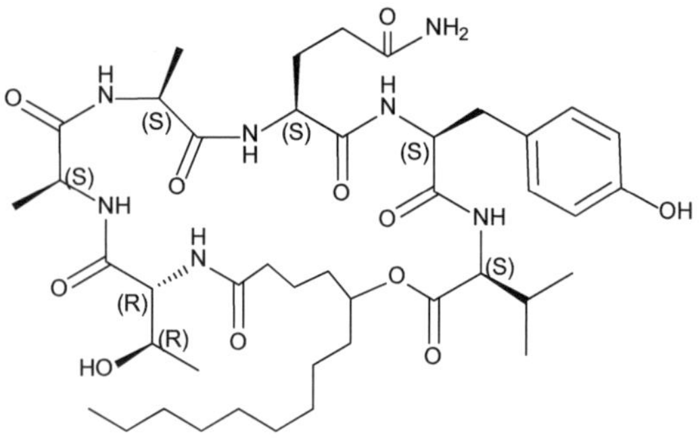

Fig. 8.12 Structure of cordycommunin

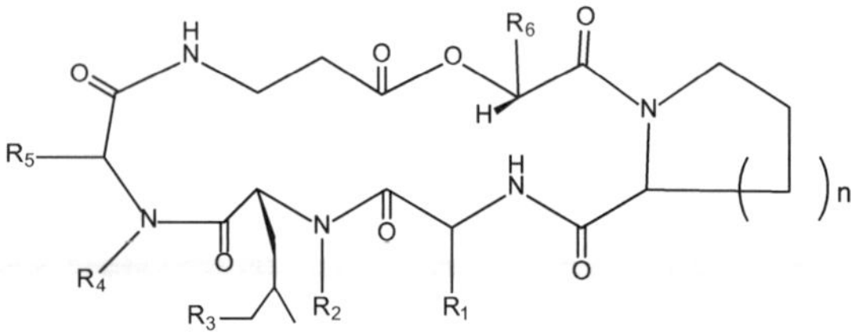

Fig. 8.13 Chemical structure of destruxins

botanical insecticides had synergistic effect (Hu et al. 2007b; Rizwan-Ul-Haq et al. 2009; Yi et al. 2012).

Destruxins damage the innate immunity of insects. Morphology and function of encapsulation and phagocytosis processes of insect hemocytes were destroyed by destruxins (Vilcinskas et al. 1997a, b; Vey et al. 2002; Fan et al. 2013). The expression of antimicrobial peptides was also reduced in *Drosophila melanogaster* treated with DA (Pal et al. 2007). Destruxin also acts as a kind of calcium ionophore and an inhibitor of V-H+-ATPase (Bandani et al. 2001; Chen et al. 2014; Vazquez et al. 2005). Other studies have reported that destruxins damage the insect organs and tissues, including the midgut, Malpighian tubules, salivary glands, and muscles (Kershaw et al. 1999; Ruiz-Sanchez et al. 2010; Ruiz-Sanchez and O'Donnell 2012; Sree and Padmaja 2008). However, the molecular mechanism of damage caused by destruxins has not yet been determined.

Destruxins were well reviewed (Pedras et al. 2002; Liu and Tzeng 2012), the authors do not repeat here.

8.3.7 Enniatins

Enniatin was first discovered in the 1940s (Gaumann et al. 1947). Enniatin analogues were produced by various species of the fungal genera *Verticillium* and *Fusarium*, etc. (Herrmann et al. 1996; Supothina et al. 2004) (Fig. 8.14 and Table 8.13). They are N-methylated cyclohexadepsipeptides, composed of three

Table 8.12 Analogues of destruxins

Destruxin analogues	n	R1	R2	R3	R4	R5	R6
Destruxin A	1	CH(Me)CH₂Me	Me	H	Me	Me	CH=CH₂
Destruxin A1	2	CH(Me)CH₂Me	Me	H	Me	Me	CH=CH₂
Destruxin A2	1	CHMe₂	Me	H	Me	Me	CH=CH₂
Destruxin A3	0	CH(Me)CH₂Me	Me	H	Me	Me	CH=CH₂
Destruxin A4	1	CH(Me)CH₂Me	Me	Me	Me	Me	CH=CH₂
Destruxin A5	1	CH(Me)CH₂Me	Me	Me	Me	Me	CH=CH₂
Destruxin A4 chlorohydrin	1	CH(Me)CH₂Me	Me	Me	Me	Me	CHOHCH₂Cl
Desmethyl destruxin A	1	CH(Me)CH₂Me	H	H	Me	Me	CH=CH₂
Dihydro-destruxin A	1	CH(Me)CH₂Me	Me	H	Me	Me	CH2Me
Pseudo-destruxin A	1	CH(Me)CH₂Me	Me	H	Me	Me	CHMe₂
Destruxin B	1	CH(Me)CH₂Me	Me	H	Me	Me	CHMe₂
Destruxin B1	2	CH(Me)CH₂Me	Me	H	Me	Me	CHMe₂
Destruxin B2	1	CHMe₂	Me	H	Me	Me	CHMe₂
Desmethyl destruxin B	1	CH(Me)CH₂Me	Me	H	Me	Me	CHMe₂
Desmethyl destruxin B2	1	CHMe₂	Me	H	Me	Me	CHMe₂
Homo-destruxin B	1	CH(Me)CH₂Me	Me	Me	Me	Me	CHMe₂
Proto-destruxin B	1	CH(Me)CH₂Me	Me	H	H	Me	CHMe₂
Hydroxy destruxin B	1	CH(Me)CH₂Me	Me	H	Me	Me	CHMe₂
Hydroxyhomo-destruxin B	1	CH(Me)CH₂Me	Me	Me	Me	Me	CHMe₂
β-ᴅ-Glucopyranosyl-hydroxyl destruxin B	1	CH(Me)CH₂Me	Me	H	Me	Me	
[Phe3, N-Me-Val5] Destruxin B	1	CH₂Ph	Me	H	Me	CHMe₂	CHMe₂
Pseudo-destruxin B	1	CH₂Ph	Me	H	Me	CH₂CHMe₂	CHMe₂
Destruxin C	1	CH(Me)CH₂Me	Me	H	Me	Me	CHMeCH₂OH
Destruxin C2	1	CHMe₂	Me	H	Me	Me	CHMeCH₂OH
Desmethyl destruxin C	1	CH(Me)CH₂Me	H	H	Me	Me	CHMeCH₂OH

(continued)

Table 8.12 (continued)

Destruxin analogues	n	R1	R2	R3	R4	R5	R6
Pseudo-destruxin C	1	CH$_2$Ph	Me	H	Me	CHMe$_2$	CHMeCH$_2$OH
Destruxin D	1	CH(Me) CH$_2$Me	Me	H	Me	Me	CHMeCOOH
Destruxin D1	2	CH(Me) CH$_2$Me	Me	H	Me	Me	CHMeCOOH
Destruxin D2	1	CHMe$_2$	Me	H	Me	Me	CHMeCOOH
Destruxin E	1	CH(Me) CH$_2$Me	Me	H	Me	Me	Oxirane
Destruxin E1	2	CH(Me) CH$_2$Me	Me	H	Me	Me	Oxirane
Destruxin E2	1	CHMe$_2$	Me	H	Me	Me	Oxirane
Destruxin E chlorohydrin	1	CH(Me) CH$_2$Me	Me	H	Me	Me	CHOHCH$_2$Cl
Destruxin E2 chlorohydrin	1	CHMe$_2$	Me	H	Me	Me	CHOHCH$_2$Cl
Destruxin E diol	1	CH(Me) CH$_2$Me	Me	H	Me	Me	CHOHCH$_2$OH
Destruxin Ed1	2	CH(Me) CH$_2$Me	H	H	Me	Me	CHOHCH$_2$OH
[β-Me-Pro] Destruxin E chlorohydrin	1	CH(Me) CH$_2$Me	Me	H	Me	Me	CHOHCH$_2$Cl
Regioisomer of destruxin E chlorohydrin	1	CH(Me) CH$_2$Me	Me	H	Me	Me	CHClCH$_2$OH
Destruxin F	1	CH(Me) CH$_2$Me	Me	H	Me	Me	CHOHMe

Notes: In destruxin A3 and destruxin A5, the proline residue is replaced by [structure] and

[structure] ·, respectively. The destruxin E chlorohydrin and [β-Me-Pro] destruxin E chlorohydrin share the same empirical formula. They are stereoisomers at R6

units each of N-methylated branched-chain L-amino acid and D-2-hydroxy acid arranged in an alternate fashion (Firakova et al. 2007).

Enniatins have been shown to act against spruce budworm larvae (*Choristoneura fumiferana*) (Strongman et al. 1988), *G. mellonella* (Mule et al. 1992), and *S. frugiperda* cell line (*Sf-9*) (Fornelli et al. 2004). Enniatins have multi-activities including antifungal, antibiotic, and cytotoxic properties. To date, 29 enniatins have been isolated and characterized, either as a single compound or mixtures of inseparable analogues. Fusafungine, a drug developed from a mixture of enniatins, is used as a topical treatment of upper respiratory tract infections by oral and/or nasal inhalation (Sy-Cordero et al. 2012).

In regard to the molecular mechanism of enniatins, researches indicated that they inhibit ABC transporters (Hiraga et al. 2005), suppress acyl-CoA:cholesterol

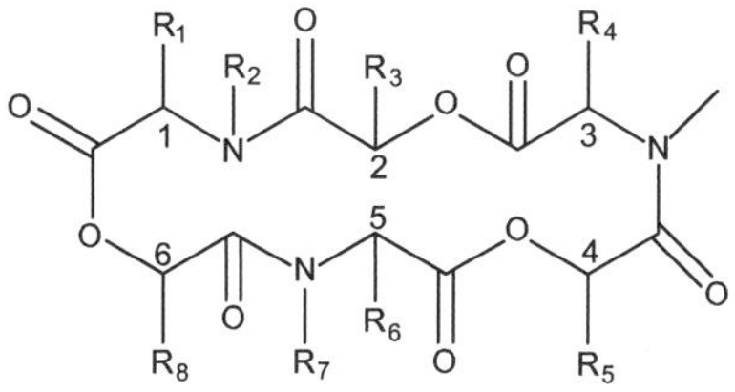

Fig. 8.14 Basic structure of enniatins

acyltransferase (ACAT) (Tomoda et al. 1992), and act as ionophores (Levy et al. 1995; Doebler 2000).

Enniatins are also a common contaminant in grain-based foods, so, research on their harmful effects on human health is gaining more importance (Feudjio et al. 2010; Santini et al. 2012).

8.3.8 Hirsutellides and Hirsutides

Hirsutellide A, a cyclohexadepsipeptide, was first isolated from a cell extract of the entomogenous fungus *Hirsutella kobayasii* BCC 1660 (Vongvanich et al. 2002) (Fig. 8.15). Hirsutellide A exhibited antimycobacterial and antimalarial activities, but was inactive toward the Vero cell line (at 50 μg/mL) (Vongvanich et al. 2002).

An entomogenous fungus *Hirsutella* spp., isolated from an infected spider, was found to produce the cyclotetrapeptide hirsutide, with the amino acid sequence of cyclo-(L-NMe-Phe-L-Phe-L-NMe-Phe-L-Val) (Lang et al. 2005) (Fig. 8.16). No other hirsutellide and hirsutide analogues have been reported yet.

8.3.9 Isariins, Isaridins, and Isarolides

Isariin was first isolated from *Isaria cretacea* (Vining and Taber 1962). Then, its analogues were isolated and structurally elucidated (Baute et al. 1981; Deffieux et al. 1981; Langenfeld et al. 2011). Isariins possess a β-hydroxyl fatty acid and five α-amino acid residues (Fig. 8.17 and Table 8.14). Isariins were originally shown to have insecticidal activity against *G. mellonella* (Baute et al. 1981). Isariins have been reported to have an inhibitory effect on the intraerythrocytic growth of *Plasmodium falciparum* (Sabareesh et al. 2007). Iso-isariin B was found to be active against the pest insects *Sitophilus* spp. with an LC_{50} value of 10 μg/mL (Langenfeld et al. 2011).

Isaridins A and B were first isolated from *Isaria* (Ravindra et al. 2004), and then, other analogues were isolated and structurally elucidated (Sabareesh et al. 2007). In contrast to isariin, they have an α-hydroxyl acid and a β-amino acid, with a preponderance of N-alkylated residues (Fig. 8.18 and Table 8.15). Isaridin also

Table 8.13 Analogues of enniatins

Analogues name	CAS registry no.	Molecular formula	Geometry configuration	R1	R2	R3	R4	R5	R6	R7	R8
Enniatin A	2503-13-1	$C_{36}H_{63}N_3O_9$	1S, 2R, 3S, 4R, 5S, 6R	Rx	Me	iPr	Rz	iPr	Rz	Me	iPr
Enniatin A1	4530-21-6	$C_{35}H_{61}N_3O_9$	1S, 2R, 3S, 4R, 5S, 6R	iPr	Me	iPr	Rz	iPr	Rz	Me	iPr
Enniatin B	917-13-5	$C_{33}H_{57}N_3O_9$	1S, 2R, 3S, 4R, 5S, 6R	iPr	Me	iPr	iPr	iPr	iPr	Me	iPr
Enniatin B1	19914-20-6	$C_{34}H_{59}N_3O_9$	1S, 2R, 3S, 4R, 5S, 6R	iPr	Me	iPr	Rz	iPr	iPr	Me	iPr
Enniatin B2	632-91-7	$C_{32}H_{55}N_3O_9$	1S, 2R, 3S, 4R, 5S, 6R	iPr	Me	iPr	iPr	iPr	iPr	H	iPr
Enniatin B3	864-99-3	$C_{31}H_{53}N_3O_9$	1S, 2R, 3S, 4R, 5S, 6R	iPr	H	iPr	iPr	iPr	iPr	H	iPr
Enniatin B5	1338161-36-6										
Enniatin C	19893-23-3	$C_{36}H_{63}N_3O_9$	1S, 2R, 3S, 4R, 5S, 6R	iBu	Me	iPr	iBu	iPr	iBu	Me	iPr
Enniatin D(B4)	19893-21-1	$C_{34}H_{59}N_3O_9$	1S, 2R, 3S, 4R, 5S, 6R	iPr	Me	iPr	iBu	iPr	iPr	Me	iPr
Enniatin E1	1450880-97-3	$C_{35}H_{61}N_3O_9$	1S, 2R, 3S, 4R, 5S, 6R	iPr	Me	iPr	iPr	Ro	iPr	Me	Rz
Enniatin E2	1450880-98-4	$C_{35}H_{61}N_3O_9$	1S, 2R, 3S, 4R, 5S, 6R	iPr	Me	iPr	Rz	iPr	iBu	Me	iPr
Enniatin F(A2)	144446-20-8	$C_{36}H_{63}N_3O_9$	–	iBu	Me	iPr	Rz	iPr	Rz	Me	iPr
Enniatin G	19893-22-2	$C_{35}H_{61}N_3O_9$	–	iBu	Me	iPr	iBu	iPr	iPr	Me	iPr
Enniatin H	561298-15-5	$C_{34}H_{59}N_3O_9$	1S, 2R, 3S, 4R, 5S, 6R	iPr	Me	Rz	iPr	iPr	iPr	Me	iPr
Enniatin I	561298-16-6	$C_{35}H_{61}N_3O_9$	1S, 2R, 3S, 4R, 5S, 6R	iPr	Me	Rz	iPr	iPr	iPr	Me	Rz
Enniatin J1	19893-15-3	$C_{31}H_{53}N_3O_9$	1S, 2R, 3S, 4R, 5S, 6R	Me	Me	iPr	iPr	iPr	iPr	Me	iPr
Enniatin J2	716318-01-3	$C_{32}H_{55}N_3O_9$	1S, 2R, 3S, 4R, 5S, 6R	Me	Me	iPr	Rz	iPr	iPr	Me	iPr
Enniatin J3	716318-02-4	$C_{32}H_{55}N_3O_9$	1S, 2R, 3S, 4R, 5S, 6R	iPr	Me	iPr	Rz	iPr	Me	Me	iPr
Enniatin K1	716318-00-2	$C_{32}H_{55}N_3O_9$	1S, 2R, 3S, 4R, 5S, 6R	Et	Me	iPr	Rz	iPr	iPr	Me	iPr
Enniatin L	791785-16-5	$C_{32}H_{55}N_3O_{10}$	1S, 2R, 3S, 4R, 5S, 6R	iPr	Me	iPr	iPr	iPr	iPr	Me	iPr
Enniatin MK1688	133869-46-2	$C_{36}H_{63}N_3O_9$	1S, 2R, 3S, 4R, 5S, 6R	iPr	Me	Rz	iPr	Rz	iPr	Me	Rz
Enniatin M1	791785-17-6	$C_{35}H_{61}N_3O_{10}$	1S, 2R, 3S, 4R, 5S, 6R	iPr	Me	iPr	iPr	Ro	iPr	Me	Rz
Enniatin M2	791785-18-7	$C_{35}H_{61}N_3O_{10}$	1S, 2R, 3S, 4R, 5S, 6R	iPr	Me	iPr	iPr	Ro	iPr	Me	iPr
Enniatin N	791785-19-8	$C_{36}H_{63}N_3O_{10}$	1S, 2R, 3S, 4R, 5S, 6R	iPr	Me	Rz	iPr	Ro	iPr	Me	Rz
Enniatin O1	847371-30-6	$C_{35}H_{61}N_3O_9$	1S, 2R, 3S, 4R, 5S, 6R	iBu	Me	Rz	iPr	iPr	iPr	Me	iPr

Enniatin O2	847371-31-7	$C_{35}H_{61}N_3O_9$	1S, 2R, 3S, 4R, 5S, 6R	iBu	Me	iPr	iPr	Rz	iPr	Me	iPr
Enniatin O3	847371-32-8	$C_{35}H_{61}N_3O_9$	1S, 2R, 3S, 4R, 5S, 6R	iPr	Me	Rz	iBu	iPr	iPr	Me	iPr
Enniatin P1	1172635-53-8	$C_{32}H_{55}N_3O_{10}$	1S, 2S, 3S, 4S, 5S, 6S	iPr	Me	iPr	Rp	iPr	iPr	Me	iPr
Enniatin P2	1172635-54-9	$C_{33}H_{57}N_3O_{10}$	1S, 2S, 3S, 4S, 5S, 6S	Rp	Me	iPr	iBu	iPr	iPr	Me	iPr

Note: Rz, ——(S, Me); Ro, ——(Et, OH); Rp, ——R(OH)

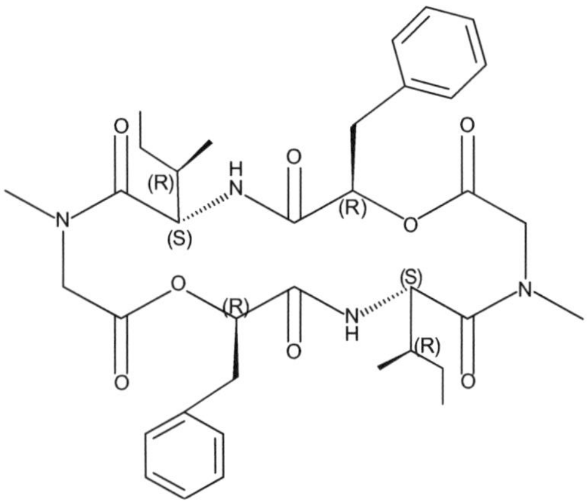

Fig. 8.15 Structure of
hirsutellide A

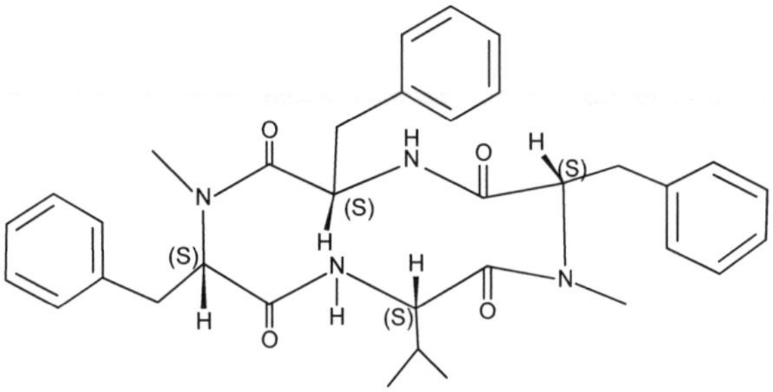

Fig. 8.16 Structure of
hirsutide

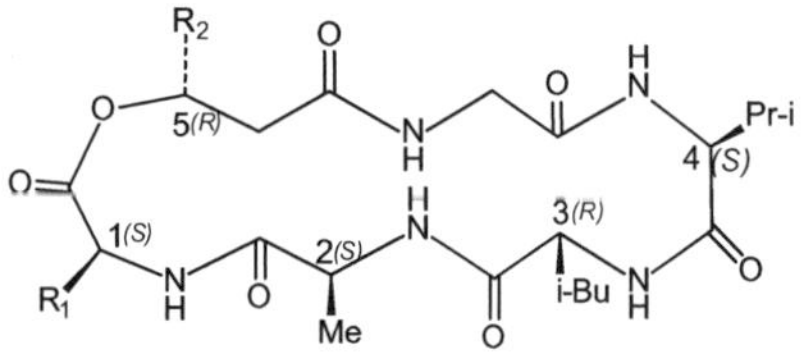

Fig. 8.17 Basic structure of
isariins

showed an inhibitory effect on the intraerythrocytic growth of *Plasmodium falciparum* (Sabareesh et al. 2007).

Isarolides A, B, and C were isolated from a new species of *Isaria*. Their structures have been elucidated (Briggs et al. 1966) (Fig. 8.19 and Table 8.16). Isarolide A is identical to beauverolide Ba, while isarolide C is identical with beauverolide Ca (Elsworth and Grove 1980).

Table 8.14 Analogues of isariins

Analogues name	CAS registry no.	Molecular formula	Geometry configuration	R1	R2
Isariin A	10409-85-5	$C_{33}H_{59}N_5O_7$	–	iPr	$-(CH_2)_8-Me$
Isariin B	80111-95-1	$C_{30}H_{53}N_5O_7$	1S, 2S, 3R, 4S, 5R	iPr	$-(CH_2)_5-Me$
Isariin C	80111-96-2	$C_{28}H_{49}N_5O_7$	1S, 2S, 3R, 4S, 5R	Me	$-(CH_2)_5-Me$
Isariin C2	944347-21-1	$C_{28}H_{49}N_5O_7$	1S, 2S, 3R, 4S, 5R	iPr	nBu
Isariin D	80111-97-3	$C_{26}H_{45}N_5O_7$	1S, 2S, 3R, 4S, 5R	Me	nBu
Isariin E	944347-11-5	$C_{27}H_{45}N_5O_7$	1S, 2S, 3R, 4S, 5R	iPr	nPr
Isariin F2	944347-18-6	$C_{29}H_{51}N_5O_7$	1S, 2S, 3R, 4S, 5R	iPr	$-(CH_2)_4-Me$
Isariin G1	944347-20-0	$C_{31}H_{55}N_5O_7$	1S, 2S, 3R, 4S, 5R	iPr	$-(CH_2)_8-Me$
Isariin G2	944347-19-7	$C_{31}H_{55}N_5O_7$	1S, 2S, 3R, 4S, 5R	Me	$-(CH_2)_6-Me$
Iso-isariin B	1290627-99-4	$C_{30}H_{53}N_5O_7$	1S, 2S, 3S, 4S	iPr	$-CH(CH_2)_3(CH_3)_2$

Note: nPr, $-CH_2CH_2CH_3$

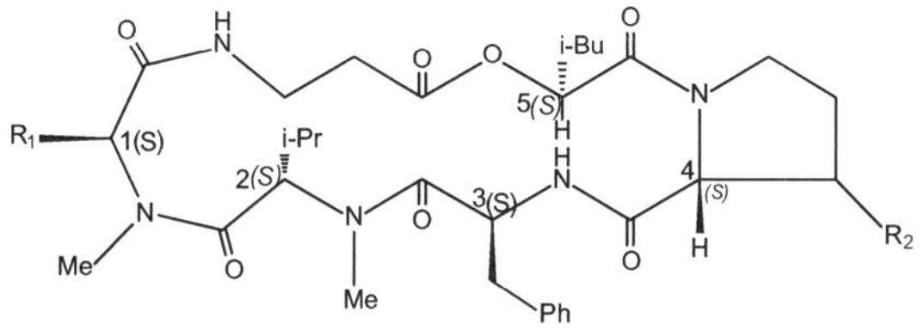

Fig. 8.18 Basic structure of isaridins

Table 8.15 Analogues of isaridins

Analogues name	CAS registry no.	Molecular formula	Geometry configuration	R1	R2
Isaridin A	780781-87-5	$C_{39}H_{53}N_5O_7$	1S, 2S, 3S, 4S, 5S	Ph	H
Isaridin B	780781-88-6	$C_{40}H_{55}N_5O_7$	1S, 2S, 3S, 4S, 5S	Ph	H
Isaridin C	342573-58-4	$C_{36}H_{55}N_5O_7$	1S, 2S, 3S, 4S, 5S	iBu	H
Isaridin C1	944346-45-6				
Isaridin C2	943896-54-6	$C_{36}H_{55}N_5O_7$	1S, 2S, 3S, 4S, 5S	iPr	Me
Isaridin D	944346-46-7				
Isaridin E	944347-22-2	$C_{35}H_{53}N_5O_7$	1S, 2S, 3S, 4S, 5S	iPr	H

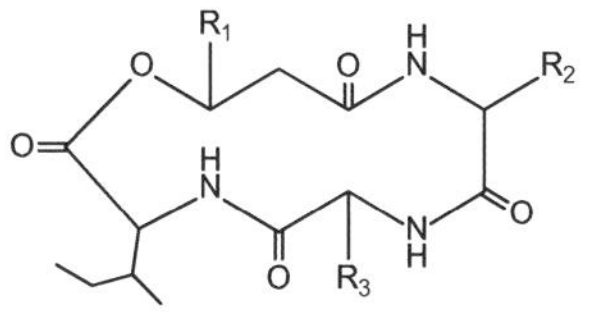

Fig. 8.19 Basic structure of isarolide

Table 8.16 Analogues of isarolides

Analogues name	CAS registry no.	Molecular formula	Geometry configuration	R1	R2	R3
Isarolide A	13594-27-9	$C_{31}H_{49}N_3O_5$	–	–CH(Me)(CH$_2$)$_5$–Me	iPr	–CH$_2$Ph
Isarolide B	13594-28-0	$C_{31}H_{49}N_3O_5$	–	–(CH$_2$)$_7$–Me	–CH$_2$Ph	iPr
Isarolide C	13594-29-1	$C_{35}H_{49}N_3O_5$	–	–CH(Me)(CH$_2$)$_5$–Me	–CH$_2$Ph	–CH$_2$Ph

Note: iPr, –CH(CH$_3$)$_2$; Ph, –C$_6$H$_5$; Me, –CH$_3$; –, no geometry configuration

8.3.10 Serinocyclins

Serinocyclins A and B were isolated from the conidia of the entomopathogenic fungus *M. anisopliae* (Krasnoff et al. 2007). The molecular structure of serinocyclin A is cyclo-(Acc-Hyp-Ser-HyLys-β-Ala-Ser-Ser), where Acc is 1-aminocyclopropane-1-carboxylic acid, Hyp is hydroxyproline, Ser is serine, HyLys is hydroxylysine, and β-Ala is β-alanine. Acc and HyLys are unusual nonprotein amino acids. Serinocyclin B contains lysine (Lys) but not hydroxylysine (HyLys). The absence of Me groups is unusual among fungal peptides and, along with the charged lysyl side chain and multiple hydroxyl groups, contributes to the polar nature of the compounds. Serinocyclin A produced a sublethal locomotory defect in mosquito larvae at an EC$_{50}$ of 59 ppm (Krasnoff et al. 2007). Figure 8.20 and Table 8.17 show the structures of serinocyclins.

8.3.11 Verticilide

Verticilide was first isolated from the fungal strain *Verticillium* spp. FKI-1033 (Omura et al. 2004). It is a 24-membered ring cyclic depsipeptide with a sequence of cyclo[(2R)-2-hydroxyheptanoyl-*N*-methyl-L-alanyl] (Omura et al. 2004; Monma et al. 2006). Another soil fungal strain, *Verticillium* spp. FKI-2679, was found to produce inhibitors of acyl-CoA:cholesterol acyltransferase (ACAT) in a cell-based assay using ACAT1- and ACAT2-expressing CHO cells. Three new verticilide analogues, verticilides A2, A3, and B1, were isolated from the fermentation broth. Their chemical structures have been elucidated (Figs. 8.21 and 8.22, Table 8.18). Verticilides A1, A2, A3, and B1 showed a high degree of selectivity toward ACAT2, with IC$_{50}$s 8.5–11-fold more potency than that observed against ACAT1 (Ohshiro et al. 2012).

Verticilides inhibited ryanodine binding to ryanodine receptors in the cockroach at an IC$_{50}$ value of 4.2 μM, whereas inhibition against mouse ryanodine receptors was weak with an IC$_{50}$ value of 53.9 μM (Monma et al. 2006; Shiomi et al. 2010).

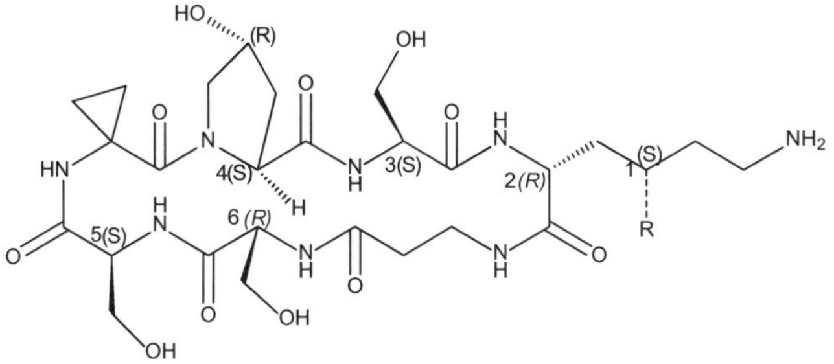

Fig. 8.20   Basic structure of serinocyclins

Table 8.17   Analogues of serinocyclins

Analogues name	CAS registry no.	Molecular formula	Geometry configuration	R
Serinocyclin A	1001897-42-2	$C_{27}H_{44}N_8O_{12}$	1S, 2R, 3S, 4S, 5S, 6R	OH
Serinocyclin B	1001897-43-3	$C_{27}H_{44}N_8O_{11}$	2R, 3S, 4S, 5S, 6R	H

Fig. 8.21   Basic structure of verticilides A1–A3

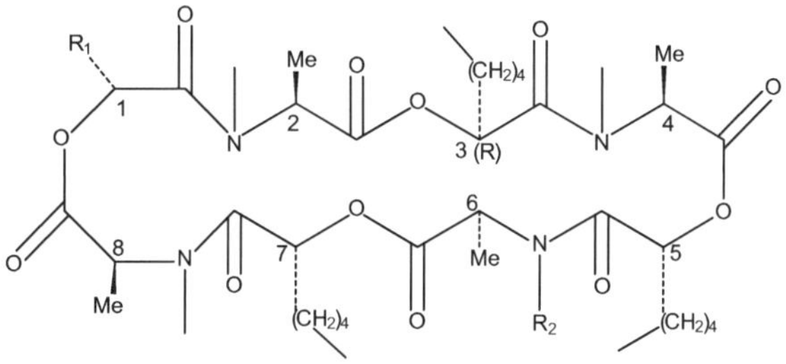

Fig. 8.22   Structure of verticilide B1

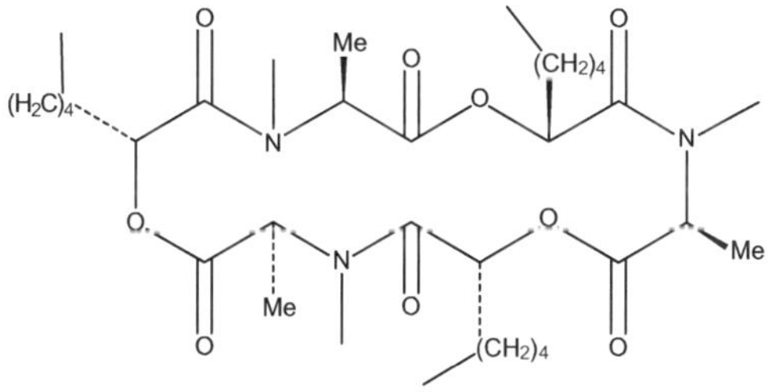

8.4   Cyclopeptides

8.4.1   Cyclosporines

Cyclosporines, or cyclosporins, discovered in the 1970s, were a series of cyclo-undecapeptide secreted by *Trichoderma polysporum* and *Cylindrocarpon lucidum* (Borel et al. 1976; Dreyfuss et al. 1976) (Fig. 8.23 and Table 8.19). Certain species

Table 8.18 Analogues of verticilides

Analogues name	CAS registry no.	Molecular formula	Geometry configuration	R1	R2
Verticilide A1	693778-57-3	$C_{44}H_{76}N_4O_{12}$	1R, 2S, 3R, 4S, 5R, 6S, 7R, 8S	–$(CH_2)_4$–Me	Me
Verticilide A2	1380602-01-6	$C_{43}H_{74}N_4O_{12}$	1R, 2S, 3R, 4S, 5R, 6S, 7R, 8S	–$(CH_2)_4$–Me	H
Verticilide A3	1380601-94-4	$C_{46}H_{80}N_4O_{12}$	1R, 2S, 3R, 4S, 5R, 6S, 7R, 8S	–$(CH_2)_6$–Me	Me
Verticilide B1	1380750-89-9	$C_{33}H_{57}N_3O_9$	–	–	–

of *Beauveria*, *Verticillium*, and *Tolypocladium* could also produce cyclosporines (Jegorov et al. 1990).

Cyclosporines have insecticidal activities. It was reported that cyclosporines are effective against mosquito larvae (Weiser and Matha 1988; Podsiadlowski et al. 1998). Cyclosporin A (CsA) has the immunosuppressive effect on insect humoral immune response (Fiolka 2008) and cellular immune response (Vilcinskas et al. 1999). CsA is an important immunosuppressant used for human organic transplantation and other disease treatments (Muellenhoff and Koo 2012; Kovarik 2013). In regard to the mechanism of action as immunosuppressant, CsA specifically binds to the receptor, cyclophilin A (CypA), which is located in the cytoplasm of T lymphocytes, and consequently inhibits CypA-mediated immune signaling pathways (Ryffel et al. 1980; Wiesinger and Borel 1980; Kallen et al. 1991; Liu et al. 1991).

8.4.2 Cordyheptapeptide

Cordyheptapeptide A was first isolated from the entomogenous fungal strain *Cordyceps* spp. BCC 1788. Its amino acid sequence was determined as cyclo-(D-NMePhe-L-Leu-L-Ile-L-NMeTyr-L-Phe-NMeGly-L-Pro)   (Rukachaisirikul et al. 2006). Then, the analogues, cordyheptapeptides C–E, were isolated from the marine-derived fungus *Acremonium persicinum* SCSIO 115 (Chen et al. 2012) (Fig. 8.24 and Table 8.20).

Cordyheptapeptide A exhibited antimalarial activity against *Plasmodium falciparum* K1 and cytotoxicity to Vero cell lines with IC_{50} values of 5.35 and >56.88 µM, respectively (Rukachaisirikul et al. 2006). The cytotoxicities of cordyheptapeptides C–E were evaluated using human glioblastoma (SF-268), human breast cancer (MCF-7), and human lung cancer (NCI-H460) cell lines (Chen et al. 2012). Cordyheptapeptide E demonstrated cytotoxicity against all three cell lines, with IC_{50} values of 3.2, 2.7, and 4.5 µM, respectively. Cordyheptapeptide C was found to possess cytotoxicity against SF-268 and MCF-7 cells with IC_{50} values of 3.7 and 3.0 µM, respectively, and weaker cytotoxicity against the

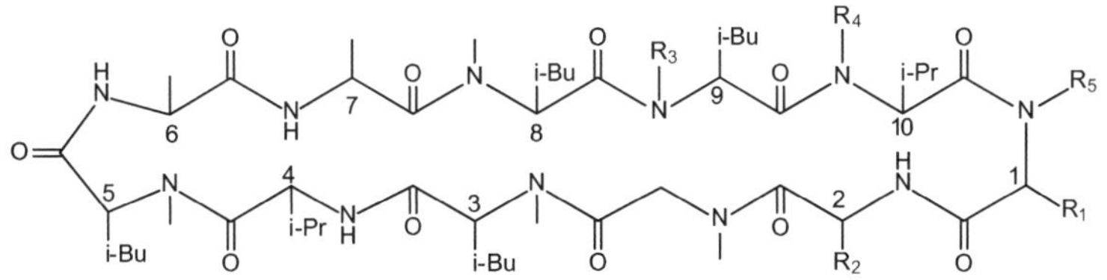

Fig. 8.23 Basic structure of cyclosporines

NCI-H460 cell line. The most polar compound, cordyheptapeptide D, displayed no activity against all three cell lines (Chen et al. 2012).

Cordyheptapeptides A and B were reported to possess cytotoxicities against KB (oral human epidermoid carcinoma), BC (human breast cancer), NCI-H187 (human small cell lung cancer), and Vero (African green monkey kidney fibroblasts) cell lines with IC_{50} values of 0.78, 0.20, 0.18, and 14 μM and 2.0, 0.66, 3.1, and 1.6 μM, respectively (Isaka et al. 2007c).

8.5   Prospects of Entomogenous Fungal NRPs

8.5.1   Prospect of NRPs for Biocontrol of Lepidopteran Pests

The insect order Lepidoptera includes very important crop pests. Use of chemical insecticides is the main method for the control of these insects. Meanwhile, chemical insecticides have been discouraged because of their side effects. More and more attention is being paid to entomogenous fungi for biocontrol of these pests. As an important pathogenic factor, NRPs will be considered more by researchers and users. Many reports have shown that NRPs can kill lepidopteran pest species, although NRPs have not been applied in fields yet. It is a rational prediction that NRPs will contribute more to pest control in the near future because of the technical development and advancement.

Advancement in technology makes the discovery of NRPs more easy and rapid. Since enniatins were discovered in the 1940s, destruxins and beauvericins in the 1960s, beauverolides, bassianolides, cyclosporins, and efrapeptins in the 1970s, etc., novel NRPs such as neoefrapeptins, cordyheptapeptides, verticilides, serinocyclins, isaridins, hirsutellides, hirsutides, cordycommunin, conoideo-crellides, and paecilodepsipeptides were successively found in this century. Meanwhile, large numbers of new analogues of NRPs were isolated and identified, for example, since destruxin A was discovered in 1961, more than 30 kinds of analogues have been successively found in the last 50 years. Because of the abundant diversity of entomogenous fungi and their NRPSs, along with more and more new fungal strains (e.g., endophytic and ocean species) being used as research materials, in addition to the application of new sensitive technologies of separation and

Table 8.19 Analogues of cyclosporines

Analogues name	CAS registry no.	Molecular formula	Geometry configuration	R1	R2	R3	R4	R5
Cyclosporin A	59865-13-3	$C_{62}H_{111}N_{11}O_{12}$	1S, 2S, 3S, 4S, 5S, 6S, 7R, 8S, 9S, 10S	Rx	Et	Me	Me	Me
Cyclosporin B	63775-95-1	$C_{61}H_{109}N_{11}O_{12}$	1S, 2S, 3S, 4S, 5S, 6S, 7R, 8S, 9S, 10S	Rx	Me	Me	Me	Me
Cyclosporin C	59787-61-0	$C_{62}H_{111}N_{11}O_{13}$	1S, 2S, 3S, 4S, 5S, 6S, 7R, 8S, 9S, 10S	Rx	CH(OH)CH$_3$	Me	Me	Me
Cyclosporin D	63775-96-2	$C_{63}H_{113}N_{11}O_{12}$	1S, 2S, 3S, 4S, 5S, 6S, 7R, 8S, 9S, 10S	Rx	iPr	Me	Me	Me
Cyclosporin E	63798-73-2	$C_{61}H_{109}N_{11}O_{12}$	1S, 2S, 3S, 4S, 5S, 6S, 7R, 8S, 9S, 10S	Rx	Et	Me	H	Me
Cyclosporin F	83574-28-1	$C_{62}H_{111}N_{11}O_{11}$	–	–CH$_2$CH(CH$_3$)CH$_2$CH=CHCH$_3$	Et	Me	Me	Me
Cyclosporin G	74436-00-3	$C_{63}H_{113}N_{11}O_{12}$	1S, 2S, 3S, 4S, 5S, 6S, 7R, 8S, 9S, 10S	Rx	nPr	Me	Me	Me
Cyclosporin H	83602-39-5	$C_{62}H_{111}N_{11}O_{12}$	1S, 2S, 3S, 4S, 5S, 6S, 7R, 8S, 9S, 10R	Rx	Et	Me	Me	Me
Cyclosporin I	83563-93-3	$C_{62}H_{111}N_{11}O_{12}$	–	–CH(OH)CH(CH$_3$)CH$_2$CH=CHCH$_3$	iPr	H	Me	Me
Cyclosporin J	121604-28-2	$C_{59}H_{107}N_{11}O_{11}$	1S, 2S, 3S, 4S, 5S, 6S, 7R, 8S, 9S, 10S	iBu	Et	Me	Me	Me
Cyclosporin K	108027-38-9	$C_{63}H_{113}N_{11}O_{11}$	–	–CH$_2$CH(CH$_3$)CH$_2$CH=CHCH$_3$	iPr	Me	Me	Me
Cyclosporin L	108027-39-0	$C_{61}H_{109}N_{11}O_{12}$	–	–CH(OH)CH(CH$_3$)CH$_2$CH=CHCH$_3$	Et	Me	Me	H

Note: nPr, –CH$_2$CH$_2$CH$_3$

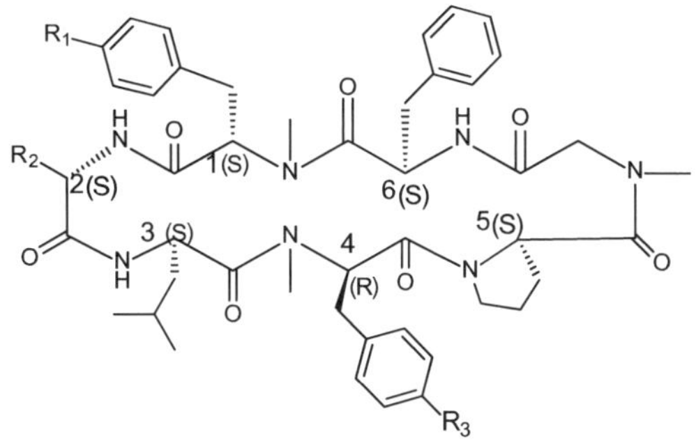

Fig. 8.24 Basic structure of cordyheptapeptide

Table 8.20 Analogues of cordyheptapeptide

Analogues name	CAS registry no.	Molecular formula	Geometry configuration	R1	R2	R3
Cordyheptapeptide A	877776-11-9	$C_{49}H_{65}N_7O_8$	1S, 2S, 3S, 4R, 5S, 6S	OH	sBu	H
Cordyheptapeptide B	957768-12-6	$C_{49}H_{65}N_7O_7$	1S, 2S, 3S, 4R, 5S, 6S	H	sBu	H
Cordyheptapeptide C	1377420-73-9	$C_{48}H_{63}N_7O_8$	1S, 2S, 3S, 4R, 5S, 6S	OH	H	H
Cordyheptapeptide D	1377420-75-1	$C_{48}H_{63}N_7O_9$	1S, 2S, 3S, 4R, 5S, 6S	OH	OH	H
Cordyheptapeptide E	1377420-77-3	$C_{49}H_{66}N_7O_9$	1S, 2S, 3S, 4R, 5S, 6S	OH	OH	Me

detection of chemical compounds, it is a rational prediction that more novel NRPs will be discovered at an easier and faster pace.

8.5.2 Molecular Mechanism of NRPs

Entomogenous fungi invade host insects mainly from the cuticle and must break through the hemolymph immunity barrier after entering the insect hemocoel to cause mycosis in insects. NRPs are the evolutionary products of fungi in order to adapt to their host insects; therefore, NRPs play a very important role in pathogenicity. Many NRPs such as destruxin and beauvericin inhibit insect immunity, some NRPs have other toxic effects, but the mechanisms of action of NRPs are not clear except for that of cyclosporins.

Slow insecticidal efficacy limits myco-insecticides to a small-scale application. In order to solve the problem, it is necessary that fundamental research on the pathogenesis of entomogenous fungi should be paid more attention. Pathogenic process of entomogenous fungi generally includes a lot of steps: conidial attachment to the insect cuticle, germination of conidium, formation of appressoria and invading structure, penetration of the cuticle, entering the hemocoel, overcoming insect's innate immunity, capturing nutrition, mass proliferation, etc. Previous studies mainly focus on the processes before fungus enters into the hemocoel (i.e., from fungal attachment to insect surface to penetrating the cuticle). Researchers paid less attention to how fungi can break through the host insect's immunity barrier after entering the hemocoel. Once the fundamental problems such as how entomogenous fungi secrete NRPs in hemocoel and how NRPs act are elucidated, it will provide a new direction to improve the efficacy of myco-insecticides.

8.5.3   NRPs as Novel Immunosuppressant Insecticides

Inventing a new pesticide is an important measure of effectively controlling agricultural pests and of developing the pesticide industry. Generally, discovery of new pesticide includes a variety of ways such as random screening, analogue synthesis, bioactive natural product model, and biorational design. Among them, bioactive natural product model is the most important. In the system of bioactive natural product model, researchers find new active natural products and optimize their structures to get lead compounds, then they determine the new target of action (target enzyme or receptor) of the leads, and finally, a new pesticide is discovered by means of computer-aided molecular design, oriented synthesis, and high-throughput screening.

Most of the insecticides in the current market act on the insect's nervous system. Their main target proteins include acetylcholinesterase (organophosphorus), acetylcholine receptor (carbamates, neonicotinoids), sodium ion channel (pyrethroids), γ-aminobutyric acid receptor (abamectin), etc. Numerous insecticides having the same target protein lead to the development of pesticide resistance in insects.

8.6   Conclusion

Insect immune system is expected to become a significant and a potential target of the new-generation pesticides. The first reason is that the immune system has different structure and function from the nervous system, muscular system, respiratory system, and endocrine system. So, the rationale is that there will be no cross resistance between immunosuppressants and the existing pesticides. The second reason is that the insect immune system is different from that of mammals. It will benefit in developing highly selective immunosuppressant insecticides. A large number of NRPs have been reported to have natural immunosuppressant properties, and new pesticide targets could be found via further study which will provide new insights into future research and development of new immunosuppressant insecticides.

Acknowledgements The authors cordially thank Dr. Zhou Zuoqiang and his MS student Liu Kai (College of Science, South China Agricultural University) for checking the chemical structures of NRPs. Ms. Chen Xiurun partly contributes to this MS. This research is supported by the National High-Technology Research and Development Program ("863" Program) of China (2011AA10A204-2).

References

Amiri-Besheli B, Khambay B, Cameron S, Deadman ML, Butt TM (2000) Inter- and intra-specific variation in destruxin production by insect pathogenic *Metarhizium* spp., and its significance to pathogenesis. Mycol Res 104:447–452. doi:10.1017/s095375629900146x

Bandani AR (2004) Effect of entomopathogenic fungus *Tolypocladium* species metabolite efrapeptin on *Galleria mellonella* agglutinin. Commun Agric Appl Biol Sci 69:165–169

Bandani AR, Amiri B, Butt TM, Gordon-Weeks R (2001) Effects of efrapeptin and destruxin, metabolites of entomogenous fungi, on the hydrolytic activity of a vacuolar type ATPase identified on the brush border membrane vesicles of *Galleria mellonella* midgut and on plant membrane bound hydrolytic enzymes. Biochim Biophys Acta Biomembr 1510:367–377. doi:10.1016/s0005-2736(00)00370-9

Baute R, Deffieux G, Merlet D, Baute MA, Neveu A (1981) New insecticidal cyclodepsipeptides from the fungus *Isaria felina*. I. Production, isolation and insecticidal properties of isariins B, C and D. J Antibiot 34:1261–1265. doi:10.7164/antibiotics.34.1261

Bernardini M, Carilli A, Pacioni G, Santurbano B (1975) Isolation of beauvericin from *Paecilomyces fumosoroseus*. Phytochemistry 14:1865. doi:10.1016/0031-9422(75)85311-8

Boettger D, Hertweck C (2013) Molecular diversity sculpted by fungal PKS-NRPS hybrids. Chembiochem 14:28–42. doi:10.1002/cbic.201200624

Boot CM, Amagata T, Tenney K, Compton JE, Pietraszkiewicz H, Valeriote FA, Crews P (2007) Four classes of structurally unusual peptides from two marine-derived fungi: structures and bioactivities. Tetrahedron 63:9903–9914. doi:10.1016/j.tet.2007.06.034

Borel JF, Feurer C, Gubler HU, Staehelin H (1976) Biological effects of cyclosporin A: a new antilymphocytic agent. Agents Actions 6:468–475. doi:10.1007/bf01973261

Briggs LH, Fergus BJ, Shannon JS (1966) Chemistry of fungi. IV. Cyclodepsipeptides from a new species of Isaria. Tetrahedron 22:269–278

Brousseau C, Charpentier G, Belloncik S (1996) Susceptibility of spruce budworm, *Choristoneura fumiferana* Clemens, to destruxins, cyclodepsipeptidic mycotoxins of *Metarhizium anisopliae*. J Invertebr Pathol 68:180–182. doi:10.1006/jipa.1996.0079

Chen Z, Song Y, Chen Y, Huang H, Zhang W, Ju J (2012) Cyclic heptapeptides, cordyheptapeptides C-E, from the marine-derived fungus *Acremonium persicinum* SCSIO 115 and their cytotoxic activities. J Nat Prod 75:1215–1219. doi:10.1021/np300152d

Chen XR, Hu QB, Yu XQ, Ren SX (2014) Effects of destruxins on free calcium and hydrogen ions in insect hemocytes. Insect Sci 2:31–38

De Zotti M, Biondi B, Crisma M, Hjorringgaard CU, Berg A, Bruckner H, Toniolo C (2012) Isovaline in naturally occurring peptides: a nondestructive methodology for configurational assignment. Biopolymers 98:36–49. doi:10.1002/bip.21679

Deffieux G, Merlet D, Baute R, Bourgeois G, Baute MA, Neveu A (1981) New insecticidal cyclodepsipeptides from the fungus *Isaria felina*. II. Structure elucidation of isariins B, C and D. J Antibiot 34:1266–1270. doi:10.7164/antibiotics.34.1266

Doebler JA (2000) Effects of neutral ionophores on membrane electrical characteristics of NG108-15 cells. Toxicol Lett 114:27–38. doi:10.1016/s0378-4274(99)00193-9

Dreyfuss M, Haerri E, Hofmann H, Kobel H, Pache W, Tscherter H (1976) Cyclosporin A and C. New metabolites from *Trichoderma polysporum* (Link ex Pers.) Rifai. Eur J Appl Microbiol 3:125–133

Dumas C, Robert P, Pais M, Vey A, Quiot J-M (1994) Insecticidal and cytotoxic effects of natural and hemisynthetic destruxins. Comp Biochem Physiol Pharmacol Toxicol Endocrinol 108:195–203

Elsworth JF, Grove JF (1977) Cyclodepsipeptides from *Beauveria bassiana* Bals. Part 1. Beauverolides H and I. J Chem Soc Perkin Trans 1:270–273

Elsworth JF, Grove JF (1980) Cyclodepsipeptides from *Beauveria bassiana*. Part 2. Beauverolides A to F and their relationship to isarolide. J Chem Soc Perkin Trans 1:1795–1799

Fan JQ, Chen XR, Hu QB (2013) Effects of destruxin A on hemocytes morphology of *Bombyx mori*. J Integr Agric 12:1042–1048. doi:10.1016/s2095-3119(13)60324-x

Feudjio FT, Dornetshuber R, Lemmens M, Hoffmann O, Lemmens-Gruber R, Berger W (2010) Beauvericin and enniatin: emerging toxins and/or remedies? World Mycotoxin J 3:415–430. doi:10.3920/wmj2010.1245

Fiolka MJ (2008) Immunosuppressive effect of cyclosporin A on insect humoral immune response. J Invertebr Pathol 98:287–292. doi:10.1016/j.jip.2008.03.015

Firakova S, Proksa B, Sturdikova M (2007) Biosynthesis and biological activity of enniatins. Pharmazie 62:563–568. doi:10.1691/ph.2007.8.7600

Fornelli F, Minervini F, Logrieco A (2004) Cytotoxicity of fungal metabolites to lepidopteran (*Spodoptera frugiperda*) cell line (SF-9). J Invertebr Pathol 85:74–79. doi:10.1016/j.jip.2004.01.002

Fredenhagen A, Molleyres L-P, Bohlendorf B, Laue G (2006) Structure determination of neoefrapeptins A to N: peptides with insecticidal activity produced by the fungus *Geotrichum candidum*. J Antibiot 59:267–280. doi:10.1038/ja.2006.38

Gaumann E, Roth S, Ettlinger L, Plattner PA, Nager U (1947) Enniatin, a new antibiotic active against *Mycobacteria*. Experientia 3:202–203

Grove JF (1980) Cyclodepsipeptides from *Beauveria bassiana*. Part 3. The isolation of beauverolides Ba, Ca, Ja, and Ka. J Chem Soc Perkin Trans 1:2878–2880

Gupta S, Krasnoff SB, Underwood NL, Renwick JAA, Roberts DW (1991) Isolation of beauvericin as an insect toxin from *Fusarium semitectum* and *Fusarium moniliforme* var. *subglutinans*. Mycopathologia 115:185–189. doi:10.1007/bf00462223

Gupta S, Montllor C, Hwang Y-S (1995) Isolation of novel beauvericin analogs from the fungus *Beauveria bassiana*. J Nat Prod 58:733–738. doi:10.1021/np50119a012

Hamill RL, Higgens CE, Boaz HE, Gorman M (1969) Structure of beauvericin, a new depsipeptide antibiotic toxic to *Artemia salina*. Tetrahedron Lett 49:4255–4258

Haritakun R, Sappan M, Suvannakad R, Tasanathai K, Isaka M (2010) An antimycobacterial cyclodepsipeptide from the entomopathogenic fungus *Ophiocordyceps communis* BCC 16475. J Nat Prod 73:75–78. doi:10.1021/np900520b

Hayakawa Y, Hattori Y, Kawasaki T, Kanoh K, Adachi K, Shizuri Y, Shin-ya K (2008) Efrapeptin J, a new down-regulator of the molecular chaperone GRP78 from a marine *Tolypocladium* sp. J Antibiot 61:365–371. doi:10.1038/ja.2008.51

He H, Janso JE, Yang HY, Bernan VS, Lin SL, Yu K (2006) Culicinin D, an antitumor peptaibol produced by the fungus *Culicinomyces clavisporus*, strain LL-12I252. J Nat Prod 69:736–741. doi:10.1021/np058133r

Herrmann M, Zocher R, Haese A (1996) Enniatin production by *Fusarium* strains and its effect on potato tuber tissue. Appl Environ Microbiol 62:393–398

Hiraga K, Yamamoto S, Fukuda H, Hamanaka N, Oda K (2005) Enniatin has a new function as an inhibitor of Pdr5p, one of the ABC transporters in *Saccharomyces cerevisiae*. Biochem Biophys Res Commun 328:1119–1125

Hu QB, Ren SX, Liu SY (2007a) Purification of destruxins produced by *Metarhizium anisopliae* and bioassay of their insecticidal activities against grubs. Kunchong Xuebao 50:461–466

Hu QB, Ren SX, An XC, Qian MH (2007b) Insecticidal activity influence of destruxins on the pathogenicity of *Paecilomyces javanicus* against *Spodoptera litura*. J Appl Entomol 131:262–268. doi:10.1111/j.1439-0418.2007.01159.x

Hur GH, Vickery CR, Burkart MD (2012) Explorations of catalytic domains in non-ribosomal peptide synthetase enzymology. Nat Prod Rep 29:1074–1098. doi:10.1039/c2np20025b

Isaka M, Palasarn S, Kocharin K, Hywel-Jones NL (2007a) Comparison of the bioactive secondary metabolites from the scale insect pathogens, anamorph *Paecilomyces cinnamomeus* and teleomorph *Torrubiella luteorostrata*. J Antibiot 60:577–581. doi:10.1038/ja.2007.73

Isaka M, Palasarn S, Lapanun S, Sriklung K (2007b) Paecilodepsipeptide A, an antimalarial and antitumor cyclohexadepsipeptide from the insect pathogenic fungus *Paecilomyces cinnamomeus* BCC 9616. J Nat Prod 70:675–678. doi:10.1021/np060602h

Isaka M, Srisanoh U, Lartpornmatulee N, Boonruangprapa T (2007c) ES-242 derivatives and cycloheptapeptides from *Cordyceps* sp. strains BCC 16173 and BCC 16176. J Nat Prod 70:1601–1604. doi:10.1021/np070357h

Isaka M, Palasarn S, Supothina S, Komwijit S, Luangsa-ard JJ (2011) Bioactive compounds from the scale insect pathogenic fungus *Conoideocrella tenuis* BCC 18627. J Nat Prod 74:782–789. doi:10.1021/np100849x

Jackson CG, Linnett PE, Beechey RB, Henderson PJF (1979) Purification and preliminary structure analysis of the efrapeptins, a group of antibiotics that inhibit the mitochondrial adenosine triphosphatase. Biochem Soc Trans 7:224–226

Jegorov A, Matha V, Weiser J (1990) Production of cyclosporins by entomopathogenic fungi. Microbios Lett 45:65–69

Jegorov A, Matha V, Hradec H (1992) Detoxification of destruxins in *Galleria mellonella* L. larvae. Comp Biochem Physiol C 103:227–229

Jegorov A, Sedmera P, Matha V, Simek P, Zahradnickova H, Landa Z, Eyal J (1994) Beauverolides L and La from *Beauveria tenella* and *Paecilomyces fumosoroseus*. Phytochemistry 37:1301–1303

Jestoi M, Rokka M, Peltonen K (2007) An integrated sample preparation to determine coccidiostats and emerging *Fusarium*-mycotoxins in various poultry tissues with LC-MS/MS. Mol Nutr Food Res 51:625–637. doi:10.1002/mnfr.200600232

Juan C, Raiola A, Manes J, Ritieni A (2014) Presence of mycotoxin in commercial infant formulas and baby foods from Italian market. Food Control 39:227–236. doi:10.1016/j.foodcont.2013.10.036

Kallen J, Spitzfaden C, Zurini MGM, Wider G, Widmer H, Wuthrich K, Walkinshaw MD (1991) Structure of human cyclophilin and its binding site for cyclosporin A determined by X-ray crystallography and NMR spectroscopy. Nature 353:276–279

Kanaoka M, Isogai A, Murakoshi S, Ichinoe M, Suzuki A, Tamura S (1978) Bassianolide, a new insecticidal cyclodepsipeptide from *Beauveria bassiana* and *Verticillium lecanii*. Agric Biol Chem 42:629–635. doi:10.1271/bbb1961.42.629

Kershaw MJ, Moorhouse ER, Bateman R, Reynolds SE, Charnley AK (1999) The role of destruxins in the pathogenicity of *Metarhizium anisopliae* for three species of insect. J Invertebr Pathol 74:213–223. doi:10.1006/jipa.1999.4884

Kodaira Y (1961) Toxic substances to insects, produced by *Aspergillus ochraceus* and *Oospora destructor*. Agric Biol Chem 25:261–262. doi:10.1271/bbb1961.25.261

Kodaira Y (1962) Studies on the new toxic substances to insects, destruxin A and B, produced by *Oospora destructor*. I. Isolation and purification of destruxin A and B. Agric Biol Chem 26:36–42. doi:10.1271/bbb1961.26.36

Kovarik J (2013) From immunosuppression to immunomodulation: current principles and future strategies. Pathobiology 80:275–281. doi:10.1159/000346960

Krasnoff SB, Gupta S (1991) Identification and directed biosynthesis of efrapeptins in the fungus *Tolypocladium geodes* Gams (Deuteromycotina: Hyphomycetes). J Chem Ecol 17:1953–1962. doi:10.1007/bf00992580

Krasnoff SB, Gupta S, St. Leger RJ, Renwick JAA, Roberts DW (1991) Antifungal and insecticidal properties of the efrapeptins: metabolites of the fungus *Tolypocladium niveum*. J Invertebr Pathol 58:180–188. doi:10.1016/0022-2011(91)90062-u

Krasnoff SB, Reategui RF, Wagenaar MM, Gloer JB, Gibson DM (2005) Cicadapeptins I and II: new Aib-containing peptides from the entomopathogenic fungus *Cordyceps heteropoda*. J Nat Prod 68:50–55. doi:10.1021/np0497189

Krasnoff SB, Keresztes I, Gillilan RE, Szebenyi DME, Donzelli BGG, Churchill ACL, Gibson DM (2007) Serinocyclins A and B, cyclic heptapeptides from *Metarhizium anisopliae*. J Nat Prod 70:1919–1924. doi:10.1021/np070407i

Lang G, Blunt JW, Cummings NJ, Cole ALJ, Munro MHG (2005) Hirsutide, a cyclic tetrapeptide from a spider-derived entomopathogenic fungus, *Hirsutella* sp. J Nat Prod 68:1303–1305. doi:10.1021/np0501536

Langenfeld A, Blond A, Gueye S, Herson P, Nay B, Dupont J, Prado S (2011) Insecticidal cyclodepsipeptides from *Beauveria felina*. J Nat Prod 74:825–830. doi:10.1021/np100890n

Lardy H, Reed P, Lin C-HC (1975) Antibiotic inhibitors of mitochondrial ATP synthesis. Fed Proc Fed Am Soc Exp Biol 34:1707–1710

Levy D, Bluzat A, Seigneuret M, Rigaud J-L (1995) Alkali cation transport through liposomes by the antimicrobial fusafungine and its constitutive enniatins. Biochem Pharmacol 50:2105–2107. doi:10.1016/0006-2952(95)02045-4

Liu B-L, Tzeng Y-M (2012) Development and applications of destruxins: a review. Biotechnol Adv 30:1242–1254. doi:10.1016/j.biotechadv.2011.10.006

Liu J, Farmer JD Jr, Lane WS, Friedman J, Weissman I, Schreiber SL (1991) Calcineurin is a common target of cyclophilin-cyclosporin A and FKBP-FK506 complexes. Cell 66:807–815

Luciano FB, Meca G, Manyes L, Manes J (2014) A chemical approach for the reduction of beauvericin in a solution model and in food systems. Food Chem Toxicol 64:270–274. doi:10.1016/j.fct.2013.11.021

Makrlik E, Toman P, Vanura P (2013a) Complexation of Pb2+ with beauvericin: an experimental and theoretical study. Monatsh Chem 144:1461–1465. doi:10.1007/s00706-013-1054-z

Makrlik E, Toman P, Vanura P (2013b) Experimental and DFT study on the complexation of Ba2+ with beauvericin. J Radioanal Nucl Chem 295:1887–1891. doi:10.1007/s10967-012-2107-1

Makrlik E, Toman P, Vanura P (2013c) Experimental and theoretical study on the complexation of Ca2+ with beauvericin. J Radioanal Nucl Chem 298:195–200. doi:10.1007/s10967-012-2335-4

Makrlik E, Toman P, Vanura P (2013d) Extraction and DFT study on the complexation of Zn2+ with beauvericin. Acta Chim Slov 60:884–888

Matsuda D, Namatame I, Tomoda H, Kobayashi S, Zocher R, Kleinkauf H, Omura S (2004) New beauveriolides produced by amino acid-supplemented fermentation of *Beauveria* sp. FO-6979. J Antibiot 57:1–9. doi:10.7164/antibiotics.57.1

Mochizuki K, Ohmori K, Tamura H, Shizuri Y, Nishiyama S, Miyoshi E, Yamamura S (1993a) The structures of bioactive cyclodepsipeptides, beauveriolides I and II, metabolites of entomopathogenic fungi *Beauveria* sp. Bull Chem Soc Jpn 66:3041–3046. doi:10.1246/bcsj.66.3041

Mochizuki K, Ohomori K, Tamura H, Shizuri Y, Nishiyama S, Miyoshi E, Yamamura S (1993b) The structures of bioactive cyclodepsipeptides, Beauveriollides I and II, metabolites of entomopathogenic fungi *Beauveria* sp. Bull Chem Soc Jpn 66:3041–3046

Molleyres L-P, Fredenhagen A, Schuez TC, Boehlendorf B, Neff S, Huang Y (2004) Production of neoefrapeptins for use as insecticides. DE10361201A1

Momose I, Ohba S-i, Tatsuda D, Kawada M, Masuda T, Tsujiuchi G, Yamori T, Esumi H, Ikeda D (2010) Mitochondrial inhibitors show preferential cytotoxicity to human pancreatic cancer PANC-1 cells under glucose-deprived conditions. Biochem Biophys Res Commun 392:460–466. doi:10.1016/j.bbrc.2010.01.050

Monma S, Sunazuka T, Nagai K, Arai T, Shiomi K, Matsui R, Omura S (2006) Verticilide: elucidation of absolute configuration and total synthesis. Org Lett 8:5601–5604. doi:10.1021/ol0623365

Moretti A, Mule G, Ritieni A, Logrieco A (2007) Further data on the production of beauvericin, enniatins and fusaproliferin and toxicity to *Artemia salina* by *Fusarium* species of *Gibberella fujikuroi* species complex. Int J Food Microbiol 118:158–163. doi:10.1016/j.ijfoodmicro.2007.07.004

Muellenhoff MW, Koo JY (2012) Cyclosporine and skin cancer: an international dermatologic perspective over 25 years of experience. A comprehensive review and pursuit to define safe use of cyclosporine in dermatology. J Dermatol Treat 23:290–304. doi:10.3109/09546634.2011.590792

Mule G, D'Ambrosio A, Logrieco A, Bottalico A (1992) Toxicity of mycotoxins of *Fusarium sambucinum* for feeding in *Galleria mellonella*. Entomol Exp Appl 62:17–22. doi:10.1111/j.1570-7458.1992.tb00636.x

Nagaoka Y, Hata K, Fukata N, Uesato S, Ueda J, Higashi K, Uchida S, Fujita T, Sasaki S, Tachikawa E (2006) Cicadapeptins from a fungus *Isaria sinclairii* inhibit acetylcholine-evoked secretion of catecholamines from bovine adrenal chromaffin cells. Pept Sci 43:248

Nakajyo S, Shimizu K, Kometani A, Kato K, Kamizaki J, Isogai A, Urakawa N (1982) Inhibitory effect of bassianolide, a cyclodepsipeptide, on drug-induced contractions of isolated smooth muscle preparations. Jpn J Pharmacol 32:55–64. doi:10.1254/jjp.32.55

Nakajyo S, Shimizu K, Kometani A, Suzuki A, Ozaki H, Urakawa N (1983) On the inhibitory mechanism of bassianolide, a cyclodepsipeptide, in acetylcholine-induced contraction in guinea-pig taenia coli. Jpn J Pharmacol 33:573–582. doi:10.1254/jjp.33.573

Nakaya S, Mizuno S, Ishigami H, Yamakawa Y, Kawagishi H, Ushimaru T (2012) New rapid screening method for anti-aging compounds using budding yeast and identification of beauveriolide I as a potent active compound. Biosci Biotechnol Biochem 76:1226–1228. doi:10.1271/bbb.110872

Namatame I, Tomoda H, Ishibashi S, Omura S (2004) Antiatherogenic activity of fungal beauveriolides, inhibitors of lipid droplet accumulation in macrophages. Proc Natl Acad Sci U S A 101:737–742. doi:10.1073/pnas.0307757100

Ohshiro T, Matsuda D, Nagai K, Doi T, Sunazuka T, Takahashi T, Rudel LL, Omura S, Tomoda H (2009) The selectivity of beauveriolide derivatives in inhibition toward the two isozymes of acyl-CoA: cholesterol acyltransferase. Chem Pharm Bull 57:377–381. doi:10.1248/cpb.57.377

Ohshiro T, Matsuda D, Kazuhiro T, Uchida R, Nonaka K, Masuma R, Tomoda H (2012) New verticilides, inhibitors of acyl-CoA:cholesterol acyltransferase, produced by *Verticillium* sp. FKI-2679. J Antibiot (Tokyo) 65:255–262. doi:ja201212 [pii]10.1038/ja.2012.12 [doi]

Ohyama M, Okada Y, Takahashi M, Sakanaka O, Matsumoto M, Atsumi K (2011) Structure-activity relationship of anthelmintic cyclooctadepsipeptides. Biosci Biotechnol Biochem 75:1354 1363. doi:10.1271/bbb.110129

Omura S, Tomoda H (2002) Selection culture media for beauveriolide I or beauveriolide III manufacture with *Beauveria*. WO2002077203A1

Omura S, Shiomi K, Masuma R (2004) Novel substance FKI-1033 manufacture with *Verticillium* as insecticide. WO2004044214A1

Pal S, St. Leger RJ, Wu LP (2007) Fungal peptide destruxin a plays a specific role in suppressing the innate immune response in *Drosophila melanogaster*. J Biol Chem 282:8969–8977

Papathanassiu AE, MacDonald NJ, Bencsura A, Vu HA (2006) F1F0-ATP synthase functions as a co-chaperone of Hsp90-substrate protein complexes. Biochem Biophys Res Commun 345:419–429. doi:10.1016/j.bbrc.2006.04.104

Papathanassiu AE, MacDonald NJ, Emlet DR, Vu HA (2011) Antitumor activity of efrapeptins, alone or in combination with 2-deoxyglucose, in breast cancer in vitro and in vivo. Cell Stress Chaperones 16:181–193. doi:10.1007/s12192-010-0231-9

Pedras MSC, Irina Zaharia L, Ward DE (2002) The destruxins: synthesis, biosynthesis, biotrans-formation, and biological activity. Phytochemistry 59:579–596. doi:10.1016/s0031-9422(02) 00016-x

Plattner RD, Nelson PE (1994) Production of beauvericin by a strain of Fusarium proliferatum isolated from corn fodder for swine. Appl Environ Microbiol 60:3894–3896

Podsiadlowski L, Matha V, Vilcinskas A (1998) Detection of a P-glycoprotein related pump in Chironomus larvae and its inhibition by verapamil and cyclosporin A. Comp Biochem Phys B 121:443–450

Prosperini A, Juan-Garcia A, Font G, Ruiz MJ (2013) Beauvericin-induced cytotoxicity via ROS production and mitochondrial damage in Caco-2 cells. Toxicol Lett 222:204–211. doi:10.1016/j.toxlet.2013.07.005

Putri SP, Ishido K-i, Kinoshita H, Kitani S, Ihara F, Sakihama Y, Igarashi Y, Nihira T (2013) Production of antioomycete compounds active against the phytopathogens Phytophthora sojae and Aphanomyces cochlioides by clavicipitoid entomopathogenic fungi. J Biosci Bioeng. doi:10.1016/j.jbiosc.2013.10.014

Ravindra G, Ranganayaki RS, Raghothama S, Srinivasan MC, Gilardi RD, Karle IL, Balaram P (2004) Two novel hexadepsipeptides with several modified amino acid residues isolated from the fungus Isaria. Chem Biodivers 1:489–504. doi:10.1002/cbdv.200490043

Rivas F, Howell G, Floyd Z (2008) Orgn 441- Progress toward the synthesis of two antibacterial heptapeptides: Cicadapeptins I and II. Abstracts of papers of the American Chemical Society, 236th National meeting of the American-Chemical-Society, Philadelphia

Rizwan-Ul-Haq M, Hu QB, Hu MY, Zhong G, Weng Q (2009) Study of destruxin B and tea saponin, their interaction and synergism activities with Bacillus thuringiensis kurstaki against Spodoptera exigua (Hubner) (Lepidoptera: Noctuidae). Appl Entomol Zool 44:419–428. doi:10.1303/aez.2009.419

Ruiz-Sanchez E, O'Donnell MJ (2012) Effects of the microbial metabolite destruxin a on ion transport by the gut and renal epithelia of Drosophila melanogaster. Arch Insect Biochem Physiol 80:109–122. doi:10.1002/arch.21023

Ruiz-Sanchez E, Orchard I, Lange AB (2010) Effects of the cyclopeptide mycotoxin destruxin A on the Malpighian tubules of Rhodnius prolixus (Stal). Toxicon 55:1162–1170. doi:10.1016/j.toxicon.2010.01.006

Rukachaisirikul V, Chantaruk S, Tansakul C, Saithong S, Chaicharernwimonkoon L, Pakawatchai C, Isaka M, Intereya K (2006) A cyclopeptide from the insect pathogenic fungus Cordyceps sp. BCC 1788. J Nat Prod 69:305–307. doi:10.1021/np0504331

Ryffel B, Donatsch P, Goetz U, Tschopp M (1980) Cyclosporin receptor on mouse lymphocytes. Immunology 41:913–919

Sabareesh V, Ranganayaki RS, Raghothama S, Bopanna MP, Balaram H, Srinivasan MC, Balaram P (2007) Identification and characterization of a library of microheterogeneous cyclohexadep-sipeptides from the fungus Isaria. J Nat Prod 70:715–729. doi:10.1021/np060532e

Santini A, Meca G, Uhlig S, Ritieni A (2012) Fusaproliferin, beauvericin and enniatins: occur-rence in food – a review. World Mycotoxin J 5:71–81. doi:10.3920/wmj2011.1331

Shiomi K, Matsui R, Kakei A, Yamaguchi Y, Masuma R, Hatano H, Arai N, Isozaki M, Tanaka H, Kobayashi S, Turberg A, Omura S (2010) Verticilide, a new ryanodine-binding inhibitor, produced by Verticillium sp. FKI-1033. J Antibiot (Tokyo) 63:77–82. doi:ja2009126 [pii] 10.1038/ja.2009.126 [doi]

Sree KS, Padmaja V (2008) Oxidative stress induced by destruxin from Metarhizium anisopliae (Metch.) involves changes in glutathione and ascorbate metabolism and instigates ultrastruc-tural changes in the salivary glands of Spodoptera litura (Fab.) larvae. Toxicon 51:1140–1150. doi:10.1016/j.toxicon.2008.01.012

Steinrauf LK (1985) Beauvericin and the other enniatins. Met Ions Biol Syst 19:139–171

Strongman DB, Strunz GM, Giguere P, Yu CM, Calhoun L (1988) Enniatins from Fusarium avenaceum isolated from balsam fir foliage and their toxicity to spruce budworm larvae,

Choristoneura fumiferana (Clem.) (Lepidoptera: tortricidae). J Chem Ecol 14:753–764. doi:10.1007/bf01018770

Supothina S, Isaka M, Kirtikara K, Tanticharoen M, Thebtaranonth Y (2004) Enniatin production by the entomopathogenic fungus *Verticillium hemipterigenum* BCC 1449. J Antibiot 57:732–738

Suzuki A, Kanaoka M, Isogai A, Murakoshi S, Ichinoe M, Tamura S (1977) Bassianolide, a new insecticidal cyclodepsipeptide from *Beauveria bassiana* and *Verticillium lecanii*. Tetrahedron Lett 18:2167–2170. doi:10.1016/s0040-4039(01)83709-6

Sy-Cordero AA, Pearce CJ, Oberlies NH (2012) Revisiting the enniatins: a review of their isolation, biosynthesis, structure determination and biological activities. J Antibiot 65:541–549. doi:10.1038/ja.2012.71

Tomoda H, Omura S (2007) Potential therapeutics for obesity and atherosclerosis: inhibitors of neutral lipid metabolism from microorganisms. Pharmacol Ther 115:375–389. doi:10.1016/j.pharmthera.2007.05.008

Tomoda H, Huang XH, Cao J, Nishida H, Nagao R, Okuda S, Tanaka H, Omura S, Arai H, Inoue K (1992) Inhibition of acyl-CoA: cholesterol acyltransferase activity by cyclodepsipeptide antibiotics. J Antibiot 45:1626–1632. doi:10.7164/antibiotics.45.1626

Tomoda H, Omura S, Takahashi T, Doi T, Hijikuro I, Imuta S, Matsuda D, Oki T (2010) Beauveriolide having ACAT-2 inhibitory activity. JP2010126454A

Vaclavik L, Vaclavikova M, Begley TH, Krynitsky AJ, Rader JI (2013) Determination of multiple mycotoxins in dietary supplements containing green coffee bean extracts using ultrahigh-performance liquid chromatography-tandem mass spectrometry (UHPLC-MS/MS). J Agric Food Chem 61:4822–4830. doi:10.1021/jf401139u

Vazquez MJ, Albarran MI, Espada A, Rivera-Sagredo A, Diez E, Hueso-Rodriguez JA (2005) A new destruxin as inhibitor of vacuolar-type H+-ATPase of Saccharomyces cerevisiae. Chem Biodivers 2:123–130

Vey A, Matha V, Dumas C (2002) Effects of the peptide mycotoxin destruxin E on insect hemocytes and on dynamics and efficiency of the multicellular immune reaction. J Invertebr Pathol 80:177–187. doi:10.1016/s0022-2011(02)00104-0

Vilcinskas A, Matha V, Gotz P (1997a) Effects of the entomopathogenic fungus *Metarhizium anisopliae* and its secondary metabolites on morphology and cytoskeleton of plasmatocytes isolated from the greater wax moth, *Galleria mellonella*. J Insect Physiol 43:1149–1159. doi:10.1016/s0022-1910(97)00066-8

Vilcinskas A, Matha V, Gotz P (1997b) Inhibition of phagocytic activity of plasmatocytes isolated from Galleria mellonella by entomogenous fungi and their secondary metabolites. J Insect Physiol 43:475–483. doi:10.1016/s0022-1910(96)00120-5

Vilcinskas A, Jegorov A, Landa Z, Goatz P, Matha V (1999) Effects of beauverolide L and cyclosporin A on humoral and cellular immune response of the greater wax moth, *Galleria mellonella*. Comp Biochem Physiol C 122:83–92

Vining LC, Taber WA (1962) Isariin, a new depsipeptide from *Isaria cretacea*. Can J Chem 40:1579–1584. doi:10.1139/v62-239

Vongvanich N, Kittakoop P, Isaka M, Trakulnaleamsai S, Vimuttipong S, Tanticharoen M, Thebtaranonth Y (2002) Hirsutellide A, a new antimycobacterial cyclohexadepsipeptide from the entomopathogenic fungus *Hirsutella kobayasii*. J Nat Prod 65:1346–1348. doi:10.1021/np020055+

Wang Q, Xu L (2012) Beauvericin, a bioactive compound produced by fungi: a short review. Molecules 17:2367–2377. doi:10.3390/molecules17032367

Weigelt S, Huber T, Hofmann F, Jost M, Ritzefeld M, Luy B, Freudenberger C, Majer Z, Vass E, Greie J-C, Panella L, Kaptein B, Broxterman QB, Kessler H, Altendorf K, Hollosi M, Sewald N (2012) Synthesis and conformational analysis of efrapeptins. Chemistry 18:478–487

Weiser J, Matha V (1988) The insecticidal activity of cyclosporines on mosquito larvae. J Invertebr Pathol 51:92–93. doi:10.1016/0022-2011(88)90092-4

Wiesinger D, Borel JF (1980) Studies on the mechanism of action of cyclosporin A. Immunobiology (Stuttgart) 156:454–463. doi:10.1016/s0171-2985(80)80078-7

Witter DP, Chen Y, Rogel JK, Boldt GE, Wentworth P Jr (2009) The natural products Beauveriolide I and III: a new class of β-amyloid-lowering compounds. Chembiochem 10:1344–1347. doi:10.1002/cbic.200900139

Xu Y, Zhan J, Wijeratne EMK, Burns AM, Gunatilaka AAL, Molnar I (2007) Cytotoxic and antihaptotactic beauvericin analogues from precursor-directed biosynthesis with the insect pathogen *Beauveria bassiana* ATCC 7159. J Nat Prod 70:1467–1471. doi:10.1021/np070262f

Xu Y, Rozco R, Wijeratne EMK, Espinosa-Artiles P, Gunatilaka AAL, Stock SP, Molnar I (2009) Biosynthesis of the cyclooligomer depsipeptide bassianolide, an insecticidal virulence factor of *Beauveria bassiana*. Fungal Genet Biol 46:353–364. doi:10.1016/j.fgb.2009.03.001

Yang M, Wang Y, Liu X, Wu J (2013) Synthesis and anti-tumor activity of cyclodepsipeptides paecilodepsipeptide A. Adv Mater Res 643:92–95. doi:10.4028/www.scientific.net/AMR.643.92

Yi F, Zou C, Hu Q, Hu M (2012) The joint action of destruxins and botanical insecticides (rotenone, azadirachtin and paeonolum) against the cotton aphid, *Aphis gossypii* Glover. Molecules 17:7533–7542. doi:10.3390/molecules17067533

Yu D, Xu F, Gage D, Zhan J (2013a) Functional dissection and module swapping of fungal cyclooligomer depsipeptide synthetases. Chem Commun 49:6176–6178. doi:10.1039/c3cc42425a

Yu D, Xu F, Zi J, Wang S, Gage D, Zeng J, Zhan J (2013b) Engineered production of fungal anticancer cyclooligomer depsipeptides in *Saccharomyces cerevisiae*. Metab Eng 18:60–68. doi:10.1016/j.ymben.2013.04.001

Zhang L, Yan K, Zhang Y, Huang R, Bian J, Zheng C, Sun H, Chen Z, Sun N, An R, Min F, Zhao W, Zhuo Y, You J, Song Y, Yu Z, Liu Z, Yang K, Gao H, Dai H, Zhang X, Wang J, Fu C, Pei G, Liu J, Si Z, Goodfellow M, Jiang Y, Kuai J, Zhou G, Chen X (2007) High-throughput synergy screening identifies microbial metabolites as combination agents for the treatment of fungal infections. Proc Natl Acad Sci U S A 104:4606–4611. doi:10.1073/pnas.0609370104

Zhang W, Sun TT, Mei D, Wang JF, Li YX (2008) Synthesis of protected (2S,4R)-2-amino-4-methyldecanoic acid, a proposed component of culicinins. Chin Chem Lett 19:1068–1070. doi:10.1016/j.cclet.2008.06.027

Zhang W, Sun T-T, Li Y-X (2009) Synthesis of the C-terminal pentapeptide of the peptaibol culicinins. J Pept Sci 15:366–368. doi:10.1002/psc.1124

Zhang W, Ding N, Li Y (2011) An improved synthesis of (2S,4S)- and (2S,4R)-2-amino-4-methyldecanoic acids: assignment of the stereochemistry of culicinins. J Pept Sci 17:576–580. doi:10.1002/psc.1376

Zhang D, Li S, Lu R, Li K, Luo F, Peng F, Hu F (2012) Influence of incubation time on metabolites in mycelia of *Paecilomyces militaris*. Weishengwu Xuebao 52:1477–1488

Chapter 9
Optimization of the Cyclodepsipeptidic Destruxin Recovery from Broth Culture of *Metarhizium anisopliae* and Its Augmentation by Precursor Supplementation

K. Sowjanya Sree and V. Padmaja

9.1 Introduction

Cyclic hexadepsipeptidic secondary metabolite, destruxin (dtx), is a major class of mycotoxins produced by the entomopathogenic fungus *Metarhizium anisopliae* (Metch.). This cyclic metabolite comprises of five amino acids, viz., proline (Pro), isoleucine (Ile), methyl valine (MeVal), methyl alanine (MeAla), and β alanine (βAla), and a hydroxyl acid, viz., sodium acetate or pipecolic acid (Pedras et al. 2002). Individual dtxs differ in the hydroxyl acid, *N*-methylation, and R group of the amino acid residues (Pedras et al. 2002). The genes responsible for the biosynthesis of destruxin from *Metarhizium* species were revealed in a recent study by Wang et al. (2012). Studies on the isotopically labeled precursors (^{13}C) administered to *M. anisopliae* cultures showed that L-methionine is the methyl group donor in the process of *N*-methylation of MeAla and MeVal residues of dtx (Jegorov et al. 1993).

Dtx A and E are the most predominant forms in the cultures of *M. anisopliae* and possess insecticidal properties against lepidopterans (Sree et al. 2008) and also against a large variety of insects as reviewed by Pedras et al. (2002). Dtxs cause an initial tetanic paralysis in the treated insect larvae, which is attributed to muscle depolarization by direct opening of the Ca^{2+} channels in the membrane (Samuels et al. 1988). It is reported that dtxs mediate specific downregulation of anti-microbial peptides (AMPs) by means of targeting the insect's innate immune signaling pathway (Pal et al. 2007). Impressive changes in the status of antioxidants, viz., catalase, total peroxidase, ascorbate peroxidase, superoxide dismutase, glutathione,

K.S. Sree (✉)
Amity Institute of Microbial Technology, Amity University Uttar Pradesh, Noida 201303, Uttar Pradesh, India
e-mail: ksowsree@gmail.com

V. Padmaja
Department of Botany, Andhra University, Visakhapatnam 530003, India

© Springer International Publishing Switzerland 2015   207
K.S. Sree, A. Varma (eds.), *Biocontrol of Lepidopteran Pests*, Soil Biology 43,
DOI 10.1007/978-3-319-14499-3_9

ascorbate, levels of lipid peroxidation, and lipoxygenase enzyme, and the ultra-structural changes in the salivary gland membrane have been shown in the 9-day-old larvae of *Spodoptera litura* upon crude dtx treatment (Sree and Padmaja 2008a, b).

It is well documented that the recovery of dtxs is influenced by the component type and ratio, usually carbon and nitrogen, in the culture medium (Liu et al. 2000; Wang et al. 2004) and also by other factors like pH and temperature (Hu et al. 2006). Optimization of media components and culture conditions through factorial (central composite design, CCD) and response surface method (RSM) is a common practice in biotechnology (Chen 1996; Prakash et al. 2008; Rao et al. 1993). In order to obtain optimum yield of a secondary metabolite, destruxin as in the present study, development of a suitable medium and culture conditions is obligatory. Statistical optimization not only allows quick screening of a large experimental domain but also reflects the role of each of the components. Basically, this optimization process involves three major steps: performing the statistically designed experiments, estimating the coefficients in a mathematical model, and predicting the response and adequacy of the model. Using the mathematical model, the levels of variables giving maximum response can be determined.

The aim of the present study was to optimize the conditions involved in the recovery of dtx A and E in Czapek–Dox (CZ) liquid medium by using CCD and RSM and also to evaluate the influence of nutrient supplements involved in dtx biosynthesis like amino acid and hydroxyl acid precursors on dtx A and E recovery. The biochemical profile of *M. anisopliae* grown on supplemented nutrient medium in terms of acid phosphatase, alkaline phosphatase, and esterase was understood in relation to the recovery of the secondary metabolite.

9.2 Materials and Methods

9.2.1 Chemicals

All the solvents used for secondary metabolite extraction, detection, and quantification were of high performance liquid chromatography (HPLC) grade from Qualigens Fine Chemicals, India. The components of fungal culture medium were from HiMedia (India) Ltd. The chemicals used for enzyme analysis were from Merck (India) Ltd. and Sigma-Aldrich Corporation (India).

9.2.2 Fungal Strains

The fungal strain M-19 of *M. anisopliae* was an ARSEF, Ithaca, collection (ARSEF-1080) isolated from the cadavers of *Heliothis* spp. Sree et al. (2008) have shown that the recovery and insecticidal activity of destruxin from this strain is very high. The propagules of these fungal strains were preserved in 30 % glycerol under low temperature for long-term storage. The cultures were grown on SDAY at 25 °C in the dark. Slants fully covered with spores, usually after 9 days of inoculation, were used for experimentation. Germination of spores on SDAY tested prior to experimentation revealed more than 95 % value.

9.2.3 Optimization of Variables Using RSM

The RSM was used to optimize the medium components and culture conditions for the growth of M-19 strain of *M. anisopliae*. Based on the earlier reports (Liu et al. 2000; Wang et al. 2004; Hu et al. 2006) and our observations, carbon and nitrogen ratio in the medium, initial pH, and temperature were selected as independent variables in the CCD experiment. Each factor had five coded levels ($-\alpha$, -1, 0, $+1$, $+$) (Table 9.1). The response was the amount of dtx A and E recovered from the culture filtrate after 7 days of incubation. The CCD contained 18 runs as shown in Table 9.2.

9.2.4 Statistical Analysis and Modeling

The results of RSM were used to fit a second-order polynomial equation.

$$Y = \beta_0 + \beta_1 X_1 + \beta_2 X_2 + \beta_3 X_3 + \beta_1\beta_1 X_1^2 + \beta_2\beta_2 X_2^2 + \beta_3\beta_3 X_3^2 + \beta_1\beta_2 X_1 X_2 \\ + \beta_1\beta_3 X_1 X_3 + \beta_2\beta_3 X_2 X_3$$

where Y is the predicted response, β_0 is the intercept, β_1, β_2, β_3 are the linear coefficients, $\beta_1\beta_1$, $\beta_2\beta_2$, $\beta_3\beta_3$ are the squared coefficients, and $\beta_1\beta_2$, $\beta_1\beta_3$, $\beta_2\beta_3$ are the interaction coefficients. X_1, X_2, and X_3 are the independent variables as given in Table 9.1.

The statistical significance of the model equation was determined by analysis of variance (ANOVA) ($\alpha = 0.05$), and the proportion of variance explained by the model was given by the multiple coefficient of determination, R squared (R^2) value. STATISTICA version 6.0 was used for the design and analysis of this experiment.

9.2.5 In Vitro Production and Recovery of dtx A and E

The crude dtx produced in vitro was extracted according to Wang et al. (2003). Briefly, 1.5 ml of spore suspension (2×10^7 conidia/ml) of *M. anisopliae* was introduced into 150 ml conical flasks charged with 50 ml of CZ liquid medium

Table 9.1 Range of values of the independent variables for response surface methodology

Independent variable	Levels				
	$-\alpha$	-1	0	$+1$	$+\alpha$
X_1—CN ratio	0.636	2.0	4.0	6.0	7.364
X_2—pH	5.318	6.0	7.0	8.0	8.682
X_3—temperature	16.591	20.0	25.0	30.0	33.409

Table 9.2 Experimental design and results of CCD of response surface method

Expt. runs	CN ratio	pH	Temperature (°C)	Dtx A (mg/l)		Dtx E (mg/l)	
				Observed	Predicted	Observed	Predicted
1	2.0	6.0	20.0	5.99	5.47	5.13	4.65
2	2.0	6.0	30.0	5.97	5.57	5.06	4.71
3	2.0	8.0	20.0	6.16	5.53	5.24	4.70
4	2.0	8.0	30.0	6.19	5.66	5.09	4.63
5	6.0	6.0	20.0	2.10	3.34	1.15	2.40
6	6.0	6.0	30.0	2.01	3.34	1.26	2.60
7	6.0	8.0	20.0	2.15	3.26	1.36	2.50
8	6.0	8.0	30.0	2.05	3.28	1.28	2.55
9	0.6	7.0	25.0	1.8	3.38	1.01	2.50
10	7.3	7.0	25.0	2.16	0.41	1.46	1.13
11	4.0	5.3	25.0	6.75	6.11	5.96	5.29
12	4.0	8.6	25.0	6.48	6.12	5.75	5.30
13	4.0	7.0	16.5	6.57	6.20	5.67	5.24
14	4.0	7.0	33.4	6.93	6.30	6.03	5.34
15	4.0	7.0	25.0	7.55	7.71	6.26	6.38
16	4.0	7.0	25.0	7.67	7.71	6.37	6.38
17	4.0	7.0	25.0	7.79	7.71	6.41	6.38
18	4.0	7.0	25.0	7.66	7.71	6.3	6.38

along with the differing amounts of glucose and peptone in order to make up to the required CN ratio (Wang et al. 2004). After incubation at varying combinations of pH and temperature (Table 9.2) for 7 days at 200 rpm in a rotary shaker, the cultures were filtered through three layers of filter papers and then through Whatman no. 1 filter paper. The culture filtrate was extracted two times for 12 h with dichloromethane–ethyl acetate (1:1, v/v). The solvent was evaporated in vacuo, and the crude extract was dissolved in 1 ml absolute methanol for HPLC analysis. Each experiment was repeated three times with three replicates each.

The reverse phase HPLC (RP-HPLC) was carried out according to Kershaw et al. (1999). A Shimadzu HPLC apparatus equipped with a Spherisorb C18 column was used. The mobile phase, i.e., 50 % acetonitrile, was run isocratically at a flow rate of 1 ml/min. Absorbance peaks obtained in the UV range at 210 nm were identified by comparison with the chromatogram published by Kershaw et al. (1999), and dtx A and E were quantified by peak area measurement with the help of the inbuilt software in HPLC system.

9.2.6 Destruxin Biosynthesis Precursors as Supplements to the Optimized Medium

The influence of different dtx biosynthesis precursors on dtx recovery from *M. anisopliae* was investigated by the addition of 0.02 % of amino acid (β-alanine, isoleucine, proline, valine, methionine) and hydroxyl acid (sodium acetate) supplements to the optimized medium. The dtx A and E were recovered and quantified as mentioned above. Each experiment was repeated three times with three replicates each.

9.2.7 Biochemical Profile of M. anisopliae Grown on Supplemented Nutrient Medium

The activity of acid phosphatase, alkaline phosphatase, and esterase were studied in *M. anisopliae* grown on optimized medium supplemented with various dtx biosynthesis precursors along with the control which was grown on optimized medium without any supplements. Each experiment was repeated three times with three replicates each.

9.2.7.1 Acid Phosphatase

The acid phosphatase enzyme was extracted by homogenizing 1 g of the fresh fungal pellets in 10 ml of ice-cold 50 mM citrate buffer (pH 5.3) in a prechilled mortar and pestle. The homogenate was centrifuged at $10,000 \times g$ for 10 min, and the supernatant thus obtained was used as the enzyme source.

Activity of the enzyme was assayed spectrophotometrically by adding 0.5 ml of the enzyme extract to 3 ml of the substrate solution (1.49 mM EDTA, 0.84 mM citric acid, and 0.03 mM *p*-nitrophenylphosphate in distilled water, pH 5.3). This was incubated at 37 °C for 15 min, and then 0.5 ml of this mixture was added to 9.5 ml of 0.085 N sodium hydroxide solution. The absorbance of the resulting solution was read at 405 nm. The activity was measured using a standard curve of *p*-nitrophenol (Lowry et al. 1954).

9.2.7.2 Alkaline Phosphatase

The alkaline phosphatase enzyme was extracted by homogenizing 1 g of the fresh fungal pellets in 10 ml of ice-cold 50 mM glycine NaOH buffer (pH 10.4) in a prechilled mortar and pestle. The homogenate was centrifuged at $10,000 \times g$ for 10 min, and the supernatant thus obtained was used as the enzyme source.

Activity of the enzyme was assayed spectrophotometrically by adding 0.5 ml of the enzyme extract to 3 ml of the substrate solution (0.15 mM glycine, 0.004 mM magnesium chloride, and 0.069 mM p-nitrophenylphosphate in 0.1 N sodium hydroxide, pH 10.4). This mixture was incubated at 37 °C for 15 min, and then 0.5 ml of this mixture was added to 9.5 ml of 0.085 N sodium hydroxide solution. The absorbance of the resulting solution was read at 405 nm. The activity was measured using a standard curve of p-nitrophenol (Lowry et al. 1954).

9.2.7.3 Esterase

The enzyme extract was prepared by homogenizing 1 g of the fresh pellets in 5 ml of the solution containing 10 mM sodium phosphate buffer (pH 9.5), 1 mM EDTA, and 1 mM β-mercaptoethanol. The homogenate was centrifuged at $10,000 \times g$ for 10 min, and the supernatant was further used as the enzyme source.

Spectrophotometric assay of esterase was carried out according to Sparks et al. (1979) with slight modifications. Initially, 0.5 ml of the enzyme extract was allowed to hydrolyze 1 ml of 0.001 M α-naphthyl acetate in 10 mM sodium phosphate buffer (pH 9.5). The mixture was incubated at 37 °C for 15 min, and then the reaction was stopped by the addition of 0.25 ml of stop solution (3.4 % SDS along with 0.8 % Fast Blue B salt solution). As a result, a Fast blue B dye complex was formed with α-naphthol. The absorbance of the resultant was read at 620 nm. The activity was calculated using a standard graph of α-naphthol prepared in the same way.

9.3 Results

9.3.1 Optimal Conditions for dtx A and E Recovery

The results of CCD experiments for studying the effect of three independent variables are presented along with the observed and predicted responses in Table 9.2. The 3D response surface plots (Figs. 9.1 and 9.2) depict the interaction between the independent variables and their effect on the amount of dtx recovery. In these figures, the relative effect of any of the two variables is shown keeping the third one constant. The mid value of each of the variable was taken as a constant.

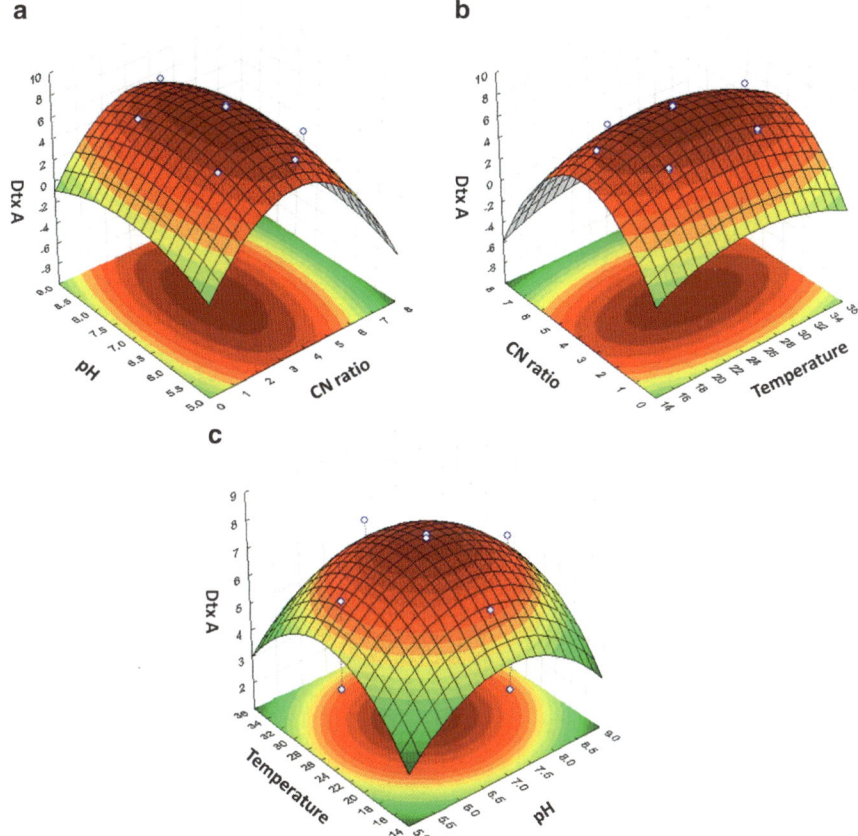

Fig. 9.1 Response surface plots indicating the effect of (**a**) CN ratio and pH, (**b**) temperature and CN ratio, and (**c**) pH and temperature on the recovery of dtx A from the culture filtrates of *Metarhizium anisopliae*

The results of ANOVA are given in Table 9.3. The regression equations obtained after the ANOVA gave the level of dtx A and E recovery as a function of the initial values of CN ratio, pH, and temperature.

The critical optimized conditions for dtx A recovery were CN ratio, 3.48; pH, 0.01; and temperature, 25.18 °C and for dtx E recovery were CN ratio, 3.46; pH, 6.99; and temperature, 25.14 °C. In both Figs. 9.1 and 9.2, the highest contour corresponded to the dtx A and E recovered under the optimum values of CN ratio, pH, and temperature. At the optimum values for dtx A, 8.91 mg/l of dtx A was recovered and at that of dtx E, 8.1 mg/l of dtx E was recovered. These values are much higher than the different combinations designed (Table 9.2). Dtx A being the most predominant form of dtx in the culture filtrate, the medium, and culture conditions optimized for dtx A recovery was used for further experimentation.

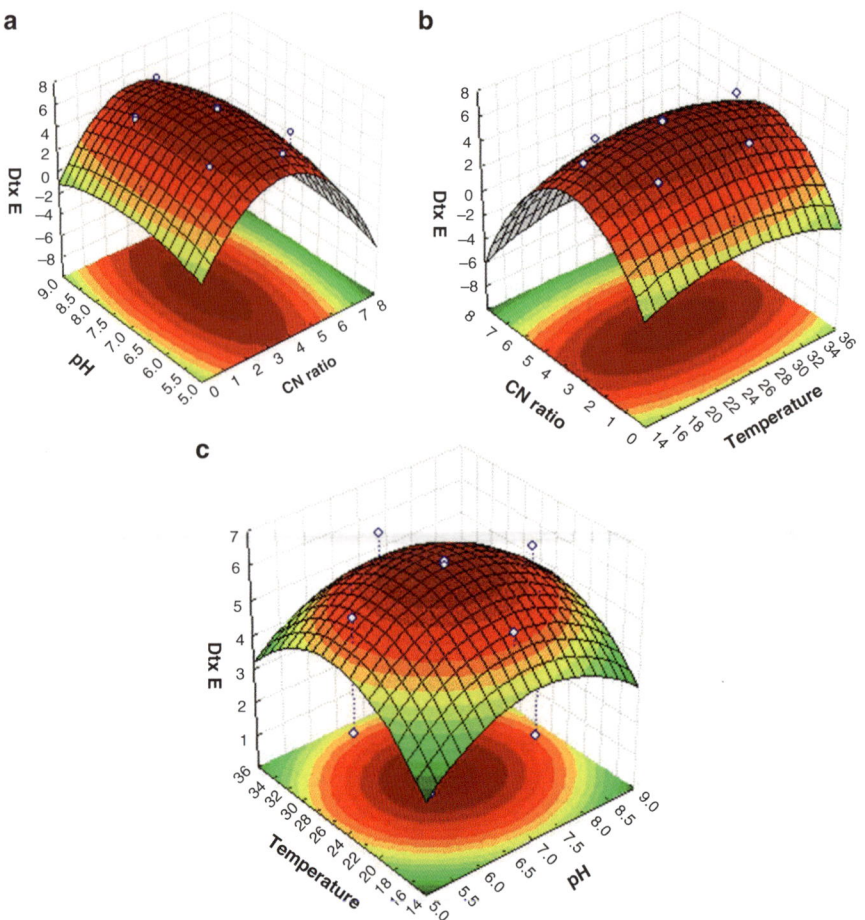

Fig. 9.2 Response surface plots indicating the effect of (**a**) CN ratio and pH, (**b**) temperature and CN ratio, and (**c**) pH and temperature on the recovery of dtx E from the culture filtrates of *Metarhizium anisopliae*

Table 9.3 ANOVA for the CCD of dtx A and E recovery

Dtx	Source	Sum of square	Degrees of freedom	Mean square	F-value	P-value
Dtx A	Model	86.10	9	86.10	39.69	0.05
	Error	1.73	8	0.22		
	Total	96.24	17			
Dtx E	Model	71.14	9	71.14	32.7	0.05
	Error	1.74	8	0.21		
	Total	84.82	17			

Dtx A: $R^2 = 0.819$; R^2 Adj $= 0.616$
Dtx E: $R^2 = 0.794$; R^2 Adj $= 0.564$

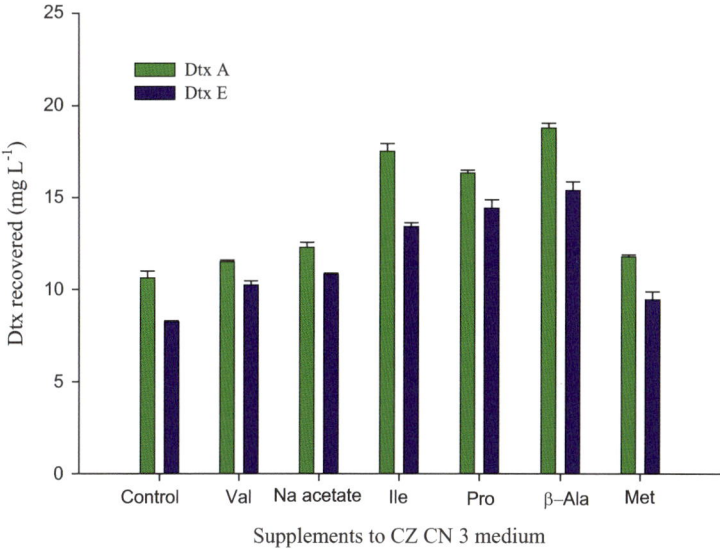

Fig. 9.3 Influence of amino acid and hydroxyl acid supplements on recovery of dtx A and E from *M. anisopliae* grown in optimized CZ culture medium and conditions (CN ratio 3.48, pH 7.01, temperature 25.18 °C). The error bar corresponds to the standard error

9.3.2 Destruxin Biosynthesis Precursors as Supplements to the Optimized Medium

The dtx biosynthesis precursors used in the present study significantly enhanced the production and recovery of both dtx A and E. The maximum recovery of dtx A and E were from the optimized medium supplemented with β-alanine. These values (dtx A, 18.9 mg/l, and dtx E, 15.3 mg/l) were twofold higher than the controls which were grown on unsupplemented CZ (CN ratio 3.48) medium. The addition of isoleucine or proline to the optimized medium also showed better values compared to controls. On the other hand, methionine supplementation recorded least influence on dtx A and E recovery (Fig. 9.3).

9.3.3 Biochemical Profile of M-19 Strain Grown on Supplemented Medium

The activity of acid phosphatase, alkaline phosphatase, and esterase in the fungal pellets derived from the media supplemented with different amino acids showed significant variation. Activity of these enzymes in the fungal pellets was directly proportional to the dtx recovery from the respective culture filtrates. The activity of

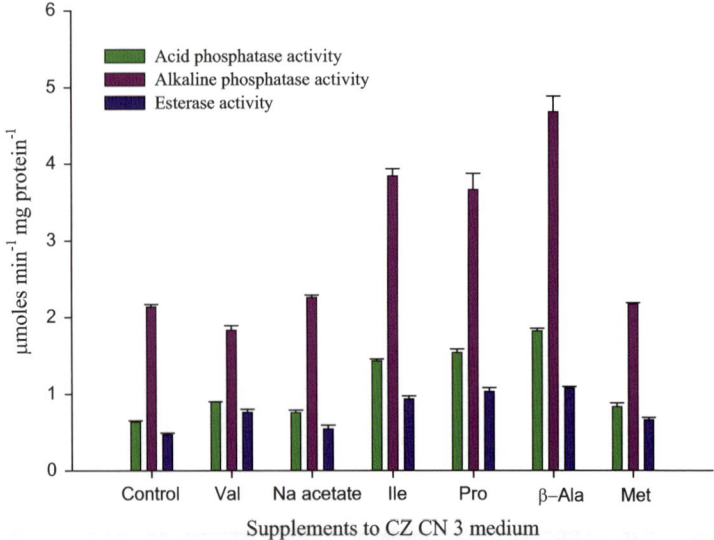

Fig. 9.4 Activity of acid phosphatase, alkaline phosphatase, and esterase enzymes in the pellets of *M. anisopliae* grown on amino acid and hydroxyl acid supplemented optimized CZ culture medium and conditions (CN ratio 3.48, pH 7.01, temperature 25.18 °C). The error bar corresponds to the standard error

all the three enzymes was maximum in the fungal pellets grown in optimized medium supplemented with β-alanine (Fig. 9.4).

9.4 Discussion

The RSM used in the current investigation revealed the optimized levels of the three independent variables—CN ratio, pH, and temperature—at which the recovery of dtx A and E was appreciably high. 81.9 % similarity in case of dtx A and 79.4 % similarity in case of dtx E were observed between the predicted and the observed experimental results from the R^2 values which reflected the accuracy and applicability of RSM to optimize the process of dtx A and E recovery.

A significant positive correlation between the recovery of dtx A and E and the culture conditions has been shown in the present study. This is a useful parameter as detection of one dtx is enough to screen for the dtx productive strains. This result is in agreement with that of Hu et al. (2006) with respect to recovery of dtx A and B. This suggests that the encoded genes for dtx biosynthesis are clustered in the genome of *M. anisopliae* (Wang et al. 2012) and are upregulated at the same level (Wang et al. 2004). This concept of toxin gene clustering has also been explained with respect to other plant toxins (Seo et al. 2001).

The addition of the five constituent amino acids and the hydroxyl acid, sodium acetate, which are the building blocks of this cyclic depsipeptide increased dtx production in the optimized culture medium and conditions. A maximum of twofold increase over the control upon supplementation with β-alanine is in accordance with its reported major role in the ring formation step of the dtx biosynthesis process (Pedras et al. 2002). Apart from this, Pro-Ile amide bond formation in the requisite linear hexadepsipeptide precursor for cyclization is also an important step in dtx biosynthesis (Pedras et al. 2002). As reported by Jegorov et al. (1993), the cyclic structure of dtx is responsible for its biological activity, and this depsipeptide is inactivated upon ring opening. Hence, the addition of β-alanine, proline, and isoleucine showed a significant positive influence on recovery of the cyclodepsipeptides, dtx A and E.

As reported in our earlier studies, increase in biomass of the M-19 strain of *M. anisopliae* showed an increment in dtx A and E production (Sree et al. 2008), suggesting positive relationship between the overall metabolism of the fungus and dtx recovery. The enzymes, viz., acid phosphatase, alkaline phosphatase, and esterase, are involved actively in cell cycle and cell differentiation which are the major processes involved in the increase in biomass of an organism (Wyckoff 1987; Vincent et al. 1992). As these enzymes play a key role in the metabolism of a cell, the increase in the activity of these enzymes suggested a corresponding increment in metabolism of the fungus which in turn augmented the production of dtx A and E. This hike was maximum with respect to β-alanine supplementation, which is involved in the key cyclization step of dtx biosynthesis.

Findings of the present study on the dtx A and E recovery from the cultures of *M. anisopliae* by manipulating substrate profile and production strategies would be of use in mycopesticide development for effective pest management.

Acknowledgments The authors acknowledge Dr. Richard A. Humber, USDA-ARS, ARSEF, Ithaca, for providing the isolate of *M. anisopliae* (Metch.) and Prof. YLN Murthy, School of Chemistry, Andhra University, for his help in the HPLC work.

References

Chen HC (1996) Optimizing the concentrations of carbon, nitrogen and phosphorus in a citric acid fermentation with response surface method. Food Biotechnol 10:13–27

Hu QB, Ren SX, Wu JH, Chang JM, Musa PD (2006) Investigation of destruxin A and B from 80 *Metarhizium strains* in China and the optimization of culture conditions for the strain MaQ10. Toxicon 48:491–498

Jegorov A, Sedmera P, Matha V (1993) Biosynthesis of destruxins. Phytochemistry 33:1403–1405

Kershaw MJ, Moorhouse ER, Bateman RP, Reynolds SE, Charnley AK (1999) The role of destruxins in the pathogenicity of *Metarhizium anisopliae* for three species of insect. J Invertebr Pathol 74:213–223

Liu BL, Chen JW, Tzeng YM (2000) Production of cyclodepsipeptides destruxin A and B from *Metarhizium anisopliae*. Biotechnol Prog 16:993–999

Lowry OH, Roberts NR, Wu MR, Hixon WS, Crawford EJ (1954) The quantitative histochemistry of brain. II. Enzyme measurements. J Biol Chem 207:19–37

Pal S, St. Leger RJ, Wu LP (2007) Fungal peptide destruxin A plays a specific role in suppressing the innate immune response in *Drosophila melanogaster*. J Biol Chem 282:8969–8977

Pedras MSC, Irina ZL, Ward DE (2002) The destruxins: synthesis, biosynthesis, biotransformation and biological activity. Phytochemistry 59:579–596

Prakash GVSB, Padmaja V, Kiran RRS (2008) Statistical optimization of process variables for the large-scale production of *Metarhizium anisopliae* conidiospores in solid-state fermentation. Bioresour Technol 99:1530–1537

Rao PV, Jayaraman K, Lakshmanan CM (1993) Production of lipase by *Candida rugosa* in solid state fermentation. 2: Medium optimization and effect of aeration. Process Biochem 28:391–395

Samuels RI, Reynolds SE, Charnley AK (1988) Calcium channel activation of insect muscle by destruxins, insecticidal compounds produced by the entomopathogenic fungus *Metarhizium anisopliae*. Comp Biochem Physiol 90:403–412

Seo JA, Proctor RH, Plattner RD (2001) Characterization of four clustered and coregulated genes associated with fumonisin biosynthesis in *Fusarium verticillioides*. Fungal Genet Biol 34: 155–165

Sparks TC, Willis WS, Shorey HH, Hammock BD (1979) Hemolymph juvenile hormone esterase activity in synchronous last instar larvae of the cabbage looper, *Trichoplusia ni*. J Insect Physiol 25:125–132

Sree KS, Padmaja V (2008a) Destruxin from *Metarhizium anisopliae* induces oxidative stress effecting larval mortality of the polyphagous pest *Spodoptera litura*. J Appl Entomol 132: 68–78

Sree KS, Padmaja V (2008b) Oxidative stress induced by destruxin from *Metarhizium anisopliae* (Metch.) involves changes in glutathione and ascorbate metabolism and instigates ultrastructural changes in the salivary glands of *Spodoptera litura* (Fab.) larvae. Toxicon 51:1140–1150

Sree KS, Padmaja V, Murthy YLN (2008) Insecticidal activity of destruxin, a mycotoxin from *Metarhizium anisopliae* (Hypocreales) strains against *Spodoptera litura* (Lepidoptera: Noctuidae) larval stages. Pest Manag Sci 64:119–125

Vincent JB, Crowder MW, Averill BA (1992) Hydrolysis of phosphate monoesters: a biological problem with multiple chemical solutions. Trends Biochem Sci 17:105–110

Wang CS, Skrobek A, Butt TM (2003) Concurrence of losing a chromosome and the ability to produce destruxins in a mutant of *Metarhizium anisopliae*. FEMS Microbiol Lett 226:373–378

Wang CS, Skrobek A, Butt TM (2004) Investigations on the destruxin production of the entomopathogenic fungus *Metarhizium anisopliae*. J Invertebr Pathol 85:168–174

Wang B, Kang Q, Lu Y, Bai L, Wang C (2012) Unveiling the biosynthetic puzzle of destruxins in *Metarhizium* species. Proc Natl Acad Sci USA 109:1287–1292

Wyckoff HW (1987) Structure of *Escherichia coli* alkaline phosphatase determined by X-ray diffraction. In: Torriani-Gorini A, Rothman FG, Silver S, Wright A, Yagil E (eds) Phosphatase metabolism and cellular regulation in microorganisms. America Society for Microbiology, Washington, DC, pp 118–126

Chapter 10
Beauveria bassiana: Biocontrol Beyond Lepidopteran Pests

H.B. Singh, Chetan Keswani, Shatrupa Ray, S.K. Yadav, S.P. Singh, S. Singh, and B.K. Sarma

10.1 Introduction

Beauveria bassiana, the anamorph stage of *Cordyceps bassiana*, is a facultative cosmopolitan entomopathogen with an extremely broad host range. First discovered by Agostino Bassi de Lodi (Keswani et al. 2013) in larval silkworms, the fungus grows as a white (hyaline) mold producing single-celled, haploid, and hydrophobic conidia. RNA-sequence transcriptomic studies have revealed the startling ability of this fungus to adapt to varied environmental niches including survival and interactions outside the insect host (Xiao et al. 2012). Thus, the ecological habitat of this entomopathogen extends from the simple insect–host interaction to a broader perspective including plant rhizosphere with a well-equipped growth-promotion attribute (Bruck 2010). A diverse array of plant groups, including the agronomic, medicinal, and cash crops, serve as a host for the endophytic form of this fungus (Ownley et al. 2008; Gurulingappa et al. 2010).

Microbial control of plant pathogens and insect pests not only reduces the dependence on chemical pesticides but also increases sustainability of agriculture. Although the number of registered microbial products has increased in recent years, many potential biopesticides either have not been developed for

H.B. Singh (✉) • B.K. Sarma
Department of Mycology and Plant Pathology, Institute of Agricultural Sciences, Banaras Hindu University, Varanasi 221005, India
e-mail: hbs1@rediffmail.com

C. Keswani • S.P. Singh
Faculty of Science, Department of Biochemistry, Banaras Hindu University, Varanasi 221005, India

S. Ray • S.K. Yadav • S. Singh
Faculty of Science, Department of Botany, Banaras Hindu University, Varanasi 221005, India

© Springer International Publishing Switzerland 2015   219
K.S. Sree, A. Varma (eds.), *Biocontrol of Lepidopteran Pests*, Soil Biology 43,
DOI 10.1007/978-3-319-14499-3_10

Table 10.1 Various host ranges of *B. bassiana*

Fungi	Bacteria	Pests
Rhizoctonia solani	*Xanthomonas axonopodis* pv. *malvacearum*	**(Coleoptera order)** *Lathrobium brunnipes, Calvia quattuordecimguttata, Phytodecta olivacea, Otiorhynchus sulcatus, Sitona lineatus, S. sulcifrons, S. macularius, S. hispidulus, Anthonomus pomorum, Hylastes ater*
Pythium ultimum, P. debaryanum, P. myriotylum	*Clostridium perfringens*	**(Hymenoptera order)** *Ichneumonidae, Lasius fuliginosus, Vespula* spp., *Bombus pratorum*
Sclerotinia sclerotiorum	*Listeria monocytogenes*	**(Heteroptera order)** *Picromerus bidens, Anthocoris nemorum*
Alternaria solani, A. tenuis	*Yersinia enterocolitica*	**(Homoptera order)** *Eulecanium* spp.
Colletotrichum gloeosporioides	*Salmonella enterica*	**(Diptera order)** *Leria serrata*
Fusarium oxysporum, F. moniliforme, F. graminearum, F. avenaceum	*Shigella dysenteriae*	**(Lepidoptera order)** *Hepialus* spp., *Hypocrita jacobaea, Cydia nigricana*
Aspergillus niger, A. parasiticus	*Staphylococcus aureus*	
Septoria nodorum	*Enterococcus faecium*	

commercial use or have had limited success due to their pathogen or pest specificity. Moreover, its inconsistent performance across environments, or a lack of understanding of the mechanism(s) of biocontrol, results into ineffective use. Hypocrean fungi, such as *B. bassiana*, offer a quick respite to this problem as they become established as epiphytes and endophytes, thereby enhancing the induced resistance of the plant or direct disease suppression by antibiosis, competition, or mycoparasitism (Ownley et al. 2010). The detailed host range of *B. bassiana* has been shown in Table 10.1.

B. bassiana has been included in a schedule as an amendment in the Insecticide Act, 1968, for commercial production as biopesticide and published in the Gazette of India dated 26th March 1999 (Keswani et al. 2013). However, its availability is still limited to some selected states in our country (Singh 2013). The present chapter focuses on the crucial attributes of *B. bassiana* that are responsible for its biocontrol activity. In addition, certain strain improvement techniques have also been taken under consideration that would augment the in situ action of the fungus. Besides, the heavy-metal remediation by *B. bassiana* has also been discussed.

10.1.1 Entomopathogenic vs. Endophytic Nature of *B. bassiana*

B. bassiana is the most appreciated endophytic fungal entomopathogen to date, with a widespread commercial availability as a potent mycopesticide. The fungus establishes itself as an endophyte either naturally, e.g., by stomatal penetration, or with the aid of inoculation methods such as soil drenches, seed coatings and immersions, radicle dressings, root and rhizome immersions, stem injection, and foliar and flower sprays. Hence, it is widely acknowledged as a success in a variety of plants such as grasses; agricultural crops, viz., tomato, cotton, corn, and potato; the medicinal group including opium poppy, cocoa, and coffee; and trees such as *Carpinus caroliniana* and western white pine (Vega et al. 2008; Ownley et al. 2008). The most preferred protocol for endophytic colonization of *B. bassiana* includes the use of formulations augmented with solid inert carriers or diluents such as diatomaceous earth, talc, clay, vermiculite, corn cob grits, and alginate gels (Wagner and Lewis 2000; Parsa et al. 2013). Irrespective of the multifarious inoculation techniques, the extent of colonization depends on the type of plant part evaluated, the inoculation method used, and the initial spore density on the plant tissue (Posada et al. 2007; Ownley et al. 2008). While leaves respond best to foliar sprays, roots colonize only to drench inoculation with stems responding equally to both forms. A plausible yet untested hypothesis suggests that the extent of endophytic colonization correlates positively with the extent of endophyte-mediated resistance by the host.

The insecticidal attribute of *B. bassiana* recommends its use in biopesticide industry with particular reference to the malaria-causing mosquito vector. The spores of *B. bassiana* have a high affinity for the female *Anopheles* mosquito. However, for proper infestation, spore sufficiency and direct contact with the insect is required so that the fungus can efficiently penetrate and germinate the insect cuticle. Female *Anopheles stephensi*, the causal agent of human malaria in Asia, relishes on the dead and dying caterpillars heavily infested with *B. bassiana*, thereby letting themselves be infested in turn with this fungus. George et al. (2013) premiered a caterpillar sans innovative technique of using oil-formulated dried spores sprayed on a cloth that resulted in the fungal infestation efficiency of 95 % in the mosquitoes. However, the utility of *B. bassiana* larvicide can be good when we have regulated resistance in bacterial *Bacillus sphaericus* and chemical larvicides.

10.2 Formulation Types of *B. bassiana*

Novel isolates of *B. bassiana* have been recommended for developing an efficient bioformulation of biopesticide and mycopesticide. Primarily, three basic types of formulations have been proposed for *B. bassiana*.

10.2.1   Conidia Mixed with Metabolites

Eyal (1993) developed a novel formulation using a saprophytic isolate producing oosporin in high yield and evaluated its efficacy as biopesticide. This bioformulation comprising of the fungal conidia (the biologically active form of the fungus) treated with oosporin (produced by submerged culture) provides an effective pest control measure particularly against Aphididae, Delphacidae, Cicadellidae, Cercopidae, Aleyrodidae, Coccidae, Coleoptera, and Lepidoptera. Moreover, it also showed potent application against mealybugs, spider mites, and other foliar insects. The mode of application involves prilling of the mycelia as it assures the retention of the biological activity of the product until application. At the time of application, the dried prill is reactivated post-wetting and further used for treating soil, seed, root, or plant (Eyal et al. 1994). In addition, the amendment with food attractants or vegetable oils rich in oleic, linoleic, and linolenic acids in bioformulation has proved a robust application in stimulation of necrophagy (Jackson et al. 2010).

10.2.2   Extracted Protein

An alternative mode of application suggests the use of finely emulated beauverial protein extract, weighing about 5 kDa, either as granule, wettable powder, or dust combined with inert materials, such as inorganic or botanical, or in liquid form such as aerosol, foam, gel, suspension, or emulsifiable concentrates. The suggested protein content in the bioformulation ranges from 1 to 95 % of the total weight of the pesticide in dry form, while the liquid formulation consists of 1–60 % of the total weight of solids in liquid phase (Leckie and Stewart 2006).

10.2.3   Endophytic Beauveria

Another novel isolate displaying endophytic colonization, reported by Vidal and Tefera (2011), is much cashed upon for commercialization purpose. The said isolate not only enjoys a broad prey range, including root weevils, wireworms, maggots, bugs, aphids, beetles, soil grubs, root maggots, termites, and ants, but also inhabits a diverse range of hosts, such as the vegetable crops and the cash crops. The added advantage of using endophytic isolates is that they posses tolerance against environmental stresses, such as UV, high temperature, rainfall, etc. The fungal strain grows along with the host, and its additional entomopathogenic quality renders a substantial lifelong protection to the host plants, particularly to agriculture crops. The mode of incorporation involves dispersion, spray, gel, emulsion, layer, cream, coating, dip, encapsulation, or granule. Moreover, they may also be incorporated as conventional microparticles or microcapsules.

10.3 Bioremediation of Heavy Metals: Potential Capacity of *B. bassiana* in the Biosorption of Heavy Metals

Heavy metals render a cumulative harm to the living system by denaturation of enzymes and proteins, production of ROS, displacement of essential metal ions from biomolecules leading to conformational changes, and damage of membrane integrity. Fungi are comparatively more tolerant to heavy-metal stress and thrive well in the acidified environment owing to their adaptation of several strategies of rendering the metal ions into innocuous forms. Hence, their use in bioremediation of industrial effluents and wastewaters containing heavy metals has been reported globally (Rajapaksha et al. 2004). The main principle for bioremediation is biosorption that primarily utilizes amino, carboxyl, hydroxyl, and carbonyl groups of the cell wall for metal binding as elucidated by FTIR analysis. A plethora of fungal species, such as *Aspergillus niger*, *Mucor rouxii*, *Rhizopus* spp., etc., have been reported to be utilized in heavy-metal remediation. Besides this, *B. bassiana* is reported to efficiently adsorb Cd and Pb from aqueous metal solutions. Physicochemical factors like solution pH and contact time at room temperature positively affected the rate of metal biosorption (Tomko et al. 2006).

Another strategy of heavy-metal uptake by fungal species involves immobilization followed by precipitation. Fungal metal leaching is promoted by proton efflux or metabolites with chelating properties. Metal immobilization can also occur through other processes of reductive metal precipitation such as synthesis of metallic nanoparticles that suggests its novel applications in the corrosion control of metal and stone artifacts. *B. bassiana* is reported to be endowed with the ability of heavy-metal chelation using oxalates as organic chelators. The beauverial oxalate reportedly chelates a variety of metals, involving iron (iron oxalates were directly sequestered on the fungal hyphae), copper (formation of characteristic "Liesegang rings" due to simultaneous diffusion and precipitation of copper oxalates), and silver (coprecipitation of copper and silver oxalates) (Joseph et al. 2012).

10.4 Secondary Metabolites of *B. bassiana*: A Boon in Disguise

B. bassiana produces a plethora of secondary metabolites having multifarious roles not only in pest control but in other human benefits too (Fig. 10.1). Some of the essential secondary metabolites of this fungus have been discussed below.

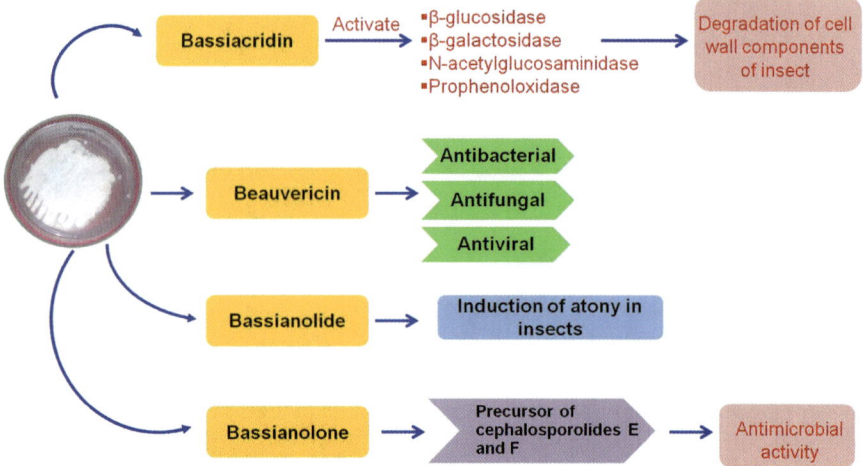

Fig. 10.1 Multifarious roles of various secondary metabolites of *B. bassiana*

10.4.1 Bassiacridin

B. bassiana is reported to secrete a protein, bassiacridin, exhibiting toxicity for locusts and comprising of about 0.1–0.3 % of the total protein content of the crude extract. The protein exhibits a monomeric structure with a molecular weight of 60 kDa and an isoelectric point of 9.5. It shows an amino acid sequence homology to the yeast chitin-binding protein, but its chitin-binding properties have not yet been determined. Being neutral for ion exchangers, bassiacridin actively participates in β-glucosidase, β-galactosidase, and *N*-acetylglucosa-minidase activities. Fourth-instar nymphs of *Locusta migratoria* injected with the protein even at low dosage expressed a mortality rate of about 50 %. The root cause of mortality as revealed by the ultrastructural studies showed melanin pigmentation in the trachea, air sacs and a formation of melanized nodules on the fat bodies leading to an alteration of the epithelial cell structure of the trachea, air-bags, and integuments, thereby turning them necrotic. However, these struc-tural changes were not reported to be caused by hemolymph pigmentation with the simultaneous production of toxic quinones. The pigmentation activity was carried out by the enzyme prophenoloxidase, activated by bassiacridin. Being novel, research needs to be targeted on the understanding of the mechanism of interaction of bassiacridin with host cells and its impact on the environment so as to include the protein as an active ingredient of biopesticides (Quesada-Moraga and Vey 2004).

10.4.2 Beauvericin

First discovered by Hamill et al. (1969), beauvericin is a cyclo-oligomeric hexadepsipeptidic ionophore produced by *B. bassiana* with a widely acknowledged insecticidal activity. The peptide binds to monovalent cations and facilitates their transport across the membrane, thereby uncoupling oxidative phosphorylation (Xu et al. 2008). Irrespective of its strong insecticidal potential, the preferential use of a potent beauvericin-producing strain as a commercial insecticide is far more advantageous than the pure compound primarily because its biosynthesis coincides with infection. During the pathogenesis phase, the developing hyphae of *B. bassiana* release extracellular hydrolytic enzymes aiding the fungal penetration of insect integument (Fan et al. 2007) and virulence factors that disables and crumbles the host immune system leading to its death. The activity of pure beauvericin in entomopathogenesis is still controversial as it was reported to be well tolerated by *Helicoverpa zea*.

Besides its role as an active insecticide, beauvericin displays a wide range of in vitro biological activities, such as antibiotic, antiviral, and cytotoxic, and an augmented antifungal activity when annexed with other antifungal agents in consortia, as mentioned below.

10.4.2.1 Antitumor Activity

The cytotoxicity of beauvericin to a human leukemia cell has been frequently reported. The peptide activates a calcium-sensitive cell apoptotic pathway. Beauvericin induces Ca^{2+} ion transport from the extracellular environment to the cytosol resulting in an augmentation of cytosolic Ca^{2+} concentration, thereby triggering the onset of an "unknown signal system" with the subsequent release of cyt c from mitochondria. The cyt c further activates the caspase inducing apoptosis (Wang and Xu 2012). Beauvericin also inhibits directional cell motility (haptotaxis) of cancer cells at sub-cytotoxic concentrations (Zhan et al. 2007)

10.4.2.2 Antibacterial Activity

Beauvericin displays a strong antibacterial activity against human, animal, and plant pathogenic bacteria with no discrimination between Gram-positive and Gram-negative bacteria. The target of action of beauvericin, contrary to other fungal antibiotics, does not involve the peptidoglycan cell wall biosynthesis, but the cell organelles and enzyme system of the bacteria. Based on its broad-spectrum antibacterial activity, beauvericin could be used to figure out the solution to drug resistance, deadly bacterial infections, and nonfood crop disease (Wang et al. 2012).

10.4.2.3 Antifungal Activity

Beauvericin, being a fungal metabolite, displays no activity as an antifungal agent. However, when used in combination with another compound, a whole new way of development and utilization of the biological activity of beauvericin is unraveled, e.g., beauvericin with ketoconazole in combination acts against *Candida parapsilosis*, one of the major culprits of neonatal mortality, while both the compounds, when applied singly, have little or no effect on this deadly fungus (Zhang et al. 2007).

10.4.2.4 Antiviral Activity

The antiviral activity of beauvericin has also been detected. Shin et al. (2009) proposed beauvericin as the most effective inhibitor of cyclic hexadepsipeptides that inhibit HIV-1 integrase. However, an effective focus on the antiviral potential of beauvericin is required for the exploration of its activity against the more fatal and epidemic disease-causing viruses, such as HBV, SARS, H1N1, and AIV (Wang et al. 2012).

10.4.3 Bassianolide

Another secondary metabolite of *B. bassiana* with insecticidal property is bassianolide, a cyclo-oligomer depsipeptide (Xu et al. 2009). Comparative sequence analysis of bassianolide synthetase (BbBSLS) suggested the presence of catalytic domains intermediate from D-2-hydroxyisovalerate and L-leucine, responsible for the iterative synthesis of dipeptidol monomer, D-hydroxyisovalerate (D-Hiv) *N*-methyl-L-leucine (*N*-Me-Leu) bound to the enzyme. Post synthesis, these domains catalyze the condensation of the monomers to form the octadepsipeptide with a 24-membered macrolactone ring. This structure further isomerizes to the final cyclic tetrameric ester form (Suzuki et al. 1977). Comparative in vitro infection studies against selected insect host establish bassianolide as a highly significant virulence factor of *B. bassiana*, inducing atony to the larvae of *H. zea* and toxicity to silkworm larvae. Knockout mutants with a targeted disruption of *bbBsls* gene were unable to produce bassianolide and showed a drastic reduction in virulence against the model insects *Galleria mellonella* and *H. zea*. However, *B. bassiana* orchestrates a definite strategy for carrying out the killing operation of insect larvae in which the metabolite production coincides with the host infection. Thus, bassianolide represents a bona fide virulence factor of the entomopathogen significantly contributing to the commercial biological insecticide preparation containing the spores of the fungus. Besides being a potent insecticide, purified bassianolide inhibits the acetylcholine-induced smooth muscle contraction, as well as displays in vitro cytotoxic, moderate antiplasmodial and antimycobacterial activities (Jirakkakul et al. 2008).

10.4.4 Bassianolone

Ackland et al. (1985) isolated two rare metabolites, cephalosporolides E and F, as a constituent of the fermentation products of the fungus, *Cephalosporium aphidicola*, growing under sulfur-limiting conditions. Soon after this serendipitous discovery, the same process could not be mimicked in laboratory conditions nor could these metabolites be isolated from elsewhere in nature. However, Oller-López et al. (2005) unexpectedly found cephalosporolides E and F together with bassianolone, an antimicrobial precursor of cephalosporolides E and F from *B. bassiana* growing under low-nitrogen conditions. Though the in vitro antimicrobial activity of cephalosporolides E and F was completely negative, bassianolone contradicted them by displaying a high antimicrobial activity with a complete inhibition of the growth of *Staphylococcus aureus* and *Candida albicans*.

10.5 Internal Machinery that Renders the Fungal Bioformulation a Success

10.5.1 Superoxide Dismutase

The basic criteria that determine the credibility of success of a particular fungal formulation are its virulence and its tolerance to environmental stresses that drastically affect the field efficacy and persistence of fungal sprays. Stress, in any form, be it UV, heat, or drought, induces the production of reactive oxygen species (ROS) which damages the essential biomolecules, like DNA, proteins, and lipids. Fungal enzymes, particularly superoxide dismutases (SODs) capable of scavenging the ROS, establish their virulence and field persistence. SODs are superoxide species scavenging metalloproteins that offer the first line of defense against superoxide damage. The SODs function in cell response, cell differentiation and infection yielded the benefit of mycelial growth, ordered conidiation rhythm, and increased sporulation potential and timely conidial germination in fungi. Fungal SODs are Cu/Zn factored and existent mainly in cytosol or mitochondria (Mn). Mn-SOD has proven to participate in the survival and resistance of aerobic fungi to multiple stresses. A new Mn-SOD has been identified from *B. bassiana* that offers multifaceted attributes to the entomopathogen, including amplification of the SOD activity by 4–10-fold, virulence to *Spodoptera litura*, and tolerance to chemical oxidation and UV-B irradiation (Xie et al. 2010).

Two isomeric forms of BbSOD, BbSOD2 and BbSOD3, characterized as cytosolic and mitochondrial isoenzymes, dominated the total SOD activity in *B. bassiana* under normal growth conditions. The biocontrol and stress retention potential of the isoenzymes was determined by constructing knockout (DBbSOD2 and DBbSOD3) and RNAi mutants with a 91–97 % suppression of the BbSOD2 and BbSOD3 transcripts. The constructs were used in conjunction with DBbSOD2/BbSOD2, DBbSOD3/BbSOD3, and wild type. Both the mutant types displayed marked phenotypic alterations, such as delayed sporulation, reduced conidial yield, and impaired conidial quality, but little change in colony morphology. The mycelia and conidia expressed greater sensitivity to menadione- or H_2O_2-induced oxidative stress but little or no response to hyperosmolarity or elevated temperature, in contrast to the yeast mitochondrial Mn-SOD which protected the yeast cells from osmotic, oxidative, and thermal stress. While disruption of either of the genes resulted in reduced conidial yield and delayed germination, the double-gene-silenced mutants expressed defective sporulation, increased sensitivity to menadione and H_2O_2, a longer delay in conidial germination, and a decreased virulence, due to a greater loss of the antioxidative capability leading to an increased level of intracellular superoxide species. This fact suggests the additive influence exertion of the two Mn-SODs mediating fungal development, antioxidative capacity, UV resistance, and virulence. However, the increased tolerance of mycelia and conidia of DBbSOD2 to the two oxidants than DBbSOD3 suggested the greater contribution of the mitochondrial BbSOD3 to the fungal antioxidative capability.

10.5.2 Hydrophobins

The entomopathogenic *B. bassiana*, existing both as a saprobe and endophyte, comprises of several infectious propagules, such as conidia (aerial and submerged), in vitro unicellular blastospores, and in vivo insect hemolymph derived cells, hyphal bodies. The process of pathogenesis commences with the adhesive interaction between the conidial surface and the insect epicuticle by nonspecific hydrophobic interactions, mediated by the spore-coat hydrophobins. *B. bassiana* comprises two genes, *hyd1* and *hyd2*, termed as the hydrophobin genes, responsible for cell-surface hydrophobicity, adhesion, virulence, spore thermotolerance, and formation of the rodlet layer, a protective spore-coat structure. Bioinformatic and phylogenetic studies classified both the proteins as class I hydrophobins with no primary sequence homology between them. N-terminal amino acid sequencing of a rodlet layer protein suggested Hyd2 as a spore-coat component. Deactivation of *hyd1*, i.e., *dhyd1*, displayed characteristic phenotypic alterations such as apparently "bald" conidia with modified surface fascicles or bundles. The spores were hydrophobic with different surface carbohydrate epitopes and β-1,3 glucan distribution. Further, their mode of dispersal was sans water, and their virulence was also lowered, but adhesion was not affected. On the contrary, adhesion was lowered in *Dhyd2*, but the mutants were unaffected in virulence. The double-gene-silenced mutants expressed a lack of bundles and rodlets and a drastically lower surface hydrophobicity, cell attachment, and virulence (Ying and Feng 2004; Zhang et al. 2011).

Apart from being a spore surface component, hydrophobins play a characteristic role in the physiology of the organism. The hydrophobin acts as a regulator that senses the environmental conditions, and if found suitable, the spore is allowed to germinate. Thus, at an elevated temperature, hyphal growth would be compromised, and conidial germination inhibition would be adaptive. The above explanation was derived from a distinctly impaired adhesion phenotype in single- and double-gene knockout mutants. *Dhyd1* mutant displayed a drastic reduction in virulence when topically assayed against *G. mellonella* larvae as compared to *Dhyd2* mutants though the latter one showed a prominent reduction in adhesion. The intrigue could be explained by the probable clumping of the *Dhyd1* conidia resulting in a decreased germination and virulence. In conclusion, *hyd1* and *hyd2* encode class I hydrophobins that are essential for spore-coat rodlet layer and fascicle formation. Besides, *Hyd1* forms the rodlet with *Hyd2* acting as an organizing partner (Zhang et al. 2011).

10.5.3 Primary Roles of Two Dehydrogenases in the Mannitol Metabolism and Multistress Tolerance of B. bassiana

Mannitol metabolism in *B. bassiana* corresponds with multistress tolerance and virulence. This attribute is of prime importance where a constant challenge from the various stress factors, such as UV and high temperature, is faced and, particularly in the case of the fungal bioformulation, exposed to the environment. *B. bassiana* hosts two enzymes, namely, mannitol-1-phosphate dehydrogenase (MPD) for reducing fructose-6-phosphate to mannitol-1-phosphate and mannitol dehydrogenase (MTD) for oxidizing mannitol to fructose. Intracellular mannitol accumulation relates to the increased conidial tolerance to various stresses. This fact was evidenced in single knockout mutants DBbMPD and DBbMTD where a drastic abatement in the mannitol content resulted in a steep decline in conidial tolerance to the various stresses as well as virulence ability as observed against *S. litura* larvae. Also, mannitol-supplemented nitrate-based minimal medium suppressed the colony growth and conidial germination of DBbMTD at a larger level as compared to DBbMPD (Wang et al. 2012). However, mannitol content decline was supported by trehalose accumulation. Trehalose accumulation is regulated by the enzyme trehalase. Trehalose acts as a carbohydrate store and offers a temporary respite during the stress conditions (Liu et al. 2009). However, owing to its compartmentalization, trehalose cannot replace mannitol to increase conidial thermotolerance. Conclusively, BbMPD and BbMTD represent the flag bearers of mannitol metabolism in *B. bassiana* contributing to conidial thermotolerance, UV-B resistance, and virulence.

10.5.4 MAP Kinase Activity

MAP kinases are widely described in eukaryotes, including fungi, and are known to play essential roles in the transduction of extracellular environmental signals, regulating development and differentiation process. Filamentous fungi harbor basically three classes of MAP kinases, Fus3/Kss1, Hog 1, and Slt2, with orthologs of each of these kinases being present in *B. bassiana*. Fus3/Kss1 MAPK deals with the regulation of infection-related development or process leading to the penetration of host tissues. The Hog1 MAPK is known to be involved in the mediation of virulence and responses against various environmental stresses (osmotic, oxidative, and thermal) and fungicides. Slt2 MAPK is likely to be crucial for the fungal survival in the environment. Bbslt2 controls growth, conidiation, cell wall integrity, response to oxidative and heat stress, heterokaryon formation, secondary metabolic pathways, and virulence in the entomopathogenic fungus. The above attributes were antipodally expressed in the knockout mutant of DBbslt2, as reduction in conidial production and viability, temperature-dependent chitin accumulation but with a simultaneous chitinal sensitivity to Congo red and fungal cell wall-degrading enzymes, and decreased conidial and hyphal hydrophobicity without alteration in the hydrophobin-encoding genes, hyd 1 and hyd 2. Besides, the cell-surface carbohydrate epitopes also expressed an alteration with a change in the content of acid-soluble, alkali-insoluble, and β-glucans as well as an attenuation in virulence as observed on topical and intra-hemocoel application. The content of trehalose was also reported to be elevated in the mutated strains as a stabilizing factor during stress conditions (Zhang et al. 2009; Luo et al. 2012).

10.5.5 Role of G-Protein Signaling (BbRGS1) in Conidiation and Conidial Thermotolerance of *B. bassiana*

The chief virulence attribute of the entomopathogenic fungus involves conidial germination leading to pathogenesis and disease transmission. As in other fungi, conidiation and germination in *B. bassiana* is regulated by regulatory G-protein signaling RGS protein encoded by *Bbrgs1*. RGS functions in a G-protein-mediated signaling pathway that responds to environmental signals by regulating the switches between vegetative growth and conidiation. Activation of the gene results in the initiation of vegetative growth and termination in conidiation, while gene repression accelerates conidiation. However, the knockout mutant exhibiting the complete loss of *Bbrgs1* gene only reduced conidial production in contrast to other fungi suggesting an alternative mechanism or other RGS proteins in addition to BbRGS1 controlling conidiation in *B. bassiana* (Fang et al. 2008). Apart from conidiation, *Bbrgs1* is actively involved in toxin synthesis, pigment production, stress tolerance, and thermotolerance. The ability to control conidiation and thermotolerance in insect pathogenic fungus as well as the ability to control the dissemination of genetically modified strains in field application has an important implication for the mass production of this biocontrol agent.

10.6 Strain Improvement in *B. bassiana*

The entomopathogenic ability of *B. bassiana* has been extensively studied and advocated in insect pest management (Roberts and St Leger 2004; Wang et al. 2004; Thomas and Read 2007). However, a major hindrance toward its commercialization and application is its slower action compared to chemical insecticides, which enables the infected insects to cause a serious damage to crops until controlled (St Leger et al. 1996). To overcome such obstacles, there have been constant efforts for the improvement of the desired strains which render their quick action and high reproducibility. Various approaches have been employed to develop enhanced *B. bassiana* strains, including the recovery of mutants after UV-light irradiation (Hegedus and Khachatourians 1995; Meirelles et al. 1997). Other reported methods involve genetic recombination (Viaud et al. 1998) and genetic constructs (Fan et al. 2007; Fang et al. 2005; Joshi et al. 1995; Sandhu et al. 2001).

Genetic engineering, involving the identification and manipulation of the virulent genes, has elevated the insecticidal efficacy and biocontrol potential of the fungus. The expression of a neurotoxin AAIT from the scorpion *Androctonus australis* and an insect cuticle-degrading protease PR1A from *Metarhizium anisopliae* has permitted a high efficiency of *B. bassiana*. When assayed against the larvae of Masson's pine caterpillar *Dendrolimus punctatus* and the wax moth *G. mellonella*, engineered strains required less spores to kill 50 % of pine caterpillars (LD_{50}) (Lu et al. 2008).

Insect cuticle is mainly composed of chitin, embedded with proteins and acts as a primary barrier against pathogen attack. *B. bassiana* produces chitinases and proteases to disintegrate insect cuticle. Two chitinases (*Bbchit1* and *Bbchit2*) have been reported in *B. bassiana* lacking a chitin-binding domain. However, hybrid chitinases were developed in which *Bbchit1* was fused to chitin-binding domains derived from plant, bacterial, or insect sources. The hybrid chitinase gene was transformed in *B. bassiana*, and the transformed strains confirmed higher levels of virulence resulting in 23 % less time to kill the targeted insects (Fang et al. 2005).

Protoplast fusion may provide as a striking method for genetic improvement of biocontrol efficacy of *B. bassiana*. Paris (1977) first described the construction of parasexual heterokaryons through a protoplast fusion between two strains of

Beauveria tenella. Till date, many intraspecific and interspecific fusions have been reported in genus *Beauveria*. The fusants obtained through protoplast fusion of *B. bassiana* with *B. sulfurescens* have been reported to possess enhanced antagonistic activity against Colorado potato beetle (*Leptinotarsa decemlineata*) and European corn borer (*Ostrinia nubilalis*) (Couteaudier et al. 1996).

T-DNA insertion mutagenesis is being used to identify and isolate the genes governing thermotolerance and osmotolerance. A pool of T-DNA inserts of *B. bassiana* have been constructed for detection of mutants deficient in thermotolerance and osmotolerance ability. Five mutants were reported which posses high conidial yield, virulence, and resistance to adverse conditions (Luo et al. 2009).

10.7 Conclusion

B. bassiana is a profusely growing saprophytic as well as endophytic fungus. However, many of its potentials remain unrealized till date. This chapter was a small attempt to enhance the knowledge of some of the beneficial attributes of *B. bassiana*. The bioformulations of this fungus will not only be realized in pathogen control but will also augment the remediation of heavy metals. These attributes of *B. bassiana* are in perfect synchronization with the environment and when optimized will lead to surplus production, thereby reducing its price and making the formulations readily available.

Acknowledgements Chetan Keswani and Shatrupa Ray are grateful to Banaras Hindu University, Varanasi, for providing the CRET-UGC fellowship.

References

Ackland MJ, Hanson JR, Hitchcock PB, Ratcliffe AH (1985) Structures of the cephalosporolides B-F, a group of C, lactones from *Cephalosporium aphidicola*. J Chem Soc Perkin Trans 1:843–847

Bruck DJ (2010) Fungal entomopathogens in the rhizosphere. In: The ecology of fungal entomopathogens. Springer, Heidelberg, pp 103–112

Couteaudier Y, Viaud M, Riba G (1996) Genetic nature, stability, and improved virulence of hybrids from protoplast fusion in *Beauveria*. Microb Ecol 32:1–10

Eyal J (1993) Novel toxin producing fungal pathogen and uses. European Patent No. 0570089A1. European Patent Office

Eyal J, Landa Z, Osborne L, Walter JF (1994) Method for production and use of pathogenic fungal preparation for pest control. U.S. Patent No. 5,360,607. U.S. Patent and Trademark Office, Washington, DC

Fan Y, Fang W, Guo S, Pei X, Zhang Y, Xiao Y, Pei Y (2007) Increased insect virulence in *Beauveria bassiana* strains overexpressing an engineered chitinase. Appl Environ Microbiol 73:295–302

Fang W, Leng B, Xiao Y, Jin K, Ma J, Fan Y, Pei Y (2005) Cloning of *Beauveria bassiana* chitinase gene Bbchit1 and its application to improve fungal strain virulence. Appl Environ Microbiol 71:363–370

Fang W, Scully LR, Zhang L, Pei Y, Bidochka MJ (2008) Implication of a regulator of G protein signalling (BbRGS1) in conidiation and conidial thermotolerance of the insect pathogenic fungus *Beauveria bassiana*. FEMS Microbiol Lett 279:146–156

George J, Jenkins NE, Blanford S, Thomas MB, Baker TC (2013) Malaria mosquitoes attracted by fatal fungus. PLoS One 8:e62632

Gurulingappa P, Sword GA, Murdoch G, McGee PA (2010) Colonization of crop plants by fungal entomopathogens and their effects on two insect pests when *in planta*. Biol Control 55:34–41

Hamill RL, Higgens CE, Boaz HE, Gorman M (1969) The structure of beauvericin, a new depsipeptide antibiotic toxic to *Artemia salina*. Tetrahedron Lett 49:4255–4258

Hegedus DD, Khachatourians GG (1995) The impact of biotechnology on hyphomycetous fungal insect biocontrol agents. Biotechnol Adv 13:455–490

Jackson MA, Dunlap CA, Jaronski ST (2010) Ecological considerations in producing and formulating fungal entomopathogens for use in insect biocontrol. In: The ecology of fungal entomopathogens. Springer, Heidelberg, pp 129–145

Jirakkakul J, Punya J, Pongpattanakitshote S, Paungmoung P, Vorapreeda N, Tachaleat A, Cheevadhanarak S (2008) Identification of the nonribosomal peptide synthetase gene responsible for bassianolide synthesis in wood-decaying fungus *Xylaria* sp. BCC1067. Microbiology 154:995–1006

Joseph E, Cario S, Simon A, Wörle M, Mazzeo R, Junier P, Job D (2012) Protection of metal artifacts with the formation of metal–oxalates complexes by *Beauveria bassiana*. Front Microbiol 2:270

Joshi L, St Leger RJ, Bidochka MJ (1995) Cloning of a cuticle-degrading protease from the entomopathogenic fungus, *Beauveria bassiana*. FEMS Microbiol Lett 125:211–217

Keswani C, Singh SP, Singh HB (2013) *Beauveria bassiana*: status, mode of action, applications and safety issues. Biotech Today 3:16–20

Leckie B, Stewart C (2006) Insecticidal compositions and methods of using the same. U.S. Patent 20070044179 A1

Liu Q, Ying SH, Feng MG, Jiang XH (2009) Physiological implication of intracellular trehalose and mannitol changes in response of entomopathogenic fungus *Beauveria bassiana* to thermal stress. Anton Leeuw Int J G 95:65–75

Lu D, Pava-Ripoll M, Li Z, Wang C (2008) Insecticidal evaluation of *Beauveria bassiana* engineered to express a scorpion neurotoxin and a cuticle degrading protease. Appl Microbiol Biotechnol 81:515–522

Luo Z, Zhang Y, Jin K, Ma J, Wang X, Pei Y (2009) Construction of *Beauveria bassiana* T-DNA insertion mutant collections and identification of thermosensitive and osmosensitive mutants. Acta Microbiol Sin 49:1301 1305

Luo X, Keyhani NO, Yu X, He Z, Luo Z, Pei Y, Zhang Y (2012) The MAP kinase Bbslt2 controls growth, conidiation, cell wall integrity, and virulence in the insect pathogenic fungus *Beauveria bassiana*. Fungal Genet Biol 49:544–555

Meirelles LDP, Boas AMV, Azevedo JL (1997) Obtention and evaluation of pathogenicity of ultra violet resistant mutants in the entomopathogenic fungus *Beauveria bassiana*. Rev Microbiol 28:121–124

Oller-López JL, Iranzo M, Mormeneo S, Oliver E, Cuerva JM, Oltra JE (2005) Bassianolone: an antimicrobial precursor of cephalosporolides E and F from the entomoparasitic fungus *Beauveria bassiana*. Org Biomol Chem 3:1172–1173

Ownley BH, Griffin MR, Klingeman WE, Gwinn KD, Moulton JK, Pereira RM (2008) *Beauveria bassiana*: endophytic colonization and plant disease control. J Invertebr Pathol 98:267–270

Ownley BH, Gwinn KD, Vega FE (2010) Endophytic fungal entomopathogens with activity against plant pathogens: ecology and evolution. In: The ecology of fungal entomopathogens. Springer, Heidelberg, pp 113–128

Paris S (1977) Heterocaryons chez *Beauveria tenella*. Mycopathologia 61:67–75

Parsa S, Ortiz V, Vega FE (2013) Establishing fungal entomopathogens as endophytes: towards endophytic biological control. J Vis Exp: JoVE 74:50360

Posada F, Aime MC, Peterson SW, Rehner SA, Vega FE (2007) Inoculation of coffee plants with the fungal entomopathogen *Beauveria bassiana* (*Ascomycota: Hypocreales*). Mycol Res 111:748–757

Quesada-Moraga E, Vey A (2004) Bassiacridin, a protein toxic for locusts secreted by the entomopathogenic fungus *Beauveria bassiana*. Mycol Res 108:441–452

Rajapaksha RMCP, Tobor-Kapłon MA, Baath E (2004) Metal toxicity affects fungal and bacterial activities in soil differently. Appl Environ Microbiol 70:2966–2973

Roberts DW, St Leger RJ (2004) *Metarhizium* spp., cosmopolitan insect-pathogenic fungi: mycological aspects. Adv Appl Microbiol 54:31–36

Sandhu SS, Kinghorn JR, Rajak RC, Unkles SE (2001) Transformation system of *Beauveria bassiana* and *Metarhizium anisopliae* using nitrate reductase gene of *Aspergillus nidulans*. Indian J Exp Biol 39:650–653

Shin CG, An DG, Song HH, Lee C (2009) Beauvericin and enniatins H, I and MK1688 are new potent inhibitors of human immunodeficiency virus type-1 integrase. J Antibiot 62:687–690

Singh HB (2013) *Beauveria bassiana* in Indian agriculture: perception, demand and promotion. In: 10th international conference of insect physiology, biochemistry and molecular biology, Nanjing, 15–19 June 2013, p 122

St Leger RJ, Joshi L, Bidochka MJ, Roberts DW (1996) Construction of an improved mycoinsecticide overexpressing a toxic protease. Proc Natl Acad Sci USA 93:6349–6354

Suzuki A, Kanaoka M, Isogai A, Tamura S, Murakoshi S, Ichinoe M (1977) Bassianolide, a new insecticidal cyclodepsipeptide from *Beauveria bassiana* and *Verticillium lecanii*. Tetrahedron Lett 18:2167–2170

Thomas MB, Read AF (2007) Can fungal biopesticides control malaria? Nat Rev Microbiol 5:377–383

Tomko J, Bačkor M, Štofko M (2006) Biosorption of heavy metals by dry fungi biomass. Acta Metallurgica Slovaca 12:447–451

Vega FE, Posada F, Catherine Aime M, Pava-Ripoll M, Infante F, Rehner SA (2008) Entomopathogenic fungal endophytes. Biol Control 46:72–82

Viaud M, Couteaudier Y, Riba G (1998) Molecular analysis of hypervirulent somatic hybrids of the entomopathogenic fungi *Beauveria bassiana* and *Beauveria sulfurescens*. Appl Environ Microbiol 64:88–93

Vidal S, Tefera T (2011) U.S. Patent Application 13/636 143

Wagner BL, Lewis LC (2000) Colonization of corn, *Zea mays*, by the entomopathogenic fungus *Beauveria bassiana*. Appl Environ Microbiol 66:3468–3473

Wang Q, Xu L (2012) Beauvericin, a bioactive compound produced by fungi: a short review. Molecules 17:2367–2377

Wang C, Fan M, Li Z, Butt TM (2004) Molecular monitoring and evaluation of the application of the insect-pathogenic fungus *Beauveria bassiana* in southeast China. J Appl Microbiol 96:861–870

Wang ZL, Lu JD, Feng MG (2012) Primary roles of two dehydrogenases in the mannitol metabolism and multi-stress tolerance of entomopathogenic fungus *Beauveria bassiana*. Environ Microbiol 14:2139–2150

Xiao G, Ying SH, Zheng P, Wang ZL, Zhang S, Xie XQ, Feng MG (2012) Genomic perspectives on the evolution of fungal entomopathogenicity in *Beauveria bassiana*. Sci Rep 2

Xie XQ, Wang J, Huang BF, Ying SH, Feng MG (2010) A new manganese superoxide dismutase identified from *Beauveria bassiana* enhances virulence and stress tolerance when overexpressed in the fungal pathogen. Appl Microbiol Biotechnol 86:1543–1553

Xu Y, Orozco R, Wijeratne EM, Gunatilaka AA, Stock SP, Molnár I (2008) Biosynthesis of the cyclooligomer depsipeptide beauvericin, a virulence factor of the entomopathogenic fungus *Beauveria bassiana*. Chem Biol 15:898–907

Xu Y, Orozco R, Kithsiri Wijeratne EM, Espinosa-Artiles P, Leslie Gunatilaka AA, Patricia Stock S, Molnár I (2009) Biosynthesis of the cyclooligomer depsipeptide bassianolide, an insecticidal virulence factor of *Beauveria bassiana*. Fungal Genet Biol 46:353–364

Ying SH, Feng MG (2004) Relationship between thermotolerance and hydrophobin-like proteins in aerial conidia of *Beauveria bassiana* and *Paecilomyces fumosoroseus* as fungal biocontrol agents. J Appl Microbiol 97:323–331

Zhan J, Burns AM, Liu MX, Faeth SH, Gunatilaka AL (2007) Search for cell motility and angiogenesis inhibitors with potential anticancer activity: beauvericin and other constituents of two endophytic strains of *Fusarium oxysporum*. J Nat Prod 70:227–232

Zhang L, Yan K, Zhang Y, Huang R, Bian J, Zheng C, Chen X (2007) High-throughput synergy screening identifies microbial metabolites as combination agents for the treatment of fungal infections. Proc Natl Acad Sci USA 104:4606–4611

Zhang Y, Zhao J, Fang W, Zhang J, Luo Z, Zhang M, Pei Y (2009) Mitogen-activated protein kinase hog1 in the entomopathogenic fungus *Beauveria bassiana* regulates environmental stress responses and virulence to insects. Appl Environ Microbiol 75:3787–3795

Zhang S, Xia YX, Kim B, Keyhani NO (2011) Two hydrophobins are involved in fungal spore coat rodlet layer assembly and each play distinct roles in surface interactions, development and pathogenesis in the entomopathogenic fungus, *Beauveria bassiana*. Mol Microbiol 80:811–826

Chapter 11
Biocontrol of Diamondback Moth, *Plutella xylostella*, with *Beauveria bassiana* and Its Metabolites

Liande Wang, Minsheng You, and Haichuan Wang

11.1 Introduction

The development of resistance to chemical insecticides and concerns over the deleterious effects of chemicals on environmental and human safety have provided a strong impetus for the development of microbial control agents for use in integrated control of insect pests. A diverse assemblage of microorganisms are currently under consideration as control agents of insects, including viruses, bacteria, protozoans, and fungi. Fungi will not be cure-alls for pest problems on all crops and in all agricultural settings, and it is unlikely that they will ever totally supplant the management of insect pests with chemical insecticides. Nevertheless, they represent a valuable management resource to be utilized within an IPM framework and will contribute significantly to reductions in chemical pesticide use (Lacey and Goettel 1995). However, the research development and final commercialization of fungal biological control agents (BCAs) continue to confront a number of obstacles, ranging from elucidating important basic biological knowledge to socioeconomic factors (Butt et al. 2001). Currently, considerable advances have been made in the infection mechanism (Wang et al. 2007, 2010). In this chapter, we present the biocontrol of diamondback moth (DBM), *Plutella xylostella* with *Beauveria bassiana* and its metabolites, with an elucidation on the infection behavior of *B. bassiana* to *P. xylostella* and virulence of the fungal isolate and its metabolites. The fungi that have received the majority of attention for *P. xylostella* control are the Deuteromycetes because of their prevalence, possibility for production on artificial media, ease of application, and relatively long shelf lives.

L. Wang (✉) • M. You
Institute of Applied Ecology, Key Laboratory of Biopesticide and Chemical Biology, MOE, Fujian Agriculture and Forestry University, Fuzhou 350002, China
e-mail: wang_liande@126.com

H. Wang
University of Nebraska-Lincoln, 312 Entomology Hall, Lincoln, NE 68583, USA

© Springer International Publishing Switzerland 2015

K.S. Sree, A. Varma (eds.), *Biocontrol of Lepidopteran Pests*, Soil Biology 43,
DOI 10.1007/978-3-319-14499-3_11

11.2   Virulence of Different *B. bassiana* Strains to *P. xylostella*

Wang (1999) bioassayed four strains of *B. bassiana* (Bb-100, Bb-71, Bb62, and Bb-38, conidial spore suspension 3×10^7 conidia/ml) to *P. xylostella* (second and third instars) in both laboratory and a small-scale field (in the campus of Fujian Agriculture and Forestry University, Fuzhou, PR China). As observed in Figs. 11.1 and 11.2, larvae became infected as they were inoculated. A time-mortality relationship was seen to exist. In the field and laboratory experiments, Bb-100 showed the strongest ability to kill the insect host. Due to imprecise mortality estimates in the field, it was difficult to quantify the effect of LT_{50}. The LT_{50} to DBM with

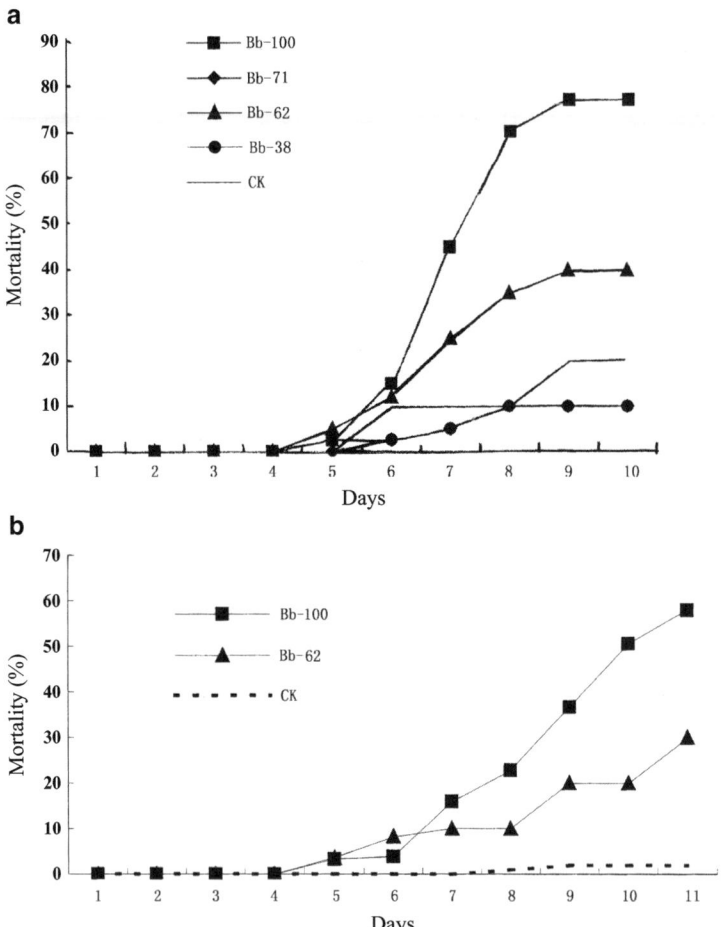

Fig. 11.1 Virulence assay of four strains of *B. bassiana* against *P. xylostella* in (**a**) laboratory, (**b**) field (cf. Wang 1999; You et al. 2004)

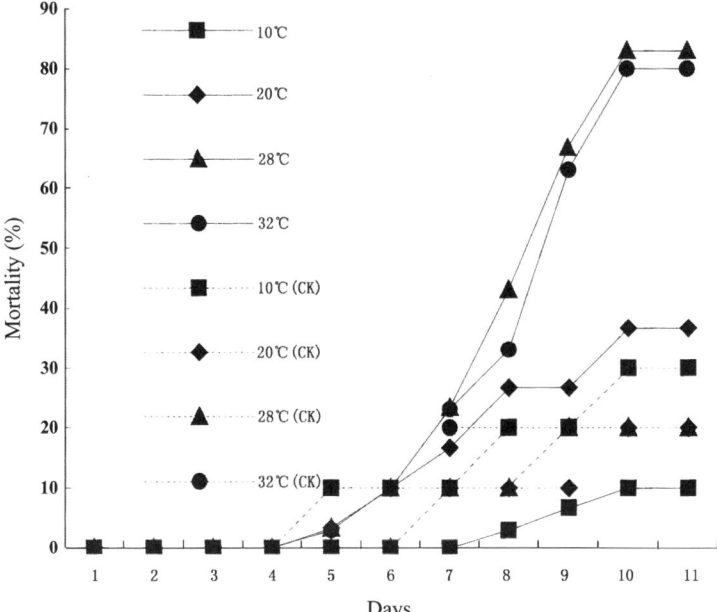

Fig. 11.2 Virulence assay of strain Bb-100 of *B. bassiana* against *P. xylostella* at different temperatures (cf. Wang 1999; You et al. 2004)

Bb-100 was 8.5 days in the laboratory; the starting time was over 5 days in the laboratory and field. Among the four strains, Bb-38 exhibited the least virulence to DBM.

To reveal environmental effect on the virulence of *B. bassiana*, a simple matrix treatment considering *P. xylostella* and Bb-100 was carried out in the laboratory. The infectious ability of Bb-100 strongly varied with temperature. Figure 11.2 indicates that the mortality approximately began 4–5 days after application. The highest level of mortality was seen at 28 °C, but the fastest time to knock down the host was at 32 °C. Low temperature delayed the activity in infection. On the other hand, it was reported that relative humidity could influence the mortality of *P. xylostella* (Masuda 1998). Undoubtedly, temperature is one of the basic factors of virulence of *B. bassiana*.

11.3   Infection Behavior of *B. bassiana* to *P. xylostella*

11.3.1   Infection Process

The process of fungal infection involves four steps: adhesion, germination, differentiation, and penetration (St. Leger et al. 1996).

11.3.1.1   Attachment

Attachment of fungal spores to the host is a prerequisite for further parasitic events and takes place together with host recognition. Conidial attachment is one of the dominant steps in the course of infection of entomopathogenic fungi. This experiment focused on the initial behavior of the conidia of *B. bassiana* on the cuticle of the third instar of *P. xylostella*. With the infection process used, conidia were commonly found on the ventral side, in segmental vicinity, and on no-sticklike area on the host. Clearly, the conidia (1.8–2.8 μm) are almost as large as the size of space among hairs on the cuticle surface of *P. xylostella*; therefore, conidia were getting a full touch with *P. xylostella* cuticle. Despite this, there were still some conidia which contacted larval epicuticle at rare hairy region (e.g., intertegumental membrane, abdomen, etc.), germinated, and breached the epicuticle (Figs. 11.3 and 11.4). The increasing evidence of conidial germination was clearly presented after 24 h. The surface of conidia possesses an outer layer of interwoven pins (Figs. 11.5 and 11.6). The structure is unique to the conidial stage, not detected on the vegetative cells (hyphal body, mycelium). The interwoven pins were arranged in groups, mainly in a vertical position, but near the top, there were several pins positioned horizontally (Fig. 11.6, arrows; Wang 1999).

This structure was rarely observed, only in conidia submerged in liquid media, and not those on dry. The superficial structure on a conidium was as similar as indicated in Fig. 11.6. The existence of such structures suggests that the topography of the conidium assists the recognition or attachment between conidium and host (Zhai and Huang 1995; Bidochka et al. 1993). After coming in contact with the cuticle of both hosts, conidium began to release some mucilage materials around the conidium (Fig. 11.7a). Obviously, at the contact point or nearby, materials cover the outer layer of conidia, which may serve as glue (Large et al. 1988). Compared with other parts of the conidium, the mucus around the part touching the cuticle was thicker. After germination, mucilage covered the outer conidium wall, but at the top

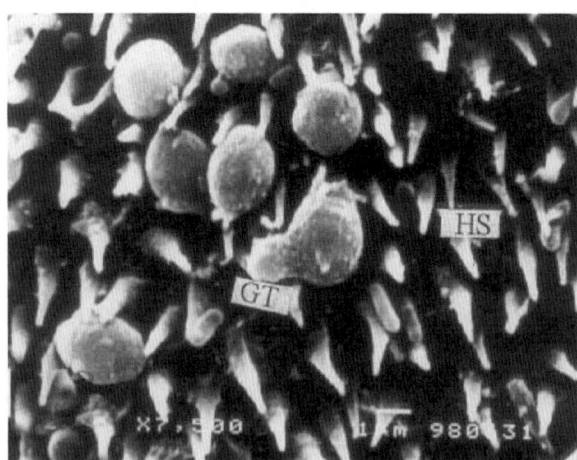

Fig. 11.3 Superficial epicuticle of *P. xylostella* infected by Bb-100 under SEM, conidia (C), germ tube (GT), hairs (HS) (cf. Wang 1999; You et al. 2004)

Fig. 11.4 Cross section of
the cuticle of *P. xylostella*
under SEM, hairs
(HS) (cf. Wang 1999; You
et al. 2004)

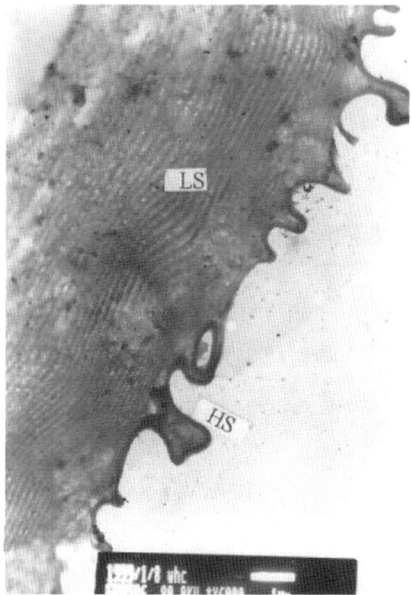

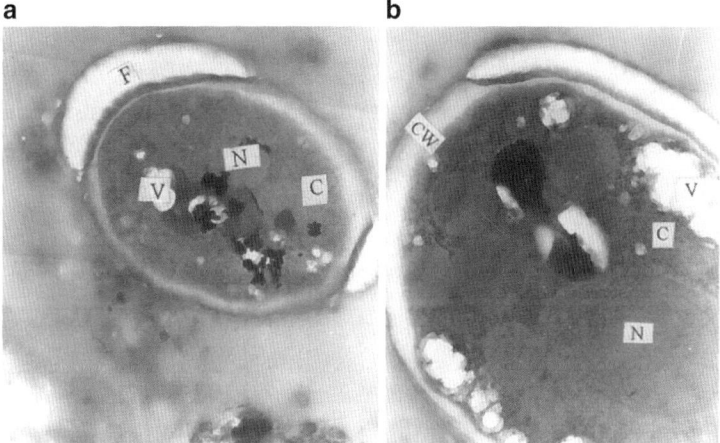

Fig. 11.5 TEM of a conidium, conidium (C), conidial wall (CW), vacuole (V), nucleus (N),
floccule (F) (cf. Wang 1999; You et al. 2004)

of the germ tube, the mucilage was much thicker than other parts (Fig. 11.7b, c
under TEM). At this stage, no septum was observed between the germ tube and
maternal conidium. Underneath, comparing with the parts nearby, the cuticle of
P. xylostella was turned to gray in color, showing a distinct deformation of the
epicuticle. It is suggested that within either part of the germ tube or contacting part
of the conidium with the cuticle would be the most metabolically active section; the

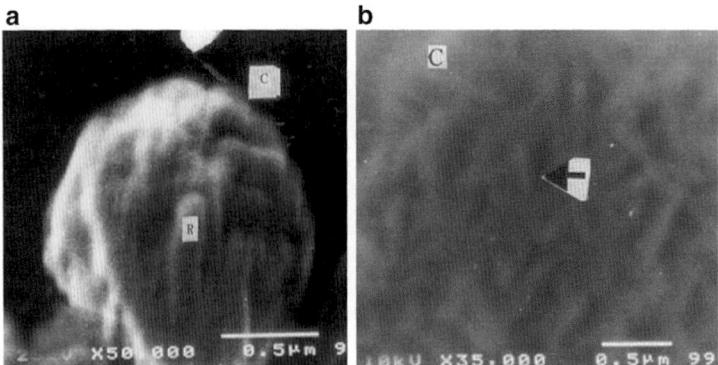

Fig. 11.6 (**a**) A view of the conidium under TEM, conidium (C), rodlet (R); (**b**) a general view of the conidium under TEM, conidium (C), rodlet-like structure (*arrow*) (cf. Wang 1999; You et al. 2004)

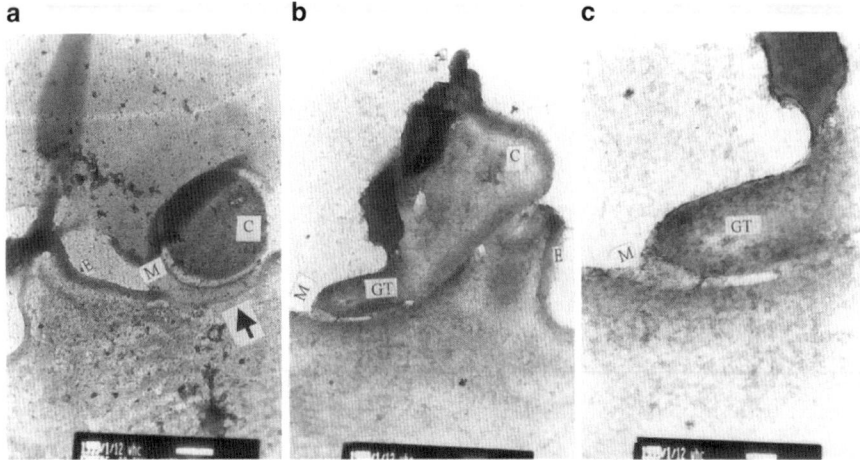

Fig. 11.7 Conidium on the cuticle of *P. xylostella* under TEM, conidium or conidia (C), mucilage (M), epicuticle (E), germ tube (GT); (**a**) conidium falling down on cuticle of *P. xylostella*, arrow indicating degraded epicuticle, bar = 500 nm; (**b** and **c**) germinating conidia on the epicuticle of *P. xylostella*, bar = 100 nm (cf. Wang 1999; You et al. 2004)

mucilage outside was secreted by the conidium and germ tube themselves (Wang 1999).

11.3.1.2 Germination

The germination tests proceeded after inoculation on different hosts in the laboratory (Table 11.1). Even within a given laboratory, the germination ratio varied with

Table 11.1 Assay of conidial germination and appressorial formation on different media

Strain	Item	Rate (%)/+0.0125 YEM			Rate (%)/+sterilized water		
		12 h	24 h	36 h	12 h	24 h	36 h
Bb-100	GC	3.5	18	25	2.0	6.6	20
	AP	1.2	3.0	5.0	1.0	2.1	3.0
Bb-62	GC	2.2	5.6	5.6	1.0	3.8	6
	AP	0.5	1.7	1.7	0.1	1.6	1.6
Bb-71	GC	1.0	4.0	5.2	0.0	4.0	4.1
	AP	0.0	1.8	2.7	0.0	2.0	2.5
Bb-38	GC	1.5	26	30	1.0	25	30
	AP	0.4	11	19	0.0	14	19

Suspension 3×10^7 conidia/ml + 0.0125 YEM, 28 °C pH = 6.5 or suspension 3×10^7 conidia/ml + sterilized water, 28 °C pH = 6.5, germinating conidia (GC), appressoria (AP), cf. Wang (1999) and You et al. (2004)

the strain and host. Clearly, the conidia could germinate on all hosts, after 24 h. With the elongation of the incubation time, the number of germinated conidia (GC) on all hosts increased linearly. In treatments adding 1.25 % YEM, the germination ratio was higher than those with only sterilized water. The result that suitable addition of nutrient could stimulate the germination of conidia was consistent with Fan and Li (1994) and St. Leger et al. (1991, 1989). Bb-38 could easily germinate on *P. xylostella* (Wang 1999).

Under the conditions of 28 °C, pH = 6, 80 % RH, in the four strains of *B. bassiana*, the amount of conidial germination increased with time. With the infection process used, the GC could be found at any part of the cuticle including the head, thorax, and elytra under SEM (Fig. 11.8).

On cuticle, although some conidia only contacted with horn-like nodules, they still appeared in taxis bending toward the cuticle (Fig. 11.8, under SEM). The germ tube of Bb-100 on the abdomen of *P. xylostella* (Fig. 11.8V) had less orientation, but on the head and other parts (Figs. 11.7 and 11.8I, II), Bb-71 showed a little taxis (Fig. 11.8III, IV). Taking the materials that exist on the epicuticle into consideration, differences in conidial orientation are probably due to components that can have prohibiting or stimulating functions on conidia (Zhai and Huang 1997; St. Leger et al. 1994). Before penetrating into *P. xylostella*, the germ tubes were 2.2–6.6 μm in length in Bb-100 and Bb-62. In Bb-71, germ tube length was around 11.2 μm. Extensive amounts of errant-growing hyphae were rare to see, but limited amounts of errant growth followed by eventual penetration over the surface were seen (Fig. 11.8VI, VII for strain Bb-100 on *P. xylostella*). At early stage about 24 h, few big vacuoles filled the whole conidium; the bud of the germ tube formed at random, into which the vacuole expanded (Fig. 11.8VIII for strain Bb-100). After 48 h, nuclear division was finished, materials in the conidium were transmitted toward the germ tube, a big vacuole was formed consequently, and the top of germ tubes sometimes tended to be thin (Fig. 11.8IX for strain Bb-100). In late stages, the septum was formed between the conidium and germ tube, and daughter nuclei moved into the germ tube (Fig. 11.8X) (Wang 1999).

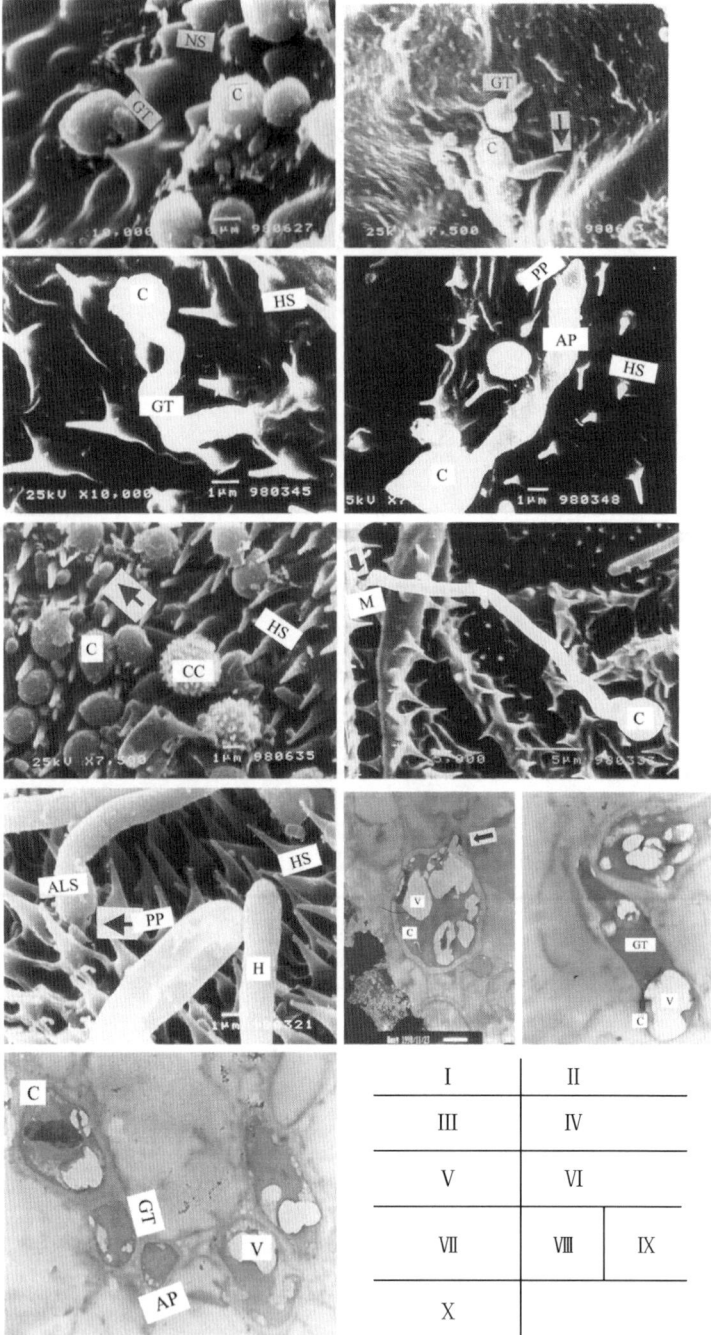

Fig. 11.8 (*I*) Superficial view of the thorax of *P. xylostella* infected with Bb-100 after 24 h under SEM, scale bar = 1 μm; (*II*) fine structure of the head of *P. xylostella* contaminated with Bb-100 after 24 h under SEM, germinating conidia (*arrow*), scale bar = 1 μm; (*III*) abdomen of *P. xylostella* contaminated by Bb-71 after over 36 h under SEM, hyphae penetrating into epicuticle

11.3.1.3 Appressoria

The second instar of *P. xylostella* was contaminated by four strains and provided a rich form of appressoria and appressoria-like structures. In Figs. 11.8IV, VII and 11.9I, II, a penetration peg for host penetration was produced, but the penetration peg was formed at a position near to germ tube in Fig. 11.9II, while it came into being at the tip of germ tube as in Figs. 11.8VII and 11.9I, II. Nearly all penetration pegs breached the cuticle. Enzyme activity was presented at the tip of the penetration peg as in Fig. 11.9II. Appressoria of *B. bassiana* were usually formed at the end of a germ tube of variable lengths and often terminated in a bulb from which a narrow hypha (penetration peg) emerged. Appressoria were occasionally formed apically on short germ tubes or directly from conidia (Fig. 11.9I, II). Evidence showed that no penetration peg or appressoria could still penetrate into the cuticle (Fig. 11.9III with Bb-71). Meanwhile, other appressoria-like structures were found (Fig. 11.8VI with Bb-100, Fig. 11.9IV with Bb-38); at the top of the germ tube, a penetration peg-like structure was produced (Wang 1999).

Under TEM, photographs of the internal structures of appressoria or appressoria-like structure were taken (Fig. 11.10I–III, appressoria-like structure, IV–IX appressoria). TEM photos presented that the pre-GT stood out, to which materials were moved from conidium (Fig. 11.10IV). There was a short stage of vigorous nuclear activity where the daughter nucleus spread inside a second time frame (Fig. 11.10V). The formation of septum was followed by the migration of the nuclei into the appressoria (Fig. 11.10IX). Inside the appressoria-like structure (Fig. 11.10I–III), conidial materials were transmitted from the conidia to the top part. The most frequent and biggest vacuoles could be found within the conidium and along the germ tube. Inside, nuclear division had been finished, and daughter nuclei could be seen in the germ tube. In late stage about 48 h, inside nuclei division had been finished; one daughter nuclei still remained in the conidium, while another moved through the germ tube and entered the forming appressorium. A septum usually was formed across the end of the germ tube, beginning to separate germ tube (appressoria) from conidia (Fig. 11.10IX). Beforehand, karyon division was

---

Fig. 11.8 (continued) (*arrow*), scale bar = 1 μm; (*IV*) superficial view of the abdomen of *P. xylostella* with Bb-71 after over 36 h under SEM, hyphae penetrating into the epicuticle (*arrow*), scale bar = 1 μm; (*V*) superficial view of the abdomen of *P. xylostella* infected with Bb-100 under SEM, conidia (*arrow*), scale bar = 1 μm; (*VI*) appressoria-like structure in Bb-100 ready to invade the epicuticle of *P. xylostella* after over 36 h under SEM, invading appressoria-like structure (*arrow*), scale bar = 1 μm; (*VII*) appressoria (Bb-38) penetrating the epicuticle of *P. xylostella* after over 36 h under SEM, a terminal swollen invading appressorium (*arrow*), scale bar = 1 μm; (*VIII*) cross section of conidia under TEM, bud of the germ tube (*arrow*), scale bar = 1 μm; (*IX*) germinated conidia under TEM, scale bar = 1,500 nm; (*X*) thin section of appressorium under TEM, scale bar = 1 μm; conidium (C), germ tube (GT), nodule structure (NS), hairs (HS), appressorium (AP), penetration peg (PP), contaminated conidia (CC), vacuole (V), nuclei (N), mucilage (M), appressoria-like structure (ALS), hyphae (H) (cf. Wang 1999; You et al. 2004)

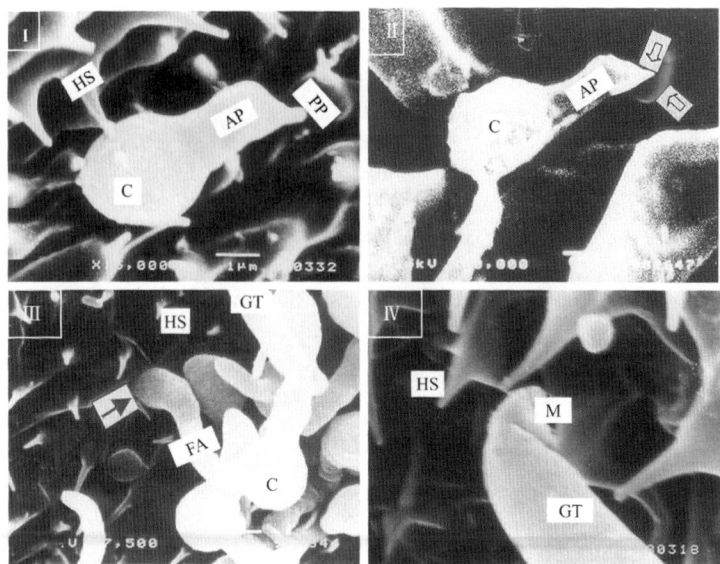

Fig. 11.9 Appressorium on the cuticle of *P. xylostella* contaminated by *B. bassiana* under SEM, conidia (C), germ tube (GT), hairs (HS), appressorium (AP), penetration peg (PP), fusiform appressorium (FA), mucilage (M); (*I*) and (*II*) contaminated by Bb-100; (*III*) contaminated by Bb-71, melanized tip of appressorium (*arrow*) during invasion; (*IV*) contaminated by Bb-38; scale bar = 1 μm (cf. Wang 1999; You et al. 2004)

processed prior to septum formation, but it is difficult to find. Clearly, the formation of septum symbolizes the finished and mature stage of appressoria. Sometimes mucus appeared at the top of appressoria (Fig. 11.10X–XII), but not in all. In Fig. 11.9III, the top of appressoria externally turned dark in color before penetration. Appressorial wall thickness was the same as conidia, but in some appressoria, a part of the top wall of appressorium was thickening (Fig. 11.10IX). Compared with appressoria, the germ tube wall was thinner than the conidia (Fig. 11.10X–XII). Inside, a great deal of vesicles could be found in maternal conidia, but in appressoria, vesicles were rare, especially at the top (Fig. 11.10IX, X) (Wang 1999).

11.3.1.4   Penetration

Penetration is another dominant step in the course of infection in entomopathogenic fungi. These experiments focused on the behavior of conidia of *B. bassiana* on the cuticle of the second instar of *P. xylostella*. The penetration was observed over 36 h contamination of host and could happen almost at any region with hole at the point of penetration of the cuticle, including the thorax (Fig. 11.8I), head (Fig. 11.8II), and abdomen (Fig. 11.8III–V). There were more penetrations at the rare hair region than regions with heavy hair. There was no errant growth of hyphae without

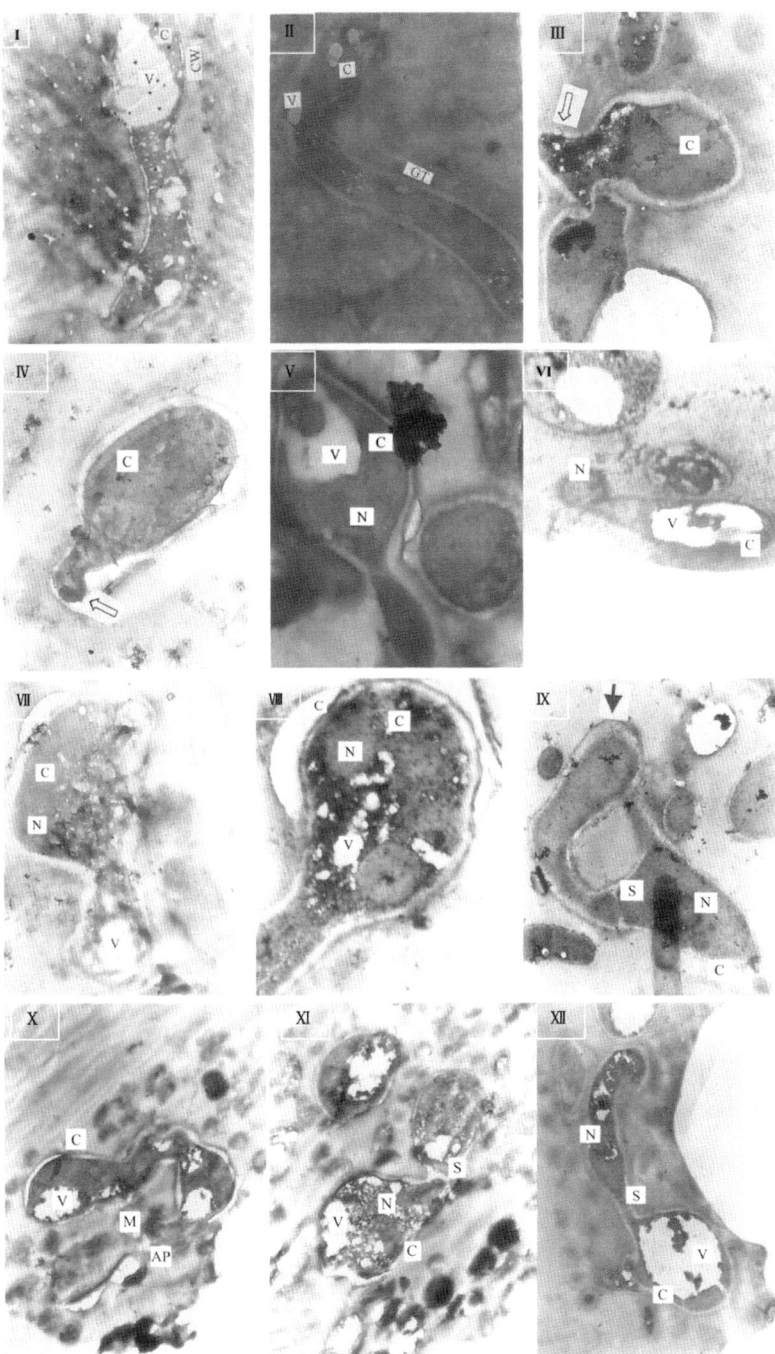

Fig. 11.10 Germ tube and appressoria under TEM (*I*) detailed germ tube, scale bar = 500 nm; (*II*) thin section of appressoria-like structure, scale bar = 500 nm; (*III*) appressoria-like structure, birdhead-like tip (*arrow*), scale bar = 500 nm; (*IV*) initial appressorial structure (*arrow*), scale bar = 200 nm; (*V*) early appressorial structure (*arrow*), scale bar = 1 μm; (*VI–VIII*) appressoria-

penetration observed. Penetration peg was formed at the tip or near the tip position of appressoria (Fig. 11.8III), which was succeeding in the cuticle. In both Bb-100 and Bb-62, most germ tubes were short and thin at the top part on *P. xylostella* when penetrating. It was also noted that terminal swelling on the end of the germ tube turned dark when breaking down the cuticle of *P. xylostella* (Fig. 11.9III with strain Bb-71). The invading hypha was quite simple, without any change at the top. No such phenomenon was found in three others (strain Ba-100, Ba-38, and Ba-62). A cross section of the cuticle of both *P. xylostella* (Fig. 11.10) was made to record the differences between them at the ultrastructural level. There is a lamellate structure under the epicuticle of the abdomen, seeing no nuance between them. Nevertheless, the gold-coloidal technique (gold-chitinase complexes) revealed some differences. Particles could be found in main parts of the cuticle in both pests. In early stages of penetration, mucilage was found at the edge of the top of the germ tube contacting with the cuticle of *P. xylostella* before breaching it (Fig. 11.7a); epicuticle deformation was also observed. A detailed micrography revealed that mucilage was double layer (Fig. 11.11I). Meanwhile, mucilage also appeared at no contacting part (at the edge of the top of the germ tube) (Fig. 11.11I–III). Under the conidium, there was a great reduction in thickness of the epicuticle of the host or even a complete absence of cuticle material (Fig. 11.11III). The deformation of epicuticle appeared under the conidium otherwise (Fig. 11.12I). These results suggest that (1) within the top part of germ tube (or appressoria) would be the most active site of enzyme metabolism and (2) results confirm the hypothesis that pathogen invasion is a combination of enzyme and mechanical pressure (Zhai and Huang 1995). On *P. xylostella*, birdhead-like appressoria were observed, but were rarely seen (Fig. 11.9I, II). Few appressoria were presented on *P. xylostella*, but their shape was the same as those generated by Bb-100. When breaching the cuticle of *P. xylostella*, mucus still covered at the top of the germ tube externally (Fig. 11.12I). After breaking down, materials inside the conidium migrated into the germ tube, which made the germ tube fully expand and the conidium was partly emptied. At the same time, daughter nuclei also moved into the germ tube (Fig. 11.12II) on *P. xylostella* with Bb-100; mucus sealed around cleavage could be found (Fig. 11.12III, IV). Without touching with any substance, mucilage could be secreted by appressoria (Fig. 11.10IX); meanwhile, septum was not found. It is proposed that mucilage material secreted by *B. bassiana* would assist fungi's infection. Near the germ tube, hyphal bodies were formed, which were full inside; nuclei were clearly seen without any vesicles (Fig. 11.12III). In later stages (after over 5–7 days on *P. xylostella*), hyphal bodies and blastospores fully filled the host body (Fig. 11.13I under TEM) and cuticle was destroyed (Fig. 11.13II under TEM).

Fig. 11.10 (continued) like structure, scale bar = 200 nm; (*IX*) appressoria in middle stage, scale bar = 1 μm; (*X–XII*) detailed appressorial structure, scale bar = 1 μm; conidia (C), conidia wall (CW), germ tube (GT), nodule structure (NS), hairs (HS), appressorium (AP), penetration peg (PP), contaminated conidia (CC), vacuole (V), nuclei (N), fusiform appressoria (FA), mucilage (M), septum (S) (cf. Wang 1999; You et al. 2004)

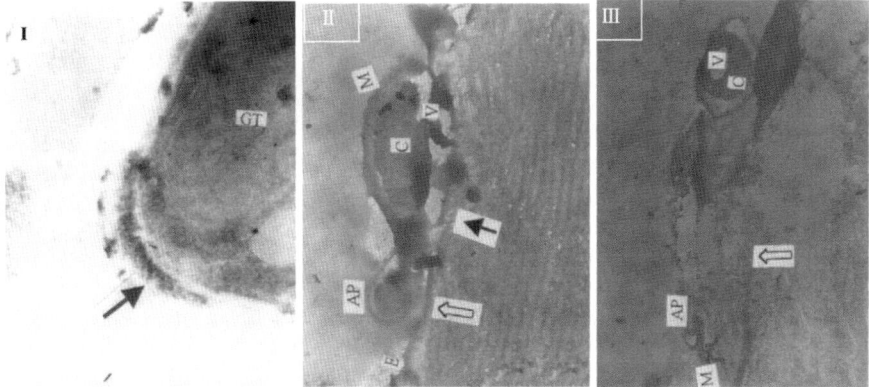

Fig. 11.11 Mucilage and appressorial structure under TEM (*I*) double mucilage cover on the tip of the germ tube (*arrow*), scale bar = 100 nm; (*II*) cross section of appressorium on epicuticle of *P. xylostella* infected by strain Bb-100, disappeared epicuticle part (*dark arrow*), scale bar = 500 nm; (*III*) appressorium on epicuticle of *P. xylostella* infected by strain Bb-62, disappeared epicuticle part (*arrow*), scale bar = 500 nm; conidia (C), germ tube (GT), appressorium (AP), vacuole (V), mucilage (M), epicuticle (E) (cf. Wang 1999; You et al. 2004)

A study of mummified larvae demonstrated that hyphae exit first at the abdomen region, but eventually appeared over the entire body (Wang 1999).

11.3.2 Enzyme Activities

11.3.2.1 Superficial Chitinase Profiles Under SEM

Bb-100 applied on the head of *P. xylostella* demonstrated an enzyme activity (arrow) (Figs. 11.8II and 11.9II on *P. xylostella*), as breaking down cuticle. Findings under SEM proposed that (1) releasing of chitinase on host depends on substrate contacted, being in agreement with reports (Coudron et al. 1984; St. Leger et al. 1996) and strain, and (2) some kinds of chitinases must be activated by proteinase before expressing activity because of their existing form of zymogen (Gooday et al. 1986). As consideration, there are some kind of substances that can induce the release of chitinase on elytra, and (3) chitinase is of potential in facilitating invasion, regardless of strain's pathogenicity.

11.3.2.2 Chitinase and Chitin Labeling Under TEM

According to the report, chitinases from different pathogens shared 66 % identity among them (St. Leger et al. 1996). For this reason, an antibody-ConA complex was applied to detect chitinase activity during invasion. Unfortunately, poor results

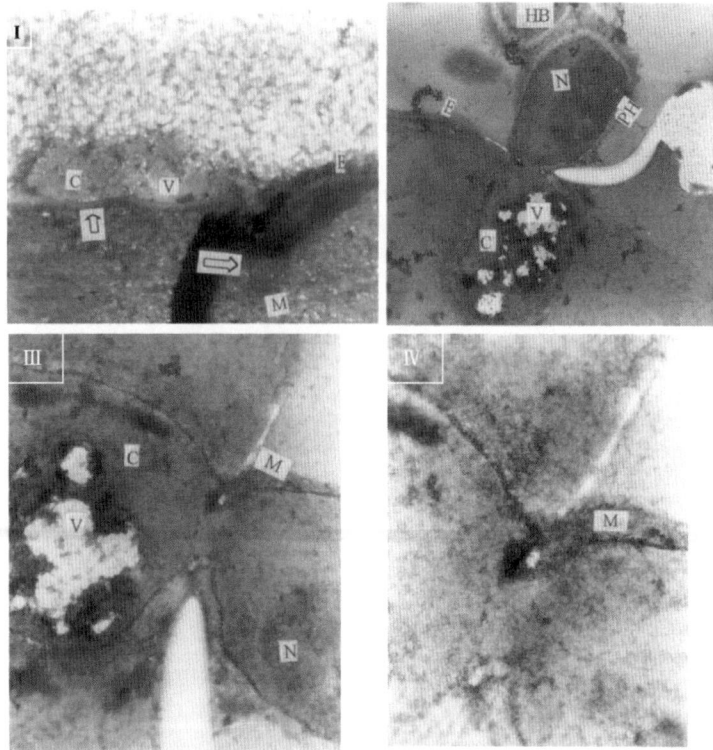

Fig. 11.12 Invasion of the cuticle of *P. xylostella* infected by Bb-100 under TEM (*I*) initial stage of invasion by the germ tube on the epicuticle after over 36 h, deformed epicuticle part due to mechanical pressure by the conidium (*arrow*), invading hyphae (*white arrow*), scale bar = 500 nm (*II–IV*) late stages of invasion of the cuticle after over 48 h, leakage of the cuticle (*arrow*), (*II*) scale bar – 200 nm, (*III*) scale bar = 200 nm, (*IV*) scale bar = 100 nm; conidia (C), appressorium (AP), vacuole (V), mucilage (M), epicuticle (E), nuclei (N), hyphal body (HB), penetrating hyphae (PH), lamellate structure (LS) (cf. Wang 1999; You et al. 2004)

were obtained. Another method was tried instead of primer with good results. Results under TEM indicated that chitinase activity could be detected through infection stage on the second instar cuticle of both *P. xylostella*, but was rare to see in later stages after penetration. Figure 11.14I–III provides a superficial show of chitinase activity before breaching *P. xylostella*. Before germination (24 h), gold particles are almost only distributed near conidia, and only a few particles were observed in conidia (Fig. 11.14III). This probably gives evidence concerning the level of chitinase activity. When germination occurred (over 36 h), large amounts of gold particles were found within the outer layer and near the germ tube. However, a few gold particles were found on the germ tube surface (Fig. 11.14II). In later penetration stage, chitinase was not labeled because hyphae had penetrated into the cuticle fully at this stage, and very little chitinase was required. Therefore,

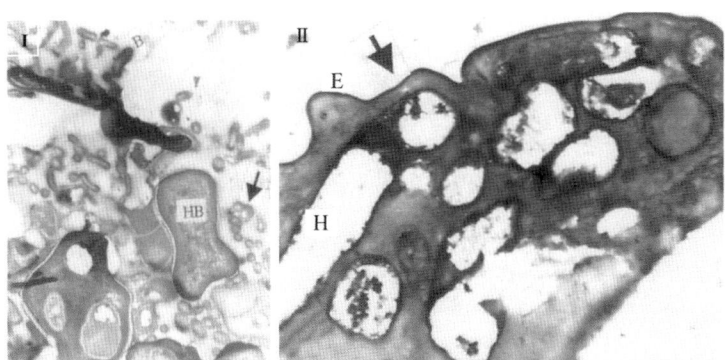

Fig. 11.13 (*I*) Rapid growth of hyphae inside the host after over 6 days under TEM, hemocyte (*arrow*), scale bar = 2 μm, (*II*) detailed view of the destroyed cuticle of *P. xylostella* after over 6 days under TEM, destroyed cuticle (*arrow*), scale bar = 1 μm; hyphae body (HB), hyphae (H), blastospore (B), epicuticle (E) (cf. Wang 1999; You et al. 2004)

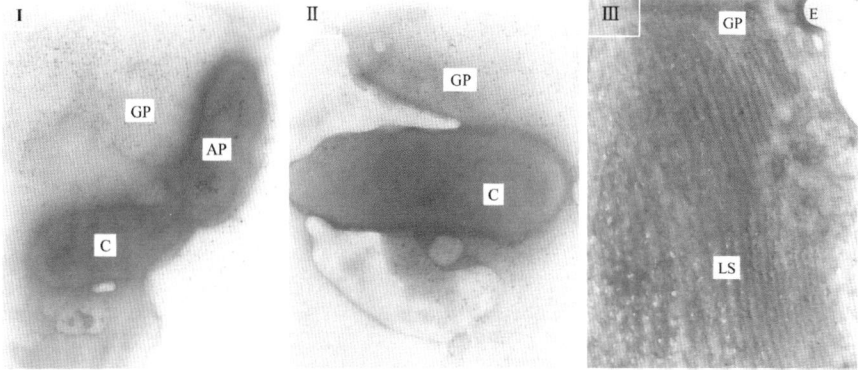

Fig. 11.14 Germinated conidia under the epicuticle of *P. xylostella* under TEM (*I*) infected by strain Bb-100, 10 nm gold particles mainly distributed around conidia, a lot of particles on the surface of appressoria (*II*) infected by strain Bb-62, 10 nm gold particles mainly distributed around conidia, a lot of particles on the wall of conidia (*III*) detailed view of the cross section of the labeled cuticle of *P. xylostella*, 30 nm gold particles distributed at random on the cuticle; conidia (C), appressoria (AP), gold particle (GP), epicuticle (E), lamellate structure (LS); scale bar = 200 nm (cf. Wang 1999; You et al. 2004)

it is hard to detect the lowest enzyme activity within conidia and germ tubes (Fig. 11.12II with strain Bb-100 on *P. xylostella*).

Chitin is an unbranched polysaccharide, composed primarily of β-1,4-linked *N*-acetylglucosamine (NAG) residue, with an occasional glucosamine residue (Brimacomge and Webber 1964). The best documented pathway for the degradation of chitin involves the sequential action of two separate hydrolases: (1) endochitinase [poly-β-1,4-(2 acetamido-2-deoxy)-D-glucoside glycanhydrolase (EC-3.2.1.14)], which produces low-molecular-weight soluble multimers of NAG, the titer from *N,N*-diacetyl chitobiose being predominant, and (2) chitobiase or exochitinase [chitobiase acetylaminodeoxyglucohydrolase (EC 3.2.1.29)], which hydrolyses the intermediates to NAG (Coudron et al. 1984). Conidia, both Bb-100 and Bb-62 germinated to form germ tubes on the surface of the third instar of *P. xylostella* and penetrated cuticles within 40 h. The thin section of the cuticle at 16, 24, 36, 48, and 60 h post-inoculation was labeled with antibody to chitinase and gold-polysaccharide complexes (dehydroxy chitin), respectively. In spite of their 66 % identity in sequence, there were still sequence differences between chitinase in *B. bassiana* and *Serratia marcescens*, whereas in experiments, just a few gold particle were found around conidia (Wang 1999). The failure to detect chitinase via antibody probably is a good explanation to this menace. Firstly, gold-polysaccharide complexes were used to show distribution of chitinase; no such thing has been reported before. The polysaccharide applied contains a variety of collection including titer of NAG, multimer of NAG, and glycol chitin, which can be taken as substrate by whatever endochitinase to degrade into NAG. Therefore, it is possible to label endochitinase in fungi by using such complexes. In addition, gold-chitinase was employed to examine chitin distribution both in fungi (Bb100, Bb62) and cuticle, but only later was labeled on *P. xylostella* (Fig. 11.14III). The failure to label chitin in hypha hints that it is just an evaluated protecting strategy for fear that fungi would be destroyed by chitinase released by themselves.

11.4 Metabolites

Numerous factors (slow action, potentially negative interactions with commonly used fungicides, limited shelf life, and dependence on favorable environmental conditions) continue to impede the commercial development and/or application of this fungus (Wang et al. 2007). *B. bassiana* requires high humidity for germination, for establishment of infection, and for sporulation and consequent epizootics, which commonly facilitates epizootics of plant disease. These factors became bottlenecks in the application of *B. bassiana* in DBM control. *B. bassiana* produces secondary metabolites with insecticidal properties during the colonization of the host tissue, which may play an important role in host mortality. Recently, the toxic substances extracted from *B. bassiana* have been found to be potent against DBM in the laboratory and at field levels. Gao et al. (2012) found that the fermentation filtrate of strain Bb-2 had bioactive and virulent properties against the larvae of *P. xylostella* with the corrected mortality of 26.7 % and 35.6 % 24 h and 48 h after treatment with fermentation filtrate.

11.5   Conclusion and Future Prospects

Future studies should be directed toward the definition of recognition factors which are involved in (1) the attachment; (2) chemical recognition between the emerging hyphal tip of germination conidia and surface of epicuticle, resulting in the hyphal tip taxis toward cuticle and penetration; (3) enzyme activity; and (4) appressoria formation. In conclusion, the application of toxic secondary metabolites from *B. bassiana* has broken through the bottlenecks in the application of *B. bassiana* for the control of diamondback moth.

References

Bidochka MJ, Miranpuri GS, Khachatourians GG (1993) Pathogenicity of *Beauveria bassiana* (Balsamo) Vuillemin toward lygus bug (Hem., Miridae). J Appl Entomol 115:313–317

Brimacomge JS, Webber JM (1964) Mucopolysaccharides: chemical structure, distribution and isolation. Elsevier, Amsterdam, pp 18–42

Butt TM, Jackson C, Magan C (2001) Fungi as biocontrol agents: progress, problems and potential. CABI Publishing, UK

Coudron TA, Kroha MJ, Ignoffo CM (1984) Levels of chitinolytic activity during development of tree entomopathogenic fungi. Comp Biochem Physiol 79B:339–348

Fan M, Li Z (1994) Impact of nutrients and culture conditions on appressorium formation of entomogenous fungi. J Aanhui Agric Univ 21:123–130

Gao P, Hu Q, Wang Y (2012) Virulence of spores and fermentation filtrates of *Beauveria bassiana* Bb-2. Hubei Agric Sci 51:3237–3239

Gooday GW, Humphreyas M, Mcintoshw H (1986) Roles of chitinases in fungal growth. In: Muzzarelli CJ, Gooday GW (eds) Chitin in nature and technology. Plenum, New York, pp 83–91

Lacey LA, Goettel M (1995) Current development in microbial control of insect pests and prospects for the early 21st century. Entomophaga 40:3–27

Large JP, Monsigny M, Prevost MC (1988) Visualization of exocellular lectins in the entomopathogenic fungus *Conidiobolus obscurus*. J Histochem Cytochem 36:1419–1424

Masuda T (1998) Microbial control of diamond back moth, *Plutella xylostella*, by entomopathogenic fungus, *Beauveria bassiana* In: Meeting Program and Abstracts VII Inter. Coll. Invertebr. Path. & Microb. Contr. IVth Inter. Confer. on *Bacillus thuringiensis*, Sapporo, 23–28 August 1998

St. Leger RJ, Butt TM, Mark S et al (1989) Production in vitro of appressoria by the entomopathogenic fungus *Metarhizium anisopliae*. Exp Mycol 13:274–288

St. Leger RJ, Roberts DW, Staples RC (1991) A model to explain differentiation of appressoria by germlings of *Metarhizium anisopliae*. J Invertebr Pathol 57:299–310

St. Leger RJ, Bidochka MJ, Donald WR (1994) Germination triggers of *Metarhizium anisopliae* conidia are related to host species. Microbial 140:1651–1660

St. Leger RJ, Joshi L, Gidochka MJ, Nancy W et al (1996) Characterization and ultrastructure localization of chitinases from *Metarhizium anisopliae*, *M. flavoviride*, and *Beauveria bassiana* during fungal invasion of host (*Manduca sexta*) cuticle. Appl Environ Microb 62(3):907–912

Wang H (1999) Ultrastructural studies on infection behaviors of *Beauveria bassiana* to *Plutella xylostella*. Ph.D. Thesis, Fujian Agricultural University, Fuzhou

Wang L, Huang J, You M, Guan X, Liu B (2007) Toxicity and feeding deterrence of crude toxin extracts of *Lecanicillium (Verticillium) lecanii* (Hyphomycetes) against sweet potato whitefly (*Bemisia tabaci*; Hom., *Aleyrodidae*). Pest Manag Sci 63:381–387

Wang L, You M, Huang J, Zhou R (2010) Diversity of entomopathogenic fungi and their application on biological control. Acta Agric Univ Jiangxiensis 32:920–927

You M, Hou Y, Yang G (2004) Control of the population system of Diamond back moth *Plutella xylostella*. Fujian Science and Technological Press, Fuzhou

Zhai J, Huang X (1995) Review on pathogenicity mechanism of entomopathogenic fungi, *Beauveria bassiana*. Mycology 22:45–48

Zhai J, Huang X (1997) Studies on behavior of germination of conidia from *Beauveria bassiana* on the cuticle of *Heliothis zea*. Acta Microbiol Sin 37:154–158

Chapter 12
Entomopathogens for Cotton Defoliators Management

K. Sahayaraj

12.1 Introduction

Cotton plants, *Gossypium hirsutum* L. (Malvaceae), are one of the most important crops cultivated in more than 80 countries (Bottrell and Adkisson 1977). China, India, the USA, Africa, Pakistan, Brazil, Uzbekistan, Australia, Turkey, Turkmenistan, Greece, Argentina, Burkina and Mexico are arguably the most important cotton-producing countries in the world (Table 12.1). It is a major world agricultural crop cultivated for the harvest of lint fibres utilised extensively in the manufacture of apparel, household and industrial goods. Although linked biologically to the production of cotton fibres, cottonseeds are more than a mere by-product of the cotton harvest. Seeds ginned from the lint fibres are processed commercially for use in animal feed, food for human consumption, and concoctions are used in the preparation of these foods as well as numerous other home and industrial products. During the years immediately following World War II, cotton pest control in the world was dominated by the organochlorine insecticides. The persistence and high toxicity of these chemicals made them effective killers of the cotton defoliators. As a result, normally only one or two treatments per season were needed to control pest satisfactory at a relatively low cost.

12.2 Bio-Limiting Factor of Cotton Production

One of the limitations of cotton production in various parts of the world is the insect pest infestation. Cotton is more attractive to pests and pathogens than practically any other plant. Worldwide entomologists have reported in excess of 1,300 species

K. Sahayaraj (✉)
Crop Protection Research Centre, St. Xavier's College, Palayamkottai 627002, India
e-mail: ksraj48@gmail.com

© Springer International Publishing Switzerland 2015
K.S. Sree, A. Varma (eds.), *Biocontrol of Lepidopteran Pests*, Soil Biology 43,
DOI 10.1007/978-3-319-14499-3_12

Table 12.1 Cotton production (million, 480 lb. bales): selected country report

Country name	2009/ 2010	2010/ 2011	2011/ 2012	2012/ 2013	2013/ 2014	2014/ 2015
China	32.0	32.5	34.0	35.0	32.0	29.5
India	24.5	27.2	29.0	28.5	29.5	28.5
United States	12.2	18.1	15.6	17.3	12.9	14.5
Pakistan	9.2	8.6	10.6	9.3	9.5	9.5
Brazil	5.5	9.0	8.7	6.0	7.5	8.3
Uzbekistan	3.9	4.1	4.2	4.5	4.2	4.2
Australia	1.8	4.2	5.5	4.6	4.1	3.1
Turkey	1.8	2.1	3.4	2.7	2.3	2.9
Turkmenistan	1.5	1.8	1.4	1.6	1.5	1.5
Greece	0.9	0.9	1.3	1.2	1.4	1.4
Argentina	1.0	1.4	1.0	0.8	1.2	1.2
Burkina	0.7	0.7	0.8	1.2	1.2	1.1
Mexico	0.5	0.7	1.2	1.0	0.9	1.1
Rest of the world	7.4	7.7	9.9	9.3	9.0	8.7
African franc zone	2.1	2.1	3.0	3.9	4.1	3.9
EU-27	1.1	1.2	1.6	1.5	1.6	1.7
World	102.8	117.0	126.6	123.0	117.1	115.5

Source: http://www.cottoninc.com/

of insects and mites inhabiting cotton, only a handful causing economic loss. In India and Pakistan, more than 130 and 145 species, respectively, of arthropod pests are known to attack cotton; the most prevalent species have been listed in Table 12.2. Any microbe or animal whose activity leads to defoliation is called as defoliator. Defoliators population is often suppressed by a complex of parasitoids, predators and pathogens under natural condition. However, their outbreaks can be triggered by application of insecticides that deplete or remove the natural enemy complex. Severe defoliator outbreaks in cotton have occurred in various parts of the world, but the factors involved in those outbreaks are poorly understood. The high volume usage of pesticides is often a negative aspect of cotton cultivation worldwide.

12.3   Pesticides in Cotton Production

In cotton-producing developing countries, cotton pesticides constitute the major part of use of agricultural chemicals (Turkey, 36 %; India, 45 %; and Egypt, 50 % of all pesticides used in agriculture) (https://www.icac.org). Some important insecticides utilised for defoliator management are aldicarb, acephate, alpha-cypermethrin, beta-cyfluthrin, chlorpyriphos, demeton, dimethoate, endosulfan, fenvalerate, phosphamidon, phosalone, malathion, methamidophos, thiram and

Table 12.2 List of cotton defoliators reported from different parts of the world and their symptoms

Pest name	Order/family	Symptoms	Country
Helicoverpa zea (Boddie) (foliage feeder)	Lepidoptera: Noctuidae	Feeding on foliage or leaves	USA
Spodoptera litura Boisd	Lepidoptera: Noctuidae	Foliage feeder, young larvae in groups skeletonise leaves and older larvae voraciously defoliate leaves	India, Mexico
Spodoptera exigua (Hübner)	Lepidoptera: Noctuidae	Leaf feeder	USA
Spodoptera littoralis	Lepidoptera: Noctuidae	Triangles cut present only at the front or rear of the body	Africa
Anomis flava Fab. (Semilooper or surveyor caterpillar)	Lepidoptera: Noctuidae	Foliage feeder called cotton semilooper causes significant loss of leaf area to young plants; larvae with looping action are seen on plant parts causing circular perforations measuring 1–3 cm in diameter in the leaves	India, Brazil, Africa
Alabama argillacea (Hübner) (cotton leafworm)	Lepidoptera: Pyraustidae	Leaf feeder	USA
Sylepta derogata Fab. (leaf roller or phyllophagous caterpillars)	Lepidoptera: Pyraustidae	Foliage roller and feeder, marginal portion of leaves eaten away, leaves are folded and larvae are seen in groups amidst faecal materials, commonly seen on leaves at the bottom of crop canopy at low infestation levels, severe infestation defoliates the whole plant	India, Africa
Acontia graellsi (Feist.)	Lepidoptera/ Noctuidae	Foliage feeding	India, China, Africa
Agrotis ipsilon (Hutnagel) (black cutworm)	Lepidoptera/ Noctuidae	Cut and feed leaves	Cosmopolitan
Melanoplus spp.	Orthoptera/ Acridoidea	Cut and feed leaves	India
Cyrtacanthacris tatarica	Orthoptera/ Acrididae	Defoliation of leaves, partial or full	India
Cyrtocanthacris ranacea	Orthoptera/ Acrididae	Feeds on leaves	
Euproctis fraterna	Lepidoptera/ Noctuidae	Initially larvae caused skeletonisation, later defoliate	India
Tarache nitidula F., *Tetranychus urticae*, *T. neocaledonicus*, and *T. falcaratus.* (red spider mites)	Trombidiformes	Feed on the underside of leaves with a necrosed appearance	India, South Africa

(continued)

Table 12.2 (continued)

Pest name	Order/family	Symptoms	Country
Polyphagotarsonemus latus	Trombidiformes	The leaves look chapped and torn as if cut by a knife	Africa
Myllocerus subfasciatus (grey weevil)	Coleoptera/ Curculionoidea, Curculionidae	Marginal notching, off of leaves	India
Gonioctena olivacea (leaf beetle)	Coleoptera	Leaf feeder	Pakistan
Nisotra dilecta, Nisotra uniformis, Podagrica decolorata (flea beetles)	Coleoptera	Make lots of holes in the leaves of young glandless cotton plants	Africa
Frankliniella occidentalis (Pergande)	Thrips	Feed on the abaxial surface of cotton cotyledons	Australia

quinalphos (Clay 2004; Ferrigno et al. 2005). Almost all of these are considered toxic enough to be classified as hazardous by the World Health Organisation (WHO).

Integrated pest management (IPM) strategies have been proposed for cotton pest management (Kogan 1988, 1998; Frisbie et al. 1989) particularly in Mexico (Williams et al. 2013) where biological control agents play an important role. Biological control is defined as the use of living organisms or their products to combat other organisms that are considered harmful. The living organisms typically used are predators, parasites, parasitoids and entomopathogens. In this chapter, we mainly focus on entomopathogens (EPs) (Butt et al. 2001) for defoliator management.

12.4 Defoliation and Its Consequences

Herbivorous arthropods, mainly insect larvae, occasionally mites and grasshoppers, represent a major challenge for plants in their natural environment. Almost all herbivores, in particular chewing insects, cause substantial injury to the site of their attack. Such mechanical wounding of plant tissues is an inevitable consequence of herbivory, although the intensity and extent of damage is different and may vary with the mode of feeding, e.g. sucking or chewing. It causes physical stress (e.g. reduction in growth) and metabolic stress (e.g. decrease in alkaloid content) which leads to loss of the leaves, indirectly reducing the yield of the crop. It varied from variety to variety despite the fact that larvae preferred feeding on specific leaves. Often, loss of either above- or below-ground tissue alters the commitment of the plant to the other. Loss of leaf material caused by herbivores above ground results in reduced root mass, while root grazing by a variety of nematodes and insect larvae leads to lower leaf mass above ground (Geiger and Servaites 1991; Mooney et al. 1991). Cotton has the capacity to sustain life during drought as well as in heavy flood. This has an impact on insect feeding. Another important feature is the production of compounds such as gossypol and tannins which are highly toxic to pests (Ted Wilson and Carter 1991).

12.4.1 Examples of Cotton Defoliators

The beet armyworm, *Spodoptera exigua* (Hübner), has been an occasional pest of cotton in the USA since the early 1900s, whereas *Spodoptera litura* (Boisd) is a major pest in many Asian countries causing damage primarily as a defoliator. *Spodoptera* spp. are considered predominately as leaf feeders. A tremendous amount of research has taken place into the biology and control of this pest. The beet armyworm, *S. exigua*, is a widely distributed polyphagous pest of numerous cultivated crops, including cotton. In the past decade, several novel insecticides have shown good activity against the beet armyworm. A partial list of these includes chlorfenapyr, tebufenozide, emamectin benzoate, indoxacarb, spinosad, etc. Another important defoliator from this genus is Egyptian cotton leafworm, *Spodoptera littoralis* (Boisd.) (El-Guindy et al. 1982).

Grasshoppers (Orthoptera: Acrididae and Romaleidae) inflict serious damage to plants throughout their development. They represent a major group of insect pests on cotton (Ribeiro et al. 2013). In India, *Chrotogonus* spp. are considered as important defoliators of cotton, whereas in the USA, Patrick (2004) reported different species in cotton field. Mites are of minor importance on cotton, and they seldom cause heavy damage. Red spider mite, *Tetranychus cinnabarinus* (Tetranychidae: Acarina); wolly mite, *Aceria gossypii*; and yellow mite/broad mite, *Polyphagotarsonemus latus* (Tarsonemidae: Acarina) are few important mites on cotton.

12.5 Entomopathogens: An Overview

The control of insect pests with EPs is unique. Naturally occurring EPs are important regulatory factors of insect populations. Many species are employed as biocontrol agents of insect pests of many economically important crops worldwide. EP is an organism [generally a bacterium, virus, protozoan, fungus, microsporidium and nematode (Vega and Kaya 2012)] causing disease in insects:

1. Entomopathogenic fungi belong to the order *Hypocreales* of the *Ascomycota* (*Beauveria*, *Metarhizium*, *Nomuraea*, *Paecilomyces* (= *Isaria*), *Hirsutella*, *Cordyceps*); others include *Entomophthora*, *Zoophthora*, *Pandora* and

Entomophaga belonging to the order *Entomophthorales* of the *Zygomycota*. Entomopathogenic fungi are a poorly exploited source of insecticidal proteins, which may be used for the development of new natural insecticides and as alternative molecules for transgenic deployment. In contrast to bacteria and viruses, fungal biocontrol agents have a unique mode of infection. They do not need to be ingested and can invade their host directly through the cuticle.

2. Entomopathogenic bacteria include *Bacillus thuringiensis*, *Clostridium bifermentans*, *Brevibacillus laterosporus*, *Chromobacterium subtsugae*, *Yersinia entomophaga*, species of *Pseudomonas*, bacteria associated with nematodes of the genus *Steinernema*, and *Heterorhabditis*, i.e., *Xenorhabdus* and *Photorhabdus* (Ruiu et al. 2013).

3. The entomopathogenic viruses include baculoviruses (nuclear polyhedrosis virus (NPV) and the granulovirus (GV), a large group of double-stranded DNA viruses), the majority of which have been isolated from a few insect orders: Lepidoptera, Diptera, Hymenoptera and Coleoptera.

4. Nematodes, which are capable of killing, sterilising or seriously hampering the development of insects and completing at least one stage of their life cycle in the host, are called entomopathogenic nematodes (EPNs). Poinar (1979) listed nine families of EPNs (Allontone-matidae, Diplogasteridae, Heterorhabditidae, Mermithidae, Neotylenchidae, Rhabditidae, Sphaerulariidae, Steinernematidae and Tetradonematidae) which attack insects and kill them, sterilise or alter host development. Out of these, only steinernematids (Steinernematidae: Rhabditida) and heterorhabditids (Heterorhabditidae: Rhabditida) have been found to be effective due to their special qualities like quick action, wide host range and wide distribution.

5. Entomopathogenic protozoans include *Mattesia grandis*, *Anthonomus grandis*, *Braconidae* (Nesema), etc.

12.6   Entomopathogens and Cotton Pest Management

12.6.1   *Alabama argillacea*

Cotton crop injured by the *A. argillacea* (Hubner, 1818) (Lepidoptera: Noctuidae) considerably reduces the yield (Bourne 1921). The cotton leafworm, *A. argillacea*, is considered to be one of the key pests in herbaceous cotton (*G. hirsutum* L. r. *latifolium*, Hutch) cropping, with constant occurrence in all cotton-growing states of Brazil (Nascimento et al. 2011). In 2000, da Silva identified entomopathogenic viruses, bacteria, fungi and protozoa from cotton fields of Brazil. Infestation of *Bacillus cereus* was reported from Colombo (Agudelo and Falcon 1977). Several insecticides and 2 formulations of *B. thuringiensis* were evaluated against the cotton leafworm, *A. argillacea*, at 0.024 and 0.032 kg a.i./ha as wettable powder, and mineral oil formulations did not give a good control of the larvae, but the reduction of the pupae was significantly 6 days after the treatment, indicating a secondary effect on the larvae before pupation (Yamamoto et al. 1990).

In Brazil, application of *B. thuringiensis* (14–21 g a.i./ha) was as good as the standard methyl parathion [parathion-methyl] (187 g a.i./ha) (Bleicher et al. 1990). Later, César Filho et al. (2002) used *Metarhizium anisopliae* (Metsch.) and *Beauveria bassiana* (Bals.) isolates for managing this pest. Results revealed that the isolate 645 of *B. bassiana* caused the highest mortality at the highest concentration, followed by isolates 634, 604 and IPA 198. The lowest lethal time for *B. bassiana* and *M. anisopliae* was achieved by the isolates 483 (4.1 days) and 1,189 (2.0 days), respectively. The isolates 1,189, 1,022 and 866 of *M. anisopliae* and 483, IPA198 and 604 of *B. bassiana*, at 10^8 and 10^9 conidia/ml are promising for use in the integrated control of *A. argillacea* larvae, but *M. anisopliae* seems more effective.

12.6.2 Anomis flava

Cotton looper *A. flava* is recorded from India, Australia, Spain, South Asian countries, Africa (Algeria, Cameroon, DR Congo, Eritrea, Ethiopia, Gabon, Gambia, Ghana, Kenya, La Reunion, Liberia, Madagascar, Morocco, Sierra Leone, Somalia, South Africa, Tanzania, Tunisia, Uganda, Zimbabwe), the Philippines, etc. Larvae feed mainly on leaves and occasionally on squares and boll surfaces. They prefer older leaves, and therefore, their damage will progress upwards on the plant. Cotton yield can be reduced if leaf tissue loss is excessive. Generally, plants are more sensitive earlier and increasingly less sensitive later. As a rough guide, leaf area loss of greater than 10–15 % could result in yield loss if it occurred before crop cut-out. After cut-out, losses of up to 15–20 % could be tolerated with low risk. According to Deutscher et al. (1999), unsprayed cotton fields in Australia have as much as 80 % of defoliation under infestation of *A. flava*.

In India, Umesh Chandra and Regupathy (2007) conducted an experiment to investigate the Cry 1Ac toxin expression and its manifestation in *A. flava* larval susceptibility, by excised leaf technique. Results revealed that toxin caused cent per cent mortality to third instar, *A. flava*. *B. thuringiensis* (Bt) products are very effective against this pest (http://www.aciar.gov.au/files/mn-157/imp11.html).

12.6.3 Helicoverpa zea

Corn earworm *H. zea* (Boddie) is a well-known pest of corn and cotton particularly in the Southern United States (Swenson et al. 2013). It attacks both non-Bt and Bt cotton. However, occurrence of larvae and ear damage on Bt corn was significantly lower than on non-Bt plants, and there were no significant differences between pure stands of Bt and 'RIB' plantings across all trials (Yang et al. 2014). *B. thuringiensis*

(Bt) toxin has reduced the use of synthetic insecticide on transgenic crops to target *Helicoverpa* spp., the major insect pest of cotton in Australia (Mensah and Austin 2012). This pest has been utilised for the mass production of entomopathogenic viruses (Reid et al. 2013), because it has susceptibility against Bt (Bailey et al. 1998).

Fungal metabolite sclerotia produced by *Aspergillus* spp. have proven to be a rich source of novel anti-insect compounds with activity against *H. zea*. Similarly, the hexane and chloroform extracts of *Eupenicillium crustaceum* displayed significant anti-insect activity in assays against this pest. Further, in dietary assay, reduction in weight gain and reduction in feeding rate were also observed in *H. zea* (Wang et al. 1995). Champlin and Grula (1979) reported that beauvericin was not toxic to *H. zea* and that bassianolide caused temporary atony.

For the first time, Pekrul and Grula (1979) reported the mechanism of action of fungi on this pest. Entomopathogenic fungus *Beauveria bassiana* (Balsamo) Vuillemin depicted direct penetration through the integument of a corn earworm, *H. zea*, without appressorial formation.

Steinernema riobravis was isolated from soil samples in corn fields near Weslaco, Texas (Cabanillas et al. 1994). Since then, successful results have been obtained with *S. riobravis* for the control of corn earworm (Cabanillas and Raulston 1994). Cabanillas and Raulston (1994) observed that prepupae of *H. zea* (Boddie) exhibited susceptibility against EPN. Later, two EPNs, *S. riobravis* and *S. carpocapsae*, were compared for their ability to parasitise corn earworm, *H. zea* (Boddie) prepupae and pupae, in corn plots at the Lower Rio Grande Valley of Texas (Cabanillas and Raulston 1996). Parasitism was higher when *S. riobravis* was applied at 200,000 IJ/m^2 through furrow irrigation (97 %) or post-irrigation (95 %) than when nematodes were sprayed onto the soil before irrigation (82 %). Parasitism of corn earworm prepupae by *S. riobravis* persisted up to 36 days after application and was higher in the treated plots (80 %) than the natural parasitism of the control plots (14 %). These results show that at high field soil temperatures, *S. riobravis* is more effective against corn earworm than *S. carpocapsae*.

12.6.4 Spodoptera *spp.*

12.6.4.1 *S. littoralis* (Boisd.)

Cotton leafworm or Egyptian leafworm, *S. littoralis* (Boisd.), a polyphagous insect herbivore, consumes large amounts of plant material in a short time (says Anderson and Alborn 1999; Shao et al. 2014). *B. bassiana* isolates have proven to be pathogenic to fall armyworm, *S. frugiperda* (J. E. Smith) (Rodrigues and Pratissoli 1989; Franca et al. 1989; Faria et al. 1992). Nuclear polyhedrosis virus (NPV) of *S. littoralis* ($1.2 \times 10^{6-8}$ polyhydra/larva), fed to third or fifth instar larvae, caused mortality of 54–100 % or 40–73 % of these larvae, respectively. It also reduced pupal weight, increased adult deformities and reduced fecundity (Abul-Nasr et al. 1979). Later in Egypt, a locally produced wettable powder formulation of an NPV [5×10^{11}, 1×10^{12} and 5×10^{12} polyhedral inclusion bodies (p.i.b.)/ha] was applied to control the Egyptian cotton leafworm, *S. littoralis*, using a knapsack sprayer fitted with a cotton tail boom. An application rate of 1×10^{12} p.i.b./ha reduced the level of this pest (Jones et al. 1994).

The crude soluble protein of *M. anisopliae*, *B. bassiana*, *B. brongniartii* and *Scopulariopsis brevicaulis* was screened against *S. littoralis* larvae. The extracts from two *M. anisopliae* and two *B. bassiana* isolates gave significant mortalities when either applied on leaf discs or incorporated into artificial diet (Quesada-Moraga et al. 2006). Further, they recorded in leaf disc assays that this extract exhibited strong dose-related toxic and antifeedant activity against the larvae. Not only the antifeeding index was dose related, but it significantly increased over time in a dose-related manner. The crude extract when exposed to higher temperature or protease treatment lost toxicity, indicating that toxicity was protein mediated. Along with the above said impact, a progressive bleeding of the midgut epithelium into the gut lumen was observed along with the lysis of the epithelium and deterioration of the microvilli. Recently, Asi et al. (2013) critically evaluated the potential of entomopathogenic fungi, *Isaria fumosorosea* (= *Paecilomyces fumosoroseus* (Wize) Brown and Smith), *B. bassiana*, *M. anisopliae* Sorokin and *Lecanicillium lecanii* (Zimmerman) (= *Verticillium lecanii*) for the biocontrol of various life stages of *S. litura* on cotton.

Chitinolytic bacterial strains induced *S. littoralis* larval mortality when combined with *B. thuringiensis* (Sneh et al. 1983) or individual Bt strains (Kalfon and De Barjac 1985; Sneh et al. 1991). This shows that we can use either bacteria or virus or their combination to reduce this pest on cotton. Previously, Moore and Navon (1973) observed that ten different laboratory strains of *B. thuringiensis* caused moderate mortality when compared to control. Later, Keller et al. (1996) explained how *B. thuringiensis* δ-endotoxins in crystalline and noncrystalline forms kill *S. littoralis*. But later, the pest developed resistance against Bt (Muller-Cohn et al. 1996). As a result, many synthetic biopesticides (e.g. methoxyfenozide, tebufenozide, spinosad, limonoids) have been on use for management. However, other entomopathogens have been utilised for this pest management. For instance, laboratory strains of *M. anisopliae*, *M. flaviviridae*, *P. farinosis*, *B. bassiana*, and *B. brongniartii* (Amer et al. 2008), *B. subtilis* strain NRC313 (*BS* NRC313) and *B. thuringiensis* strain NRC335 (*BT* NRC335) (Abd El-Salam et al. 2011), new strain CCM 8367 of *I. fumosorosea* (Hussein et al. 2013), Bt var. *kurstaki* (Btk2, Btk3 and Btk66) and Bt var. *mexicanensis* (Btm27), in addition to two reference strains (4D20 and 4AC1) (Alfazairy et al. 2013), at Egypt, were tested against this pest and showed insecticidal activity. In Saudi Arabia, using bacterial transconjugant technology, AlOtaibi (2013) showed combined effects of crystals +endospores of *B. thuringiensis* serovar kurstaki, *B. subtilis* and four of their transconjugants against this pest which showed higher mortality, and this factor was important to be considered in designing resistance management strategies.

Similarly, nematodes also have been in use for this management. In 2006, Razek recorded infectivity of nematodes against this pest. In laboratory studies, demonstrated that five native EPN species/isolates caused 100 % mortality of *S. cilium* larvae, a soil surface-feeding pest of turfgrass. At 25 infective juveniles/ cm^2 applied to sod, two selected Turkish species, *Steinernema carpocapsae* and *Heterorhabditis bacteriophora* (Sarigerme isolate), showed 77 % and 29 % average larval mortality, respectively (Gulcu et al. 2014).

12.6.4.2 *S. litura* (Fab.)

Endophytic fungi are a group of microbial plant symbionts that occur in living tissues of plants without causing visible disease symptoms. *S. litura* (F.) fed on leaves of *Nigrospora oryzae*- and *Cladosporium uredinicola*-infected plants exhibited abnormalities such as change in shape, extensive vacuolisation and necrosis in significantly higher percentage of haemocytes (Thakur et al. 2014). Entomopathogenic fungi *B. bassiana* at four different concentrations (2.4×10^7, 2.4×10^6, 2.4×10^5, 2.4×10^4 conidia/ml) were tested against *S. litura* larvae. Results revealed that the fungi reduced pupation (43.33 %) at 2.4×10^7 conidia/ ml, caused mortality, adult malformation and completely arrested fecundity (Malarvannan et al. 2010).

B. thuringiensis (Bt) is a microbial pesticide widely used to control crop pests. Its strains have good biocontrol activity against crop insect pest and, however, lack some desirable characteristics that are found in *B. subtilis*. Revathi et al. (2013), using protoplast fusion technique, fused *B. thuringiensis* with a strain of *B. subtilis*, and the fusants produced almost 95 % mortality in first instar larvae. Baculovirus infection not only disturbs moulting but also affects digestive physiology (Subrahmanyam and Ramakrishnan 1981). Similarly in another study, a transgenic *B. bassiana* strain (BbV28) expressing Vip3Aa1 (a Vip3A toxin) was created to infect the larvae of *S. litura* through conidial ingestion and cuticle adhesion and tested against this pest. Feeding reduced the LC_{50} of the transformant by 17.2 and 1.3-fold on days 3 and 7, respectively. Median lethal times (LT_{50}s) of BbV28 were shortened by 23–35 %, declining with conidial concentrations. The larvae infected by ingestion of BbV28 conidia showed typical symptoms of Vip3A action, i.e. shrinkage and palsy. However, neither LC_{50} nor LT_{50} trends differed between BbV28 and its parental strain if the infection occurred through the cuticle only. Our findings indicate that fungal conidia can be used as vectors for spreading the highly insecticidal Vip3A protein for the control of foliage feeders such as *S. litura* (Qin et al. 2010).

Though NPV alone has more impact on *S. litura* (Sahayaraj and Joseph 2003), Senthil-Nathan and Kalaivani (2005, 2006) studied the combined effect of NPV and azadirachtin and suggested this combination for this pest management. Similar concept was also proposed by Gopalakrishnan et al. (2013). They reported that extracts of Annona, *Chrysanthemum*, *Datura*, *Jatropha*, Neem, *Parthenium*, *Pongamia*, *Tridax* and *Vitex*; and plant growth-promoting (PGP) bacteria, *viz.*, *B. subtilis* (BCB-19), *B. megaterium* (SB-9), *Serratia marcescens* (HIB-28) and *Pseudomonas* spp. (SB-21), and fungus (*M. anisopliae*), were evaluated for their efficacy against *S. litura* and showed high mortality along with weight reduction of the larvae (73 % and 91 %). It was therefore concluded that the aforementioned six botanicals and five entomopathogens have great potential in the management of *S. litura*. Further, it was proposed that botanicals and microorganisms have the capability to synthesise biologically active secondary metabolites such as antibiotics, herbicides and pesticides.

12.6.5 *Acontia graellsii*

Cotton semilooper (*A. graellsii = Acontia draellsii, Xanthodes graellsii, Xanthodes fimbriata, Xanthodes innocens*) has been recorded from various parts of India and Africa (Equatorial Guinea, Ethiopia, Gambia, Ghana, La Reunion, Madagascar, Malawi, Mauritania, Nigeria, Sierra Leone, Somalia, South Africa, Sudan, Zambia, Zimbabwe). However, not much work has been carried out in Africa for this pest management. Initially, occurrence of the entomopathogenic fungi *Nomuraea rileyi* (Farlow) Samson on *A. graellsii* F. (Noctuidae: Lepidoptera) (Gopalakrishnan and Narayanan 1988) was reported.

12.6.6 *Sylepta derogata*

Cotton leaf roller, *S. derogata* (Fabricius) (Lepidoptera: Crambidae), is distributed in India, Nigeria and China. In India, chlorpyriphos, dichlorvos, fenitrothion, monocrotophos, quinalphos, phenthoate, phosalone, tetrachlorvinphos, dicrotophos and endosulfan are recommended for this pest management. In China, along with few insecticides (flufenoxuron, dichlorvos, cypermethrin and abamectin), *B. thuringiensis* has been utilised for this control and showed good results (Wu et al. 2008). In India, *P. farinosus* was shown to infect larvae of *S. derogata* (Kuruvilla and Jacob 1980). Later, Jiji et al. (2005) identified *B. bassiana* as one of the best fungi for this pest control.

12.6.7 *Agrotis ipsilon*

Black cutworm is distributed in Europe, China, India, Canada and North America. It has a wide host range. Nearly all vegetables can be consumed, and this species also feeds on alfalfa, clover, cotton, rice, sorghum, strawberry, sugar beet, tobacco, and sometimes grains and grasses. Larvae can consume over 400 cm^2 of foliage during their development, but over 80 % occurs during the terminal instar and about 10 % in the instar immediately preceding the last. Kunkel et al. (2004) reported that the black cutworm, *A. ipsilon*, is less susceptible to the EPN, *S. carpocapsae*, when it consumes the endophytic grass. Two entomopathogenic fungi, *M. anisopliae* and *B. bassiana* were tested against this insect. Results show that the damage was higher in the earlier stages of seedling growth, whereas in later stages they gave good protection to the crop (Viji and Bhagat 2001). Tripathi et al. (2003) utilised *B. thuringiensis* (Dipel 8L, subsp. *kurstaki*) against *A. ipsilon* for 2 years in Garhwal Himalayan region, Uttar Pradesh, India, and they succeeded to manage this pest under field conditions.

12.6.8 Mites

Cotton in the early stage as well as at the late stage is attacked by mites. They are generally found on the undersurface of leaves wire fine webbings. During heavy infestation, they may be found all over the leaf surface. Mites puncture the leaf tissue and the oozing plant sap is sucked. Removal of plant sap with chlorophyll and other plant pigments results in characteristic blocking with reddish bronze discoloration of leaves. Severe infestation leads to premature defoliation of leaves. Application of dicofol or wettable sulphur reduces the mite incidence and further infestation.

In China, aerial conidia of isolates of *B. bassiana* (Bb734 and Bb2860) and *M. anisopliae* (Ma456 and Ma759) was mixed with an emulsifiable oil and sprayed in block-randomised triple plots of two irrigated cotton fields for the control of summer populations of cotton spider mites, mainly *Tetranychus truncates* and *T. turkestani*. Results revealed that overall means of relative efficacies during the periods of both trials were 85.8 % (77.9–94.9 %) and 88.0 % (82.4–94.0 %) for Ma456 and 77.9 % (68.6–89.6 %) and 85.7 % (77.8–87.7 %) for Bb734 (Shi et al. 2008). Very recently, Bt has been utilised to manage cotton mites at field level in Brazil (Agostini et al. 2014).

12.6.9 Grasshoppers

Grasshoppers are considered as occasional as well as seasonal pests of cotton. Grasshoppers feed on foliage, most often on the edges of fields near pasture areas or roadsides. They seldom cause economically significant injury. Since grasshoppers are highly mobile and migratory pests, no specific entomopathogens have been recommended so far. However, in China, a new species of *Sporothrix* was isolated from cotton-dwelled grasshoppers (Huang and Zhen 1997).

12.7 Conclusion and Recommendations

The demand for safe and 'biologically' healthy foods has stimulated an increase of research on biological control of pests. Pest control through these microorganisms, naturally found in soil, is an important ecosystem service that maintains the stability of agricultural systems and has the potential to mitigate costs for control of pests. Considering the above-mentioned available literature, the following recommendations are put forth for cotton defoliator management worldwide:

1. Since not much work has been available on *A. graellsii* in Africa, an important pest of cotton, available entomopathogens can be utilised for this pest management.
2. *A. argillacea* is an important pest of cotton in Brazil. Hence, we recommend to utilise as many as entomopathogens for the management.
3. Many protozoans and nematodes can be utilised in cotton defoliators control.
4. Physical, chemical and molecular mechanism involved in the action of EPs on cotton defoliators should be studied to utilise and popularise the microbial insecticides among farmers.
5. Specificity of EPs against the cotton pest should be known.
6. Compatibility with other pest management components can be known for cotton defoliators.
7. Agro-climate-based EPs should be identified to manage the defoliators at different locations.
8. Mite infestation can be considered seriously and can be reduced using the proposed option.

Acknowledgement The author is grateful to the management, St. Xavier's College, Palayamkottai, for the laboratory support and encouragement.

References

Abd El-Salam AME, Nemat AM, Magdy A (2011) Potency of *Bacillus thuringiensis* and *Bacillus subtilis* against the cotton leafworm, *Spodoptera littoralis* (Boisd.) Larvae. Arch Phytopathol Plant Protect 44:204–215

Abul-Nasr SE, Ammar ED, Abul-Ela SM (1979) Effects of nuclear polyhedrosis virus on various developmental stages of the cotton leafworm *Spodoptera littoralis* (Boisd.). Z Angew Entomol 88:181–187. doi:10.1111/j.1439-0418.1979.tb02494.x

Agostini LT, Duarte RT, Volpe HXL, Agostini TT, de Carvalho GA, Abrahão YP, Polanczyk RA (2014) Compatibility among insecticides, acaricides, and *Bacillus thuringiensis* used to control *Tetranychus urticae* (Acari: Tetranychidae) and *Heliothis virescens* (Lepidoptera: Noctuidae) in cotton fields. Afr J Agric Res 9:941–949

Agudelo F, Falcon LA (1977) Some naturally occurring insect pathogens in Colombia. Turrialba 27:423–424

Alfazairy AA, El-Ahwan AM, Mohamed EA, Zaghloul HA, El-Helow ER (2013) Microbial control of the cotton leafworm *Spodoptera littoralis* (Boisd.) by Egyptian *Bacillus thuringiensis* isolates. Folia Microbiol 58:155–162

AlOtaibi SA (2013) Mortality responses of *Spodoptera litura* following feeding on BT-sprayed plants. J Basic Appl Sci 9:195–215

Amer MM, El-Sayed TI, Bakheit HK, Moustafa SA, El-Sayed YA (2008) Pathogenicity and genetic variability of five entomopathogenic fungi against *Spodoptera littoralis*. Res J Agric Biol Sci 4:354–367

Anderson P, Alborn H (1999) Effects on oviposition behaviour and larval development of *Spodoptera littoralis* by herbivore-induced changes in cotton plants. Entomol Exp Appl 92:45–51

Asi MR, Bashir MH, Afzal M, Zia K, Akram M (2013) Potential of entomopathogenic fungi for biocontrol of *Spodoptera litura* Fab. (Lepidoptera: Noctuidae). J Animal Plant Sci 23:913–918

Bailey WD, Zhao G, Carter LM, Gould F, Kennedy GG, Roe RM (1998) Feeding disruption bioassay for species and *Bacillus thuringiensis* resistance diagnosis for *Heliothis virescens* and *Helicoverpa zea* in cotton (Lepidoptera: Noctuidae). Crop Prot 17:591–598

Bleicher E, de Jesus FMM, de Sousa SL (1990) Use of selective insecticides to control the cotton leaf worm. Pesq Agrop Brasileira 25:277–280

Bottrell DG, Adkisson PL (1977) Cotton insect pest management. Annu Rev Entomol 22:451–481

Bourne BA (1921) Report of the assistant director of agriculture on the entomological and mycological work carried out during the season under review. Report, 1919–1920. Department of Agriculture, Barbados, pp 10–31

Butt TM, Jackson C, Magan N (2001) Fungi as biocontrol agents: progress, problems and potential. CAB International, Wallingford

Cabanillas HE, Raulston JR (1994) Pathogenicity of *Steinernema riobravis* against corn earworm, *Helicoverpa zea* (Boddie). Fundam Appl Nematol 17:219–223

Cabanillas HE, Raulston JR (1996) Evaluation of *Steinernema riobravis*, *S. carpocapsae*, and irrigation timing for the control of corn earworm, *Helicoverpa zea*. J Nematol 28:75

Cabanillas HE, Poinar GO Jr, Raulston JR (1994) *Steinernema riotrravis* n. sp. (Rhabditida: Steinernematidae) from Texas. Fundam Appl Nematol 17:123–131

César Filho E, Marques EJ, Barros R (2002) Selection of *Metarhizium anisopliae* (metsch.) and *Beauveria bassiana* (bals.) isolates to control *Alabama argillacea* (huebner) caterpillars. Sci Agric 59:457–462

Champlin FR, Grula EA (1979) Noninvolvement of beauvericin in the entomopathogenicity of *Beauveria bassiana*. Appl Environ Microbiol 37:1122–1125

Clay J (2004) World agriculture and the environment: a commodity-by-commodity guide to impacts and practices. Island Press, Washington, DC

da Silva CAD (2000) Entomopathogenic microorganisms associated with insects and mites of cotton. Documentos-Embrapa Algodão 77:45

Deutscher SL, Wilson S, Whiteside ST, Mansfield M, Rossiter WL (1999) Pest and beneficial guide. Australian Cotton Cooperative Research Centre, Narrabri, p 218

El-Guindy MA, Madi SM, Keddis ME, Issan YH, Abdel-Satto MM (1982) Development of resistance to pyrethroids in field populations of the Egyptian cotton leafworm, *Spodoptera littoralis* Boisd. Int Pest Control 24:6–11

Faria LLF, de Oliveia JV, Barros R (1992) Patogenicidade do fungo *Beauveria bassiana* (Bals.) Vuill., em lagartas de *Spodoptera frugiperda* (J. E. Smith, 1797) (Lepidoptera: Noctuidae) sob condições de laboratório. Caderno Ômega 4:207–217

Ferrigno S, Ratter SG, Ton P, Vodouhe DS, Williamson S, Wilson J (2005) Organic cotton: a new development path for African Smallholders? vol 120, Gatekeeper series. International Institute for Environment and Development, London

Franca MM, Tigano MS, Carvalho RS (1989) Suscetibilidade de *Spodoptera frugiperda* aos fungos entomopatogênicos *Beauveria bassiana* e *Nomuraea rileyi*. In: Congresso Brasileiro de Entomolgia, vol 12. Belo Horizonte, Resumos. SEB, Belo Horizonte, p 254

Frisbie RE, El Zik K, Wilson LT (1989) Integrated pest management systems and cotton production. Wiley, New York, p 437

Geiger DR, Servaites JC (1991) Carbon allocation and response to stress. In: Mooney HA, Winner WW (eds) Response to stress. Academic, New York, pp 103–127

Gopalakrishnan M, Narayanan K (1988) Occurrence of the entomopathogenic fungi *Nomuraea rileyi* (Farlow) Samson on *Acontia graellsii* F. (Noctuidae: Lepidoptera) and *Beauveria bassiana* (Balsamo) Vuill. on *Myllocerus subfaciatus* G. (Curcolionidae: Coleoptera). J Biol Control 2:58–59

Gopalakrishnan S, Rao GR, Humayun P, Rao VR, Alekhya G, Jacob S, Rupela O (2013) Efficacy of botanical extracts and entomopathogens on control of *Helicoverpa armigera* and *Spodoptera litura*. Afr J Biotechnol 10:16667–16673

Gulcu B, Ulug D, Hazir C, Karagoz M, Hazir S (2014) Biological control potential of native entomopathogenic nematodes (Steinernematidae and Heterorhabditidae) against *Spodoptera cilium* (Lepidoptera: Noctuidae) in turfgrass. Biocontrol Sci Technol 24(8):965–970

Huang BL, Zhen MZL (1997) *Sporothrix chondracris*, a new entomopathogenic fungus on cotton. Hycosena 16:88–90

Hussein HM, Zemek R, Habuštová SO, Prenerová E, Adel MM (2013) Laboratory evaluation of a new strain CCM 8367 of *Isaria fumosorosea* (syn. *Paecilomyces fumosoroseus*) on *Spodoptera littoralis* (Boisd.). Arch Phytopathol Plant Protect 46:1307–1319

Jiji T, Praveena R, Naseema A, Anitha N (2005) *Beauveria bassiana* (Bals) Vuill. on *Sylepta derogata* Fab. and its cross infectivity on other major pests of vegetables. Insect Environ 11:144

Jones KA, Irving NS, Grzywacz D, Moawad GM, Hussein AH, Fargahly A (1994) Application rate trials with a nuclear polyhedrosis virus to control *Spodoptera littoralis* (Boisd.) on cotton in Egypt. Crop Prot 13:337–340

Kalfon AR, De Barjac H (1985) Screening of the insecticidal activity of *Bacillus thuringiensis* strains against the egyptian cotton leaf worm *Spodoptera littoralis*. Entomophaga 30:177–186

Keller M, Sneh B, Strizhov N, Prudovsky E, Regev A, Koncz C, Zilberstein A (1996) Digestion of δ-endotoxin by gut proteases may explain reduced sensitivity of advanced instar larvae of *Spodoptera littoralis* to CryIC. Insect Biochem Mol Biol 26:365–373

Kogan M (1988) Ecological theory and integrated management – theory and practice. Wiley, New York, p 362

Kogan M (1998) Integrated pest management: historical perspectives and contemporary developments. Annu Rev Entomol 43:243–270

Kunkel BA, Grewal PS, Quigley MF (2004) A mechanism of acquired resistance against an entomopathogenic nematode by *Agrotis ipsilon* feeding on perennial ryegrass harboring a fungal endophyte. Biol Control 29:100–108

Kuruvilla S, Jacob A (1980) Pathogenicity of the entomogenous fungus *Paecilomyces farinosus* (Dickson ex Fries) to several insect pests. Entomon 5:175–176

Malarvannan S, Murali PD, Shanthakumar SP, Prabavathy VR, Nair S (2010) Laboratory evaluation of the entomopathogenic fungi, *Beauveria bassiana* against the Tobacco caterpillar, *Spodoptera litura* Fabricius (Noctuidae: Lepidoptera). J Biopestic 3:126–131

Mensah RK, Austin L (2012) Microbial control of cotton pests. Part I: use of the naturally occurring entomopathogenic fungus Aspergillus sp. (BC 639) in the management of Creontiades dilutus (Stal) (Hemiptera: Miridae) and beneficial insects on transgenic cotton crops. Biocontrol Sci Technol 22:567–582

Mooney HA, Winner WE, Pell EJ (eds) (1991) Response of plants to multiple stresses. Academic, New York, p 422

Moore I, Navon A (1973) Studies of the susceptibility of the cotton leafworm, *Spodoptera littoralis* (Boisduval), to various strains of *Bacillus thuringiensis*. Phytoparasitica 1:23–32

Muller-Cohn J, Chaufaux J, Buisson C, Gilois N, Sanchis V, Lereclus D (1996) *Spodoptera littoralis* (Lepidoptera: Noctuidae) resistance to CryIC and cross-resistance to other *Bacillus thuringiensis* crystal toxins. J Econ Entomol 89:791–797

Nascimento AR, Ramalho FS, Azeredo TL, Fernandes FS, Júnior JN, Silva CAD, Malaquias JB (2011) Feeding and life history of *Alabama argillacea* (Lepidoptera: Noctuidae) on cotton cultivars producing colored fibers. Ann Entomol Soc Am 104:613–619

Patrick CD (2004) Grasshoppers and their control. http://www.hdl.handle.net/1969.1/87825

Pekrul S, Grula EA (1979) Mode of infection of the corn earworm (*Heliothus zea*) by *Beauveria bassiana* as revealed by scanning electron microscopy. J Invertebr Pathol 34:238–247

Poinar GO Jr (1979) Steinernematids and heterorhabditis have been observed to infect over 200 species of insects from several orders. In: Swan JL, Kaya HK (eds) A few other arthropods are also susceptible to EPNs: a handbook of techniques, CRC Press, Inc., p 277

Qin Y, Ying SH, Chen Y, Shen ZC, Feng MG (2010) Integration of insecticidal protein Vip3Aa1 into *Beauveria bassiana* enhances fungal virulence to *Spodoptera litura* larvae by cuticle and per os infection. Appl Environ Microbiol 76:4611–4618

Quesada-Moraga E, Carrasco-Díaz JA, Santiago-Álvarez C (2006) Insecticidal and antifeedant activities of proteins secreted by entomopathogenic fungi against *Spodoptera littoralis* (Lep., Noctuidae). J Appl Entomol 130:442–452. doi:10.1111/j.1439-0418.2006.01079.x

Razek AS (2006) Infectivity prospects of both nematodes and bacterial symbionts against cotton leafworm, *Spodoptera littoralis* (Biosduval) (Lepidoptera: Noctuidae). J Pest Sci 79:11–15

Reid S, Chan L, van Oers MM (2013) Production of entomopathogenic viruses. In: Morales-Ramos JA, Rojas MG, Shapiro-Ilan DI (eds) Mass production of beneficial organisms: invertebrates and entomopathogens. Academic, London, pp 437–482

Revathi K, Chandrasekaran R, Thanigaivel A, Arunachalam Kirubakaran S, Senthil-Nathan S (2013) Biocontrol efficacy of protoplast fusants between *Bacillus thuringiensis* and *Bacillus subtilis* against *Spodoptera litura* Fabr. Arch Phytopathol Plant Protect 47(11):1365–1375

Ribeiro RC, Lemos WDP, Poderoso JCM, Pikart TG, Zanuncio JC (2013) New record of grasshopper (Orthoptera: Acrididae & Romaleidae) defoliators and population dynamics of insects on crops of *Heliconia* spp in the Amazon. Florida Entomol 96:225–228

Rodrigues C, Pratissoli D (1989) Avaliação de patogenicidade dos fungos entomógenos *Beauveria bassiana* e *Metarhizium anisopliae* sobre *Spodoptera frugiperda* (lagarta do cartucho). In: Congresso Brasileiro De Entomologia, vol 12. Belo Horizonte, Resumos. SEB, Belo Horizonte, p 223

Ruiu L, Satta A, Floris I (2013) Emerging entomopathogenic bacteria for insect pest management. Bull Insectol 66:181–186

Sahayaraj K, Joseph M (2003) Impact of (S) NPV on *Spodoptera litura* (Fab.) mortality and flora. Nat Conserv 15:43–50

Senthil-Nathan S, Kalaivani K (2005) Efficacy of nucleopolyhedrovirus (NPV) and azadirachtin on *Spodoptera litura* Fabricius (Lepidoptera: Noctuidae). Biol Control 34:93–98

Senthil-Nathan S, Kalaivani K (2006) Combined effects of azadirachtin and nucleopolyhedrovirus (SpltNPV) on *Spodoptera litura* Fabricius (Lepidoptera: Noctuidae) larvae. Biol Control 36:94–104

Shao Y, Arias-Cordero E, Guo H, Bartram S, Boland W (2014) In vivo pyro-SIP assessing active gut microbiota of the cotton leafworm, *Spodoptera littoralis*. PLoS One 9:e85948

Shi WB, Zhang LL, Feng MG (2008) Field trials of four formulations of *Beauveria bassiana* and *Metarhizium anisopliae* for control of cotton spider mites (Acari: Tetranychidae) in the Tarim Basin of China. Biol control 45:48–55

Sneh B, Schuster S, Gross S (1983) Improvement of the insecticidal activity of *Bacillus thuringiensis* var. *entomocidus* on larvae of *Spodoptera littoralis* (Lepidoptera, Noctuidae) by addition of chitinolytic bacteria, a phagostimulant and a UV-protectant. Z Angew Entomol 96:77–83. doi:10.1111/j.1439-0418.1983.tb03644.x

Sneh B, Schuster S, Broza M (1991) Insecticidal activity of *Bacillus thuringiensis* strains against the Egyptian cotton leafworm, *Spodoptera littoralis* Boisd. (Lep. Noctuidae). Entomophaga 26:179–190

Subrahmanyam B, Ramakrishnan N (1981) Influence of a baculovirus infection on molting and food consumption by *Spodoptera litura*. J Invertebr Pathol 38:161–168

Swenson SJ, Prischmann-Voldseth DA, Musser FR (2013) Corn earworms (Lepidoptera: Noctuidae) as pests of soybean. J Integr Pest Manag 4:D1–D8

Ted Wilson L, Carter F (1991) Leaf feeding insects and mites. Phys Today 2:1–8

Thakur A, Singh V, Kaur A, Kaur S (2014) Suppression of cellular immune response in *Spodoptera litura* (Lepidoptera: Noctuidae) larvae by endophytic fungi *Nigrospora oryzae* and *Cladosporium uredinicola*. Ann Entomol Soc Am 107:674–679

Tripathi DM, Bisht RS, Mishra PN (2003) Bio-efficacy of some synthetic insecticides and bio-pesticides against black cutworm, *Agrotis ipsilon* infesting potato (*Solanum tuberosum*) in Garhwal Himalaya. Indian J Entomol 65:468–473

Umesh Chandra S, Regupathy A (2007) Exploring the Bt cotton potential for the management of semilooper complex, *Anomis flava* Fab. and *Tarache nitidula* Fab. In: World cotton research conference-4, 10–14 September 2007. International Cotton Advisory Committee (ICAC), Lubbock

Vega FE, Kaya HK (2012) Insect pathology, 2nd edn. Elsevier, London

Viji CP, Bhagat RM (2001) Bioefficacy of some plant products, synthetic insecticides and entomopathogenic fungi against black cutworm, *Agrotis ipsilon* larvae on maize. Indian J Entomol 63:26–32

Wang H, Gloer JB, Wicklow DT, Dowb PF (1995) Aflvinines and other antiinesctan metabolites from the ascostroma of *Eupenicillium crustaceum* and related species. Appl Environ Microbiol 61:4429–4435

Williams T, Arredondo-Bernal HC, Rodríguez-del-Bosque LA (2013) Biological pest control in Mexico. Annu Rev Entomol 58:119–140

Wu J-H, Huang Z, Ren S-X, Zhou H-P (2008) Control of different insecticides in field on *Sylepta derogata* population. J South China Agric Univ 02:6

Yamamoto PT, Benetoli I, Fernandes OD, Gravena S (1990) Effect of *Bacillus thuringiensis* and various insecticides on the cotton leafworm (*Alabama argillacea*) and its predators. Ecossistema Faculdade de Agronomia e Zootecnia Manoel Carlos Goncalves, Brazil 15:36–44

Yang F, Kerns DL, Head GP, Leonard BR, Niu Y, Huang F (2014) Occurrence, distribution, and ear damage of *Helicoverpa zea* (Lepidoptera: Noctuidae) in mixed plantings of non-Bt and Bt corn containing Genuity SmartStax™ traits. Crop Prot 55:127–132

Chapter 13
Entomopathogenic Nematodes and Their Bacterial Symbionts as Lethal Bioagents of Lepidopteran Pests

Sharad Mohan

13.1 Introduction

Entomopathogenic nematodes (EPN) are soil-dwelling parasites of a wide range of insects. Their ubiquity and insecticidal specificity makes them ideally suited to form part of integrated pest management programmes. Found widely in natural and agricultural ecosystems across the world, they are relatively specific to target pests and non-toxic to humans and the environment. Although EPN have been described from 23 nematode families (Koppenhöfer 2007), those belonging to the families Steinernematidae (Chitwood and Chitwood 1937) and Heterorhabditidae (Poinar 1976) are the ones most commonly studied due to their significant biocontrol potential. The nematode genera *Steinernema* and *Heterorhabditis* are taxonomically, biologically and commercially the most important ones, as they carry the lethal bacteria *Xenorhabdus* spp. and *Photorhabdus* spp., respectively, which are responsible for killing insects. The symbiotic nematode–bacteria pairs exhibit a wide spectrum of insecticidal activity across several economically important orders of the class Insecta. Considerable success has been achieved under field conditions in the biocontrol of several pests belonging to the orders Lepidoptera, Coleoptera and Diptera on high-value commercial crops.

Lepidoptera is second only to Coleoptera in its range of insect diversity, which includes pests destructive to agricultural, ornamental, landscape and forest vegetation. As *Steinernema* and *Heterorhabditis* are soil-borne, they attack a wide range of insect pests, which spend at least a part of their life cycle, particularly the feeding stages, in the soil. Many lepidopteran species overwinter as the last instar larvae or pupae in the soil or under fallen leaves and other debris on the soil surface. Generally, in cryptic habitats, chemical pesticides and biological control agents other than EPN are unsuccessful in reaching the overwintering stages of larvae or

S. Mohan (✉)
Division of Nematology, Indian Agricultural Research Institute, New Delhi 110012, India
e-mail: sharad@iari.res.in

© Springer International Publishing Switzerland 2015   273
K.S. Sree, A. Varma (eds.), *Biocontrol of Lepidopteran Pests*, Soil Biology 43,
DOI 10.1007/978-3-319-14499-3_13

pupae effectively. The foraging strategies of EPN are ideally suited for hunting out these lepidopteran pests, in their various stages, and their success against some commercially important species of pests in cryptic habitats has been quite encouraging.

Glaser's (1932) path-breaking report on the infection of the Japanese beetle by *Steinernema carpocapsae* gave new directions to the control of insect pests. With the inception of several new species of EPN pathogenic to insects and other invertebrates, biocontrol of insect pests using nematodes gained credence. The ease in rearing and handling of lepidopteran pests, such as *Galleria*, *Corcyra*, *Bombyx*, *Plutella*, *Helicoverpa*, *Spodoptera*, *Pieris*, *Agrotis*, etc., has made them the perfect hosts in laboratory investigations involving biocontrol by nematodes. These insects have served to effectively illustrate a range of EPN biocontrol attributes such as their biology, reproduction, in vivo mass production, pathogenicity and insect immunity.

Notwithstanding the overwhelming efficacy of EPN in laboratory bioassays in achieving mortality against almost all major lepidopteran pests, the results have not been replicated at the field level with similar success. For example, the EPN biocontrol of cutworms—*Agrotis*, *Amathes*, *Noctua*, *Peridroma* and *Prodenia* spp. (Lepidoptera: Noctuidae)—which are voracious feeders on leaf, bud, stem and roots of several crops and turf has not been successfully implemented on a large scale in field conditions (Shapiro-Ilan et al. 2002; Ebssa and Koppenhöfer 2011). However, there have been a few instances of field-level success too as detailed in Table 13.1.

13.2   Bacterial Symbionts of EPN

Phylogenetically, the two families Steinernematidae and Heterorhabditidae are dissimilar, but they share a similar life cycle through convergent evolution (Poinar 1993). These nematodes are characterized by their ability to carry specific pathogenic symbiotic bacteria, *Photorhabdus* (Boemare et al. 1993) with Heterorhabditidae and *Xenorhabdus* (Thomas and Poinar 1979) with Steinernematidae. The two bacteria belonging to the family *Enterobacteriaceae* are medium to long motile rods with peritrichous flagella and are gram-negative facultative anaerobes. They do not have an environmentally resistant stage and have been found to exist only as nematode vectors or insect hosts. *Xenorhabdus* occurs naturally in a special intestinal vesicle of the *Steinernema* infective juveniles (IJ) (Bird and Akhurst 1983), while *Photorhabdus* is located mainly in the anterior part of the *Heterorhabditis* IJ gut (Ciche and Ensign 2003). Both the bacteria are vectored by their respective nematode hosts into the haemocoel of the insect and released within 5 h of invasion. They suppress the insect immune response by producing an array of virulence factors, which eventually kill the host. The bacteria further contribute to the symbiotic relationship by secreting several exoenzymes that stimulate macromolecular degradation, the products of which together with the

Table 13.1 Current use of *Steinernema* and *Heterorhabditis* nematodes as biological control organisms (modified from Shapiro-Ilan and Gaugler 2010)

Crops	Pest common name	Pest scientific name	Efficacious nematodes
Artichokes	Artichoke plume moth	*Platyptilia carduidactyla*	Sc
Vegetables	Armyworm	Lepidoptera: Noctuidae	Sc, Sf, Sr
Ornamentals	Banana moth	*Opogona sacchari*	Hb, Sc
Turf, vegetables	Black cutworm	*Agrotis ipsilon*	Sc
Fruit trees, ornamentals	Borer	*Synanthedon* spp. and other sesiids	Hb, Sc, Sf
Pome fruit	Codling moth	*Cydia pomonella*	Sc, Sf
Vegetables	Corn earworm	*Helicoverpa zea*	Sc, Sf, Sr
Cranberries	Cranberry girdler	*Chrysoteuchia topiaria*	Sc
Grapes	Grape root borer	*Vitacea polistiformis*	Hz, Hb
Iris	Iris borer	*Macronoctua onusta*	Hb, Sc
Nut and fruit trees	Navel orangeworm	*Amyelois transitella*	Sc

1. Nematodes listed provided at least 75 % suppression of these pests in field or greenhouse experiments
2. Nematode species are abbreviated as follows: Hb *H. bacteriophora*, Hz *H. zealandica*, Sc *S. carpocapsae*, Sf *S. feltiae* and Sr *S. riobrave*
3. Efficacy against various pest species within this group varies among nematode species

bacteria themselves are thought to provide a nutrient-rich base to their nematode partners (Waterfield et al. 2001; Cowles and Goodrich-Blair 2005; Clarke 2008). The bacteria proliferate exponentially and produce diverse antimicrobial compounds that inhibit the growth of a wide range of organisms and thus protect the cadavers from purification (Akhurst 1982; Furgani et al. 2008). The nematodes feed and reproduce inside the insect cadavers until their growing numbers and diminishing nutrition compel the IJ to re associate with the bacteria and emerge into the soil to look for new hosts (Martens et al. 2003; Snyder et al. 2007).

13.3 Foraging Strategy

The third-stage juvenile or the dauer larva is the infective stage for both *Steinernema* and *Heterorhabditis*. The IJ occur naturally as free-living forms in the soil and are resistant to the environmental conditions (Poinar and Georgis 1990). They contain carbohydrate energy reserves which help them to survive for long periods in the soil under unfavourable temperature, moisture and aeration conditions. The ambient conditions required for survival, infection and reproduction vary with the nematode species and their natural habitat.

IJ have a unique ability to locate their hosts in both soil and cryptic habitats. An understanding of this ability is fundamental to arriving at precise EPN–pest matches and thus enabling effective biocontrol. IJ adopt two modes, ambushing and cruising (Gaugler et al. 1989), to locate and infect an insect larva. An ambushing nematode stands upright on its tail, raising more than 95 % of its body off the substrate; this stance is known as nictation. The nictating nematode attaches to an insect host passing by it (Campbell et al. 1996). Ambushers target highly mobile lepidopteran pests active at the soil surface, such as cutworms and armyworms. Cruisers, as the name implies, are EPN that move through the soil in search of an insect host. They are most effective against sedentary and slow-moving insect pests found at various soil depths, such as white grubs and root weevils. Cruisers seek out their prey by sensing the carbon dioxide or other volatiles released by it: *S. carpocapsae* and *S. scapterisci* are ambushers, and *Heterorhabditis bacteriophora*, *H. megidis*, *S. kraussei* and *S. glaseri* are cruisers, while *S. riobrave* and *S. feltiae* are both (Campbell and Gaugler 1997).

13.4   Efficacy of EPN Against Lepidopteran Pests in Field Conditions

In general, under field conditions, steinernematids have performed better against lepidopteran pests as compared to heterorhabditids. The ambusher nematode, *S. carpocapsae*, has been relatively more successful followed by *S. glaseri*, *S. feltiae* and *S. riobrave* against several lepidopteran pests causing substantial damage to orchard trees and field crops. Campbell and Gaugler (1993) reported that *S. carpocapsae* was up to 43 times more effective at finding a mobile insect host as compared to *H. bacteriophora*. It also tended to search the prey more effectively along a surface than through a matrix. This section pertains to the successful field applications of EPN against some significant lepidopteran pests.

13.4.1   Orchard Trees

The control of lepidopteran pests in orchard trees is a major challenge throughout the world, due to the tendency of the final stage larvae of several species to moult into cocoon and overwinter in cryptic habitats such as under loose bark and fallen leaves, in piles of litter at the base of a tree, in woodpiles or even in fruit bins. By virtue of their foraging strategies and their symbiotic bacterial toxicity, EPN are ideally equipped to reach, infect and kill those stages of the pests that are concealed in the cryptic habitat, while chemical pesticides, and fungal, bacterial, or viral biocontrol agents have a limited reach and lethal capacity.

13.4.1.1 Apple and Pear

Apple and pear trees are attacked by a broad spectrum of lepidopteran pests that feed on the fruits, leaves and vascular tissues of the trees. The damage caused can range from huge economic losses incurred due to reduced fruit quality to a total destruction of the tree.

Apples are one of the most valuable fruit crops in the USA, and the codling moth, *Cydia pomonella*, is considered a major constraint, in not only apple orchards but also in pear, walnut and other fruit trees. The neonates bore into the fruit and spend their entire feeding time inside it until the last stage larvae leave the fruit in search of concealed spaces, beneath the bark or under the fallen leaves to pupate.

S. carpocapsae has provided an excellent control of this pest at a dose of 1–2×10^6 IJ/tree and the surrounding areas, provided the treated areas are kept moist for 8 h or more (Lacey and Unruh 1998; Lacey and Chauvin 1999; Unruh and Lacey 2001). Inasmuch as it has been observed that a dry environment is detrimental to the survival and persistence of the IJ, the active nematodes effectively kill the insect in its various stages: the final instar larvae after they exit the fruit, the cocooned prepupae and the cocoon stages that overwinter in the cryptic habitats.

A 4-year field trial was conducted in the apple and pear orchards in eastern Washington State Field with *S. carpocapsae* and *S. feltiae* to determine the effects of seasonal temperatures, adjuvants, post-application irrigation and the methods of application in controlling the cocooned codling moth larvae. EPN were applied to apple trees (Golden Delicious) with a backpack sprayer at the rate of 10^6 IJ/tree, supplemented with a wetting agent to aid the survival of the IJ. In September 1999, a high mortality of 94–95 % was observed in the treatments of both the species. However, the efficacy of *S. carpocapsae* reduced to 58 % as against 90 % in *S. feltiae* in October, owing to the low temperature. In March, cool windy conditions reduced the efficacy of *S. carpocapsae* and *S. feltiae* to 26 % and 65 %, respectively. Warmer weather in April stepped up the efficacy of *S. carpocapsae* and *S. feltiae* to 71 % and 86 %, respectively. In further tests on the same location, in mid-October 2001, *S. feltiae* when combined with a wetting agent (Silwet L77) or a humectant (STOCKOSORB) (10^6 IJ/tree) was found to be the most effective control of the sentinel codling moth larvae, cocooned on cardboard strips (80 % mortality) and logs (34–47 %). Also posttreatment, the trees were misted (wetted) for 4 h. The misting of trees, both before and after the application of *S. carpocapsae*, has been recommended by Unruh and Lacey (2001) to facilitate the IJ to enter the cryptic microhabitat of the host and penetrate the cocoon.

Cydia pomonella granulovirus (CpGV) is considered a highly effective and environmentally benign control agent for the codling moth in organic apple production. However, in Germany, this pest reportedly developed resistance against CpGV leading to an investigation of the potential of EPN in controlling the populations of the codling moth in organic fruit orchards. In two separate field tests, treatments of *S. glaseri* (cruiser) and *S. carpocapsae* (ambusher) resulted in a 90 % reduction in adult emergence from the treated stems. On-farm trials, under favourable weather conditions, showed about 50–60 % control of infestation by the codling moth in the year following the application (Kienzle et al. 2008).

In northern China, the larvae of the peach fruit borer, *Carposina niponensis*, is a major pest in apple causing monetary loss of up to 1.7 billion dollars per annum in 1 million ha of orchards. Bedding (1990) reported more than 90 % mortality in the pest by the application of *S. carpocapsae*.

13.4.1.2 Nuts

The navel orangeworm, *Amyelois transitella* Walker, is a major pest of commercially grown almonds, pistachios, figs and walnuts. In California, it poses a significant threat to the cultivation of almonds and pistachios, crops with high and increasing value. The moths oviposit during summer, and the larvae spend the winter feeding on the nuts left behind on the trees and the ground, post harvest. Thereafter, they infect the new crop of nuts the following year. Growth regulators and pyrethroid insecticides have been seen to have limited effect in controlling the overwintering larvae (Zalom et al. 1984; Bentley et al. 2008). *S. carpocapsae* was able to kill >72 % of the navel orangeworm larvae, when applied at a rate of 10^5 IJ m^{-2} by the manual spraying method (Siegel et al. 2004).

In another application in Ballico, California, *S. feltiae* Mexican strain was applied with a backpack Echo duster–mister or a pick-up mounted handgun sprayer at the rate of 3×10^6 IJ in 5 l of water per almond tree infected with the navel orangeworm. Up to 78 % mortality was recorded, and the application schedule was recommended for small-scale nematode control on almonds (Lindegren et al. 1987). The IJ were found to remain viable and infective inside the almond hulls for at least 8–10 days or more.

The filbertworm, *Cydia latiferreana*, is considered an economically important insect pest of hazelnuts in North America. *S. carpocapsae* strain All was used to manage the overwintering worms on the hazelnut orchard floor. The filbertworm larvae were treated with the nematode at 40–150 IJ/cm^2 applied in 75 or 190 ml/m^2 water, in plots with either bare soil or litter (leaves, twigs, husks and blank nuts). This resulted in a 2–11 % reduction in their population in October 2007 and at 50–78 % in May 2008 (Chambers et al. 2010).

13.4.1.3 Peach

The lesser peach tree borer, *Synanthedon pictipes* Grote and Robinson, a pest of commercially grown *Prunus* spp., is indigenous to eastern North America, especially in the southeastern peach orchards. The larvae treated with four strains of *S. carpocapsae* (All, DD136, Sal and Hybrid2) resulted in <20 % survival, whereas larval survival was always >50 % when treated with three strains of *S. riobrave* (3-8b, 7-12 and 355) or *H.* spp. Due to the simultaneous occurrence of overlapping generations of *S. pictipes* larvae in the orchards, the susceptibility of the larvae could also be mapped versus their size: large and medium larvae were more susceptible than the small ones. The above-ground application of these nematodes was more efficacious due to the moisture-retaining covers placed over *S. pictipes*-infested wounds on the peach limbs (Cottrell et al. 2011).

In China, the peach fruit moth, *Carposina niponensis*, which pupates in the soil, has been effectively controlled by *S. carpocapsae* (Wang 1993).

13.4.1.4 Strawberry

The *S. carpocapsae* strain Agriotos and the *H. bacteriophora* strain Oswego were extremely virulent to the last instar larvae of the strawberry crown moth, *Synanthedon bibionipennis*, even in the protected environment inside the strawberry crown, causing 96 % and 94 % mortality, respectively, in the laboratory. However, field applications in late fall (October) were less effective with *S. carpocapsae* and *H. bacteriophora*, resulting in 51 % and 33 % infection, respectively (Bruck et al. 2008). It was recommended that to improve field-level control, nematode applications should be made in late summer to early fall when larvae are present in the soil, and the soil temperatures are more favourable for nematode infection.

13.4.1.5 Grape

The grape root borer, *Vitacea polistiformis* Harris, is the most important insect pest of grapes in Florida. It is a difficult pest to detect and control because it spends the majority of its life cycle underground. The larvae sequester the grape roots, and thus the damage is not readily visible. The exposed roots of an infested vine show tunnels just under the cambium filled with a reddish frass and trunk girdling. An application of 5 billion/ha of *H. zealandica* achieved 70 % control of the borers. In another study, an application of 60,000 *H. zealandica* IJ/plant caused 96 % mortality, as the IJ could reach the borer even in the root pieces (Pollock 2002). Saunders and All (1985) found that *S. carpocapsae* burrowed down to the primary feeding sites of the borer in the roots and effectively killed the first larval instars. Gray and Johnson (1983) also found *S. carpocapsae* to be most effective in suppressing the borer activity at 29 °C, and their survival was directly correlated to the soil temperatures (30 °C) and soil moisture (>79.5 % RH) at which IJ lived for up to 1.5–2 years.

13.4.1.6 Litchi

S. carpocapsae was found to be highly lethal to the litchi stem borer, *Indarbela dea*, causing over 86 % mortality by spraying 1,000 IJ around the borer hole, where the larvae are usually active during night (Xu and Yang 1992).

13.4.1.7   Alder and Sycamore

The larvae of the clear-wing moth borer, *Synanthedon* spp., bore into plant tissues and feed in the galleries made in the barks and roots of trees, resulting in wood defects or structural weakness that may eventually destroy the tree. *S. culiciformis*, which bores into the heartwood of alder trees, was sprayed with *S. feltiae* with a Hudson sprayer at the rate of 6.5 or 11.5×10^6 IJ/tree or with a 1-pint squirt bottle at the rate of 18,000 or 36,000 IJ/gallery, providing 17–84 % and 86–93 % borer control, respectively, in Davis California. *S. resplendens*, a bark borer of sycamore trees, treated with 11.3×10^6 *S. feltiae* IJ/tree or 8.6×10^6 *S. bibionis* IJ/tree in Riverside, California, resulted in 61 % and 13 % population reduction, respectively (Kaya and Brown 1986). As *S. culiciformis* occurs in the moist heartwood, it makes a large gallery opening (an average of 0.28 in.), thus allowing nematodes to enter or be sprayed directly into the gallery. Therefore, the moist galleries enhanced the nematode survival and host-finding ability in alders. A comparatively poor mortality of borers in sycamore was attributed primarily to the drier galleries in the bark and the smaller size of the gallery openings (average of 0.05 in. in October and 0.17 in. in April).

13.4.2   Vegetables

In 1996 and 1999, two studies were carried out, at the experimental farm of Agriculture and Agri-Food Canada at L'Acadie, on the foliar application of *S. carpocapsae*. In 1996, *S. carpocapsae*, applied at 4 billion/ha, provided 35.3 % control of *Artogeia rapae* on Brussels sprouts and 33.0 % on broccoli, while the application of the same in 1999 resulted in 24.9 %, 19.4 % and 14.9 % control on Brussels sprouts, broccoli and cauliflower, respectively. Based on these field results, it was concluded that foliar applications of *S. carpocapsae* did not provide an acceptable level of *Artogeia rapae* control under Quebec's environmental conditions (Bélair et al. 2003).

Field inundation of *S. feltiae* strain All, *S. bacteriophora* strain Hp88 and *H. heliothidis* strain NC to control the larvae of the banana moth, *Opogona sacchari* Bojer, infesting potato and bamboo palms in Florida resulted in successful establishment of the nematodes and a 58–100 % reduction in the pest population. The residual of *H. heliothidis* strain NC reduced the new infestations more effectively than the residual of *S. feltiae* strain All, as the latter had a lower survival rate (Pena et al. 1990).

In India, field trials conducted between 2002 and 2004 to control diamondback moth, *Plutella xylostella*, infestation of cabbage through the foliar application of *S. thermophilum* caused up to 46 % mortality in the larvae. EPN treatment reduced crop damage up to 43.1 % as compared to 49.5 % in treating with the insecticide Lambda-cyhalothrin (Somvanshi et al. 2006).

13.4.3 Corn

S. riobrave was found to be highly effective as inundative control against the prepupae and pupae of the corn earworm, *Helicoverpa zea*, in the cornfields of Texas, under conditions of high soil temperature, optimal irrigation and appropriate time of application (Cabanillas and Raulston 1995, 1996). It was also successful against this pest in the Arkansas cornfields (Feaster and Steinkraus 1996).

13.4.4 Cotton

Cotton, the most important natural fibre crop in the world, harbours a large number of insect pests that are susceptible to EPN. An application of *S. riobrave* at 2.5 billion IJ/ha in cotton fields in Arizona heavily infested with the pink bollworm, *Pectinophora gossypiella*, showed a reduction in infested bolls and a 19 % increase in the yield as compared to the untreated fields. The nematodes persisted in large numbers for 19 days and were recovered up to 75 days after application (Gouge et al. 1996), while similar results were obtained with *S. riobrave* and *S. carpocapsae* in the cotton fields in Texas (Gouge et al. 1997).

13.4.5 Turf

The black cutworm, *Agrotis ipsilon*, a polyphagous pest, is a perennial problem of golf course greens. The larvae spend their life cycle in the soil and are susceptible to a number of EPN. Georgis and Poinar (1994) reported 95 % control by *S. carpocapsae*. Buhler and Gibb (1994) observed that while both *S. carpocapsae* and *S. glaseri* significantly reduced the cutworm larvae in creeping bentgrass, *S. carpocapsae* provided a slightly higher level of control than *S. glaseri*. However, the persistence of nematodes was lost 8 days after their application in the field.

13.4.6 Management of Pests in Fruit Bins

The propagation of the codling moth larvae, *Cydia pomonella*, as a diapausing stage in fruit bins, is a potential source of reinfestation of orchards. The immersion of the infested bins in suspensions of commercially produced nematodes ranging from 10 to 50 IJ/ml water, along with a wetting agent and a humectant, resulted in 45–87 % and 56–85 % mortality in the cocooned codling moth larvae for *S. feltiae* and *S. carpocapsae*, respectively (Lacey et al. 2005). The wetting agent, Silwet L77, was used to assist the penetration of EPN into the codling moth hibernacula and the humectant to prevent EPN desiccation.

In another study, the diapausing cocooned oriental fruit moth, *Grapholita molesta* Busck, larvae in fruit bins were found to be susceptible to the IJ of *S. feltiae*. The treatment of bins with suspensions of 10 or 25 *S. feltiae* IJ/ml water along with the wetting agent, Silwet L77, resulted in 33.3–59 % and 77.7–81.6 % mortality in the corner supports and cardboard strips, respectively (Riga et al. 2006).

13.4.7   EPN in Combination with Bt

To improve the efficiency and efficacy of EPN, several attempts were made to combine them with other bioagents. The application of *Bacillus thuringiensis* (*Bt*) and *S. carpocapsae*, both at half rate, to control *P. xylostella* on watercress resulted in 58 % control, which was significantly higher than that of the individual application of each at full rate (Baur et al. 1998). However, the use of additives to improve EPN persistence and efficacy could not increase the feasibility of foliar applications against *P. xylostella* on watercress (Baur et al. 1997). The combination of *S. carpocapsae* and *Bt* with a polymer, by Schroer and Ehlers (2005) and Schroer et al. (2005), provided similar results. The application of 0.5 million/m^2 of *S. carpocapsae*, along with a surfactant–polymer formulation achieved a significant reduction of *P. xylostella* per plant with >50 % control after 7 days and 45 % control after 14 days in cabbage cultivated in east Java and Indonesia. Weekly applications of *Bt* (TUREX®) or its alternate application with the nematodes achieved >80 % control. The application of both biological agents together every second week reached insignificant lower efficacy (70 %). The success of the nematodes in these trials could mainly be attributed to the high humidity in the experimental area.

The moist microhabitat and moderate temperatures during the artichoke-growing season in the fog belt region of the central coast of California were found to be ideal for the management of the artichoke plume moth, *Platyptilia carduidactyla*, which typically tunnels young vegetative shoots. Field application of *Neoaplectana carpocapsae* (=*S. feltiae*) Weiser at 2,000 IJ/ml gave 100 % control of the third and fourth instars, 15 days after treatment. Its residual effect, evaluated 24 days after treatment, was greater than that of a commonly used insecticide, methidathion. *Bt var.* kurstaki at 19.6 billion international units/ha gave significant larval control; its efficacy was not significantly lower than that of methidathion. A combination of the nematode and *Bt* var. kurstaki did not result in significantly greater control than that achieved by the nematode used alone at 1,000 IJ/ml (Barp and Kaya 1984).

13.4.7.1 Incorporation of EPN in Bt Refuge

The evolution of resistance by pests can reduce the efficacy of *Bt* transgenic crops. In conjunction with refuges of non-*Bt* host plants, fitness costs can delay the evolution of resistance and thereby prolong the efficacy of *Bt* crops. Baur et al. (1998) had suggested that EPN could serve as components of integrated pest management of *P. xylostella* and could help manage resistance to *Bt*. Gassmann et al. (2006) reiterated that EPN could increase the fitness costs of resistance to *Bt* toxin. Later, Gassmann et al. (2008) tested the effects of *S. riobrave* on insect mortality and fitness costs of resistance to *Bt* toxin Cry1Ac in the pink bollworm *Pectinophora gossypiella* Saunders, a major pest of cotton in the southwestern United States. The results indicated that EPN could act synergistically with *Bt* crops by killing the pests in non-*Bt* refuges and could delay resistance by pests to *Bt* crops. However, no effect on fitness costs was detected for the nematode *H. bacteriophora*. Hannon et al. (2010) concluded that EPN could bolster insect resistance management, but the success of this approach would depend on the selection of the appropriate species of nematode and the environment.

13.5 Symbiotic Bacteria as Biopesticide

Both *Xenorhabdus* and *Photorhabdus* can be grown as free-living organisms under standard laboratory conditions on growth media. They are known to secrete several extracellular products, including lipase, protease and lipopolysaccharides, and many broad spectrum antibiotics in the culture medium (Akhurst 1982; Forst et al. 1997). Bacterial proteins that are effective insecticides include crystal proteins (δ endotoxins) of *Bt* and *Bacillus sphaericus*, the Vip toxins of *Bt* and cholesterol oxidase of *Streptomyces* spp., all of which lyse the midgut epithelium of the insect host (Bowen et al. 1998). Similar insecticidal properties have been found in *Photorhabdus luminescens* and *Xenorhabdus nematophilus*. Most of the bacterial strains have been found to be highly toxic when injected, and a few display lethal oral toxicity as well. Recently, a 51.8 kDa protease, purified from *P. luminescens* strain 0805-P5G, showed high insecticidal activity in *G. mellonella* when injected, and it also showed high oral toxicity in the *P. xylostella* of a Taiwan field-collected strain, but low toxicity in an American strain (Chang et al. 2013).

The insecticidal toxin proteins secreted by *Photorhabdus*, during its growth, into the culture medium, have been purified (Bowen and Ensign 1998) and their genes cloned (Bowen et al. 1998). Such reports have influenced the utilization of these bacteria as biopesticides independent of their nematode hosts. A patent for the use of *Xenorhabdus nematophilus* against fire ants has been issued (Dudney 1997). Direct toxicity of *P. luminescens* to insects under natural conditions was demonstrated by the author for the first time. Within 24 h of foliar application of

P. luminescens, 100 % of the cabbage white butterfly *Pieris brassicae* larvae naturally infesting the *Nasturtium* fields were killed (Mohan et al. 2003). Foliar application of *P. luminescens* and *X. nematophilus* resulted in 60 % and 40 % mortality to the pupae of *P. xylostella*, respectively (Razek 2003).

Actively growing cells of *P. luminescens* (2.5×10^7/bead) encapsulated in sodium alginate and mixed with sterilized soil, when exposed to *Spodoptera litura* larvae, resulted in 100 % mortality in 48 h, while the use of alginate-encapsulated *H. indica* effected 40 % mortality after 72 h. The LC_{50} of *Photorhabdus* cells was estimated at 1,010 cells per larva for killing the *S. litura* 6th instar larvae in 48 h (Rajagopal et al. 2006).

In the wake of such advances, *P. luminescens* and its formulations are being explored for inclusion in pest management strategies. However, their wide-ranging entomopathogenicity gives rise to doubts regarding their ecological safety against beneficial/nontarget fauna in general and insects in particular. The ecological compatibility of *P. luminescens* cells or their secreted toxins against the hatching of two species, *Trichogramma chilonis* Ishii and *T. japonicum* Ashmead, of a commercially recommended biological control agent on the hyperparasitized eggs of the stored grain moth *Corcyra cephalonica* was studied in vitro. In 65 % of the eggs exposed to *P. luminescens* cells alone or their toxins, the *Corcyra* eggshells became flaccid, and there was up to 84 % reduction in the emergence of the *Trichogramma* adults (Mohan and Sabir 2005). Such results point to the bioecological hazards of indiscriminate use of *P. luminescens* as a biopesticide. Thus, due to its wide host range, the inclusion of *P. luminescens* in any integrated pest management programme would be suspected, until proven safe for natural enemies and nontarget organisms.

13.6 Conclusion

Despite the successes achieved in field trials, a more promising future for the EPN would be to optimize their efficacy under location-specific sites. It is crucial to identify and select more pathogenic native nematode species and strains, which can adapt to the ecology and biology of the lepidopteran host insect. An optimal combination of a nematode species and its dosage vis-a-vis a specific insect pest is imperative for field-level success of EPN. A fundamental understanding of the interaction between indigenous EPN species and isolates with the susceptible stages of the pest is important.

The contention of the biosafety of the symbiotic bacteria to beneficial organisms needs to be examined more critically as these organisms are ecologically obligate to EPN, with specific mechanisms of pathogenicity, and their existence in free form in nature is believed to last a very short while due to photo- and thermal sensitivity.

References

Akhurst RJ (1982) Antibiotic activity of *Xenorhabdus* spp., bacteria symbiotically associated with insect pathogenic nematodes of the families Heterorhabditidae and Steinernematidae. J Gen Microbiol 128:3061–3065

Barp MA, Kaya HK (1984) Evaluation of the entomogenous nematode *Neoaplectana carpocapsae* (=*S. feltiae*) Weiser (Rhabditida: *S*.tidae) and the bacterium *Bacillus thuringiensis* Berliner var. kurstaki for suppression of the Artichoke Plume Moth (Lepidoptera: Pterophoridae). J Econ Entomol 77:225–229

Baur ME, Kaya HK, Gaugler R, Tabashnik B (1997) Effects of adjuvants on entomopathogenic nematode persistence and efficacy against *Plutella xylostella*. Biocontrol Sci Technol 7: 513–525

Baur ME, Kaya HK, Tabashnik BE, Chilcutt CF (1998) Suppression of diamondback moth (Lepidoptera: Plutellidae) with an entomopathogenic nematode (Rhabditida: Steinernematidae) and *Bacillus thuringiensis* Berliner. J Econ Entomol 91:1089–1095

Bedding RA (1990) Logistic and strategies for introducing entomopathogenic nematode technology into developing countries. In: Gaugler R, Kaya HK (eds) Entomopathogenic nematodes in biological control. CRC Press, Boca Raton, pp 233–246

Bélair G, Fournier Y, Dauphinais N (2003) Efficacy of steinernematid nematodes against three insect pests of crucifers in Quebec1. J Nematol 35:259–265

Bentley W, Siegel JP, Holtz BA, Daane KM (2008) Navel orangeworm (*Amyelois transitella*) (Walker) and obliquebanded leafroller (*Choristoneura rosaceana*) (Harris) as pests of pistachio. In: Ferguson LF, Beede RH, Haviland DH, Holtz BA, Kallsen CE, Sanden BL (eds) Pistachio production manual. University of California, Davis, pp 179–191

Bird AF, Akhurst RJ (1983) The nature of the intestinal vesicle in nematodes of the family Steinernematidae. Int J Parasitol 13:599–606

Boemare NE, Akhurst RJ, Mourant RG (1993) DNA relatedness between *Xenorhabdus* species, a symbiotic bacteria from entomopathogenic nematodes, and a proposal to transfer *Xenorhabdus luminescens* to a new genus *Photorhabdus* gen. nov. Int J Syst Bacteriol 43:249–255

Bowen DJ, Ensign JC (1998) Purification and characterization of a high-molecular-weight insecticidal protein complex produced by the entomopathogenic bacterium *Photorhabdus luminescens*. Appl Environ Microbiol 64:3029–3035

Bowen DJ, Blackburn RM, Andreev O, Golubeva E, Bhartia R, ffrench-Constant RH (1998) Insecticidal toxins from bacterium *Photorhabdus luminescens*. Science 280:2129–2132

Bruck DJ, Edwards DL, Donahue KM (2008) Susceptibility of the strawberry crown moth (Lepidoptera: Sesiidae) to entomopathogenic nematodes. J Econ Entomol 101:251–255

Buhler WG, Gibb TJ (1994) Persistence of *S. carpocapsae* and *S. glaseri* as measured by their control of black cutworm larvae in bentgrass. J Econ Entomol 87:638–642

Cabanillas HE, Raulston JR (1995) Impact of *S. riobravis* (Rhabditida: Steinernematidae) on the control of *Helicoverpa zea* (Lepidoptera: Noctuidae) in corn. J Econ Entomol 88:58–64

Cabanillas HE, Raulston JR (1996) Evaluation of *S. riobravis*, *S. carpocapsae*, and irrigation timing for the control of corn earworm, *Helicoverpa zea*. J Nematol 28:75–82

Campbell JF, Gaugler R (1993) Nictation behaviour and its ecological implications in the host search strategies of entomopathogenic nematodes. Behaviour 126:155–169

Campbell JF, Gaugler R (1997) Inter-specific variation in entomopathogenic nematode foraging strategy: dichotomy or variation along a continuum? Fund Appl Nematol 20:393–398

Campbell J, Lewis E, Yoder F, Gaugler R (1996) Entomopathogenic nematode spatial distribution in turfgrass. Parasitology 113:473–482

Chambers U, Bruck DJ, Olsen J, Walton VM (2010) Control of overwintering filbertworm (Lepidoptera: Tortricidae) larvae with *S. carpocapsae*. J Econ Entomol 103:416–422

Chang YT, Hsieh C, Wu LC, Chang HC, Kao SS, Meng M, Hsieh FC (2013) Purification and properties of an insecticidal metalloprotease produced by *Photorhabdus luminescens* strain 0805-P5G, the entomopathogenic nematode symbiont. Int J Mol Sci 14:308–321

Ciche TA, Ensign JC (2003) For the insect pathogen *Photorhabdus luminescens*, which end of a nematode is out? Appl Environ Microbiol 69:1890–1897

Clarke DJ (2008) *Photorhabdus*: a model for the analysis of pathogenicity and mutualism. Cell Microbiol 10:2159–2167

Cottrell TE, Shapiro-Ilan DI, Horton DL, Mizell RF (2011) Laboratory virulence and orchard efficacy of entomopathogenic nematodes against the lesser peachtree borer (Lepidoptera: Sesiidae). J Econ Entomol 104:47–53

Cowles KN, Goodrich-Blair H (2005) Expression and activity of a *Xenorhabdus* nematophila haemolysin required for full virulence towards *Manduca sexta* insects. Cell Microbiol 7: 209–219

Dudney RA (1997) Use of *Xenorhabdus nematophilus* Im/1 and 19061/1 for fire ant control. US Patent US 5,616,318 970, 401. Application Information US 95-488,820 950,609

Ebssa L, Koppenhöfer AM (2011) Efficacy and persistence of entomopathogenic nematodes for black cutworm control in turfgrass. Biocontrol Sci Technol 21:779–796

Feaster MA, Steinkraus DC (1996) Inundative biological control of *Helicoverpa zea* (Lepidoptera: Noctuidae) with the entomopathogenic nematode *S. riobravis* (Rhabditida: Steinernematidae). Biol Control 7:38–43

Forst S, Dowds B, Boemare N, Stackebrandt E (1997) *Xenorhabdus* and *Photorhabdus* spp.: bugs that kill bugs. Annu Rev Microbiol 51:47–72

Furgani G, Boszormenyi E, Fodor A, Mathe-Fodor A, Forst S (2008) *Xenorhabdus* antibiotics: a comparative analysis and potential utility for controlling mastitis caused by bacteria. J Appl Microbiol 104:745–758

Gassmann AJ, Stock SP, Carrière Y, Tabashnik BE (2006) Effect of entomopathogenic nematodes on the fitness cost of resistance to *Bt* toxin Cry1Ac in pink bollworm (Lepidoptera: Gelechiidae). J Econ Entomol 99:920–926

Gassmann AJ, Stock SP, Sisterson MS, Carrière Y, Tabashnik BE (2008) Synergism between entomopathogenic nematodes and *Bacillus thuringiensis* crops: integrating biological control and resistance management. J Appl Ecol 45:957–966

Gaugler R, Campbell J, McGuire T (1989) Selection for host finding in *S. feltiae*. J Invertebr Pathol 54:363–372

Georgis R, Poinar GO Jr (1994) Nematodes as bioinsecticides in turf and ornamentals. In: Leslie AR (ed) Integrated pest management of turfgrass and ornamentals. Lewis Publishers, Boca Raton, pp 477–489

Glaser RW (1932) Studies on *Neoaplectana glaseri*, nematode parasite of the Japanese beetle (Popillia japonica). N J Dept Agric 211:34

Gouge DH, Reaves LL, Stoltman MM, Van Berkum JR, Burke RA, Forlow Jech LJ, Henneberry TJ (1996) Control of Pink boll-worm *Pectiniphora gossipiella* (Saunders) larvae in Arizona and Texas cotton fields using entomopathogenic nematodes *S. riobravis*. In: Ritcher DA, Armour J (eds) Proceedings of the Beltwide cotton production research conference. National Cotton Council of America, Memphis, pp 1078–1082

Gouge DH, Smith KA, Payne C, Lee LL, Van Berkum JR, Ortega D, Henneberry, TJ (1997) Control of Pink boll-worm *Pectiniphora gossipiella* (Saunders) with biocontrol and biorational agents. In: Dugger P, Ritcher DA (eds) Proceedings of the Beltwide cotton production research conference. National Cotton Council of America, Memphis, pp 1066–1072

Gray PA, Johnson DT (1983) Survival of the nematode *Neoaplectana carpocapsae* in relation to soil temperature, moisture, and time. J Georgia Entomol Soc 18:454–460

Hannon ER, Sisterson MS, Stock SP, Carrière Y, Tabashnik BE, Gassmann AJ (2010) Effects of four nematode species on fitness costs of pink bollworm resistance to *Bacillus thuringiensis* toxin Cry1Ac. J Econ Entomol 103:1821–1831

Kaya HK, Brown LR (1986) Field application of entomogenous nematodes for biological control of clear-wing moth borers in alders and sycamore trees. J Arboric 12:150–154

Kienzle J, Zimmer J, Volk F, Zebitz CPW (2008) Experiences with entomopathogenic nematodes for the control of overwintering codling moth larvae in Germany. Archived at http://orgprints. org/13712

Koppenhöfer AM (2007) Nematodes. In: Lacey LA, Kaya HK (eds) Field manual of techniques in invertebrate pathology: application and evaluation of pathogens for control of insects and other invertebrate pests, 2nd edn. Springer, Dordrecht, pp 249–264

Lacey LA, Chauvin RL (1999) Entomopathogenic nematodes for control of codling moth in fruit bins. J Econ Entomol 92:104–109

Lacey LA, Unruh TR (1998) Entomopathogenic nematodes for control of codling moth, *Cydia pomonella* (Lepidoptera: Tortricidae): effect of nematode species, concentration, temperature, and humidity. Biol Control 13:190–197

Lacey LA, Neven LG, Headrick HL, Fritts R Jr (2005) Factors affecting entomopathogenic nematodes (Steinernematidae) for control of overwintering codling moth (Lepidoptera: Tortricidae) in fruit bins. J Econ Entomol 98:1863–1869

Lindegren JE, Agudelo-Silva F, Valero KA, Curtis CE (1987) Comparative small scale field application of *S. feltiae* for navel orangeworm control. J Nematol 19:503–504

Martens EC, Heungens K, Goodrich-Blair H (2003) Early colonization events in the mutualistic association between *S. carpocapsae* nematodes and *Xenorhabdus nematophila* bacteria. J Bacteriol 185:3147–3154

Mohan S, Sabir N (2005) Biosafety concerns on the use of *Photorhabdus luminescens* as biopesticide: experimental evidence of mortality in egg parasitoid *Trichogramma* spp. Curr Sci 89:1268–1272

Mohan S, Raman R, Gaur HS (2003) Foliar application of *Photorhabdus luminescens*, symbiotic bacteria from entomopathogenic nematode *H. indica*, to kill cabbage butterfly *Pieris brassicae*. Curr Sci 84:1397

Pena JE, Schroeder WJ, Osborne LS (1990) Use of entomogenous nematodes of the families Heterorhabditidae and Steinernematidae to control banana moth (*Opogona sachari*). Nematropica 20:51–55

Poinar GO Jr (1993) Taxonomy and biology of Steinernematidae and Heterorhabditidae. In: Gaugler R, Kaya HK (eds) Entomopathogenic nematodes in biological control. Marcel Dekker, New York, pp 23–61

Poinar GO Jr, Georgis R (1990) Characterization and field application of *H. bacteriophora* strain HP88. Revue-de-Nematologie 13:387–393

Pollock C (2002) Nematodes effective against grape pests. http://fusion.ag.ohio state.edu/news/ story.asp. Accessed 18 Feb 2003

Rajagopal R, Mohan S, Bhatnagar RK (2006) Direct infection of *Spodoptera litura* by *Photorhabdus luminescens* encapsulated in alginate beads. J Invertebr Pathol 93:50–53

Razek AAS (2003) Pathogenic effects of *Xenorhabdus nematophilus* and *Photorhabdus luminescens* against pupae of the Diamondback moth, *Plutella xylostella*. J Pest Sci 76: 108–111

Riga E, Lacey LA, Guerra N, Headrick HL (2006) Control of the oriental fruit moth, *Grapholita molesta*, using entomopathogenic nematodes in laboratory and fruit bin assays. J Nematol 38: 168–171

Saunders MC, All JN (1985) Association of entomophilic rhabditoid nematode populations with natural control of first-instar larvae of the grape root borer, *Vitacea polistiformis*, in Concord grape vineyards. J Invertebr Pathol 45:147–151

Schroer S, Ehlers RU (2005) Foliar application of the entomopathogenic nematode *S. carpocapsae* for biological control of diamondback moth larvae (*Plutella xylostella*). Biol Control 33:81–86

Schroer S, Sulistydanto D, Ehlers RU (2005) Control of *Plutella xylostella* using polymer-formulated *S. carpocapsae* and *Bacillus thuringiensis* in cabbage fields. J Appl Entomol 129: 128–204

Shapiro-Ilan DI, Gaugler R (2010) Nematodes: Rhabditida: Steinernematidae & Heterorhabditidae. In: Shelton A (ed) Biological control: a guide to natural enemies in North America. Cornell University. http://www.biocontrol.entomology.cornell.edu/pathogens/nematodes.htm

Shapiro-Ilan DI, Gouge DH, Koppenho"fer AM (2002) Factors affecting commercial success: case studies in cotton, turf and citrus. In: Gaugler R (ed) Entomopathogenic nematology. CABI Publishing, Wallingford, pp 333–355

Siegel J, Lacey LA, Fritts R, Higbee BS, Noble P (2004) Use of Steinernematid nematodes for post harvest control of navel orangeworm (Lepidoptera: Pyralidae, *Amyelois transitella*) in fallen pistachios. Biol Cont 30:410–417

Snyder H, Stock SP, Kim SK, Flores-Lara Y, Forst S (2007) New insights into the colonization and release processes of *Xenorhabdus nematophila* and the morphology and ultrastructure of the bacterial receptacle of its nematode host, *S. carpocapsae*. Appl Environ Microbiol 73: 5338–5346

Somvanshi VS, Ganguly S, Paul AVN (2006) Field efficacy of the entomopathogenic nematode *S. thermophilum* Ganguly and Singh (Rhabditida: Steinernematidae) against diamondback moth (*Plutella xylostella* L.) infesting cabbage. Biol Control 37:9–15

Thomas GM, Poinar GO Jr (1979) Amended description of the genus *Xenorhabdus* Thomas and Poinar. Int J Syst Bacteriol 33:878–879

Unruh TR, Lacey LA (2001) Control of codling moth, *Cydia pomonella* (Lepidoptera: Tortricidae) with *S. carpocapsae*: effects of supplemental wetting and pupation site on infection rate. Biol Control 20:48–56

Wang J (1993) Control of the peach fruit moth, *Carposina niponensis*, using entomopathogenic nematodes. In: Bedding R, Akhurst R, Kaya H (eds) Nematodes and the biological control of insect pests. CSIRO Publications, East Melbourne, pp 59–65

Waterfield N, Bowen DJ, Fetherston JD, Perry RD, ffrench-Constant RH (2001) The toxin complex genes of *Photorhabdus*: a growing gene family. Trends Microbiol 9:185–191

Xu JL, Yang P (1992) The application of the codling moth nematode against the litchi stemborer. Acta Phytophylacica Sinica 19:217–222

Zalom FG, Barnett WW, Weakley CV (1984) Efficacy of winter sanitation for managing the navel orangeworm, *Paramyelois transitella* (Walker), in California almond orchards. Prot Ecol 7: 37–41

Chapter 14
The Management of *Helicoverpa* Species by Entomopathogenic Nematodes

M. Abid Hussain and Wasim Ahmad

14.1 Introduction

Helicoverpa spp. (Lepidoptera: Noctuidae) are known by many common names such as American bollworm, cotton bollworm, corn earworm, tobacco budworm, old world bollworm, legume/gram pod borer, etc. It is widely distributed in Asia, Africa, Australia, the Mediterranean Europe and the semiarid tropical regions of the world. It is a polyphagous pest of many agricultural, horticultural and ornamental crops (Reed and Pawar 1982; Zalucki et al. 1986; Fitt 1989; Sharma 2001). The geographic distribution of the Heliothine lepidopteran species complex (*Helicoverpa virescens*, *H. armigera*, *H. punctigera* and *H. zea*) is listed in Table 14.1. The status of *Helicoverpa* as a serious pest is due to its high mobility, polyphagy, facultative diapause as pupae, rapid generation turnover, fecundity and predilection for harvestable parts of high-value crops such as cotton, tomato, pulses and cereals (Fitt 1989). All of these life-history features contribute to make *Helicoverpa* spp. one of the world's worst pest.

The estimates of yield losses due to this pest vary from 5 to 10 % in the temperate regions and 50–100 % in the tropics (van Emden et al. 1988). The avoidable losses in food legumes at production levels of 60.45 million tonnes are predicted to be nearly 18.14 million tonnes (at an average loss of 30 %), valued at nearly US$10 billion (Sharma et al. 2008). Barwale et al. (2004) estimated about Rs. 1,200 crore (equivalent to US$273 million at exchange rate in 2004) worth of pesticides used in India to control the bollworm complex of cotton, whereas *Helicoverpa* causes an estimated loss of US$927 million in chickpea and pigeon

M.A. Hussain
Department of Plant Protection, Hamelmalo Agricultural College, Keren, Eritrea

W. Ahmad (✉)
Section of Nematology, Department of Zoology, Aligarh Muslim University, Aligarh 202002, India
e-mail: ahmadwasim57@gmail.com

© Springer International Publishing Switzerland 2015
K.S. Sree, A. Varma (eds.), *Biocontrol of Lepidopteran Pests*, Soil Biology 43,
DOI 10.1007/978-3-319-14499-3_14

Table 14.1 Geographic distribution of the *Helicoverpa* species complex (Deguine et al. 2008)

Species	Geographic distribution	Main host plants
Helicoverpa armigera	Africa, Central and Southeastern Asia, Australia, Southern Europe, India, New Zealand and many eastern Pacific Islands, Brazil[a]	Cotton, groundnut, maize, pulses, rapeseed, safflower, sorghum, soybean, sunflower, tobacco, tomato, etc.
Helicoverpa zea	North and South America	Cotton, maize, sorghum, soybean, sunflower, tomato
Helicoverpa punctigera	Australia	Chickpea, cotton, lucerne, safflower, soybean, sunflower
Helicoverpa virescens	North and South America	Cotton, soybean, sunflower, tobacco, sunflower

[a]Tay et al. (2013)

pea, and more than US$5 billion on different crops worldwide (Sharma 2005). Several factors, including increasing levels of resistance to pesticides and rise in cropping intensity, have contributed to greater importance of this pest (Shanower et al. 1998). Agronomic factors, such as high-yielding varieties, increased use of irrigation and fertilisers and large-scale production and planting of alternate crop hosts, contribute towards greater prevalence and increased severity (Reed and Pawar 1982; Fitt 1989).

The fundamental of effective integrated pest management (IPM) programmes is the development of appropriate pest management strategies and tactics that best interface with the cropping system and the pest situations. The potential of some of the control tactics to reduce the population density of *Helicoverpa* spp. in different cropping systems is evaluated by several researchers. Attempts have been made to develop integrated management approach for *Helicoverpa* spp. using host–plant resistance (Mihm 1997; Sharma et al. 2005; Sharma 2007; Kumari et al. 2010; Naseri et al. 2010; Soleimannejad et al. 2010) including transgenic Bt crops (Mendelsohn et al. 2003; Gopala Swamy et al. 2009; Sanahuja et al. 2011; James 2012; Acharjee and Sarmah 2013), biological control (predators and parasitoids) (King and Coleman 1989; Sharma 2001), interference methods including sex pheromones for population monitoring or mating disruptions (Reddy and Manjunatha 2000), biopesticides (especially commercial formulations of *Bacillus thuringiensis*) (Navon 2000; Liao et al. 2002; Sanahuja et al. 2011), cultural practices (including appropriate crop rotations, trap crops, planting date and habitat complexity) (Jallow et al. 2004) and chemical control (COPR 1983; Zalucki et al. 1986; Matthews 1989; Jackson 2014). Another IPM tool is push-pull strategy which can also be deployed to reduce pesticide input (Cook et al. 2007). Nevertheless, the pest become serious with regular outbreaks and developed resistance to almost all conventional insecticides including synthetic pyrethroids (Armes et al. 1996; Kranthi et al. 2002; Whalon et al. 2007). As a result, chemical control through the use of synthetic insecticides could not become panacea in the protection of agriculturally important crops. There are recent claims of field-evolved resistance to proteins in transgenic toxin Bt crops. For example, a survey during 2010 revealed field-evolved resistance in *H. armigera* to Cry1Ac protein in northern China (Zhang et al. 2011). Yet in another study in Pakistan, the development of 580-fold resistance ratio to Bt toxin Cry1Ac has been reported and compared with several conventional insecticides (Alvi et al. 2012). Five-year data set of Downes et al. (2010) shows a significant exponential increase in the frequency of alleles conferring Cry2Ab Bt toxin resistance in Australian field populations of *H. punctigera*.

The use of microbial pesticides has been considered a more sustainable and environmentally benign option to control *Helicoverpa* (King and Coleman 1989). Unfortunately, previous control attempts using entomopathogens were not effective to a great extent, and substantial reduction in degree of losses caused by this dreaded pest could not be achieved (Sachan and Lal 1997). The deactivation by UV light, insufficient titres ingested by larvae and lack of virulence could be the practical problems. Environmentally safe techniques such as the release of *Trichogramma* egg parasitoids, the use of *B. thuringiensis* (Bt) sprays, *H. armigera* nucleopolyhedrovirus (HearSNPV) and sex pheromones are not yet readily available in rural areas or are relatively less effective than the synthetic insecticides and, as a result, have not been widely adopted by the farmers (Kumari et al. 2010). Another potential soil-dwelling biocontrol agents are entomopathogenic nematodes (EPNs), which can be used as an important tool for *Helicoverpa* control (Richter and Fuxa 1990; Cabanillas and Raulston 1996a, b; Feaster and Steinkraus 1996; Raulston et al. 2001; Hussain et al. 2014).

For any successful IPM programme, one must have complete understanding of the factors that regulate the pest population. Every effort should be made to take advantage of weak points in the biology of the pest to devise the control measures. Therefore, a brief account of *Helicoverpa* spp. (particularly *H. armigera* and *H. zea*) as well as the potential of EPNs and its limitations is discussed in this chapter to correlate a better suite of control tactics utilising this bio-agent.

14.2   Biology of *Helicoverpa* spp.

Morphology and biology of various life stages of *Helicoverpa* spp. have been described by various workers (Hardwick 1965; Armes 1989; Matthews 1999, etc.). Sharma (2001) gave a comprehensive account of the biology of *H. armigera* which is typical of the noctuid insect. Life cycle consists of four stages, viz., egg, larva, pupa and adult. A female lays 500–3,000 eggs singly on its host plants. Oviposition period lasts for 5–24 days, and the duration of the egg incubation period varies between 2 and 5 days. The newly emerged larvae eat empty eggshell and then move around for some distance, with occasional feeding on the surface before settling down at a preferred site (flower bud or flower in cotton and pigeon pea, young leaves in chickpea, corn silks and the soft grain or young whorled leaves of cereals). Older larvae prefer flower buds and young cotton bolls, legume pods or cereal grain, although leaves are also eaten when plants are small or only a few fruiting bodies are present. Encounters between older (>third instars) larvae usually result in cannibalism, often resulting in only one large larva per flowering bud, boll, pod, whorled leaves or a panicle (Sharma 2001). The average larval period of 21 days is recorded on cotton flower buds at 21–27 °C (Reed 1965), whereas Singh and Singh (1975) reported 8–12 days on tomato and 18 days on cotton and maize.

On completion of larval development, the larvae drop or crawl to the ground and enter the soil for pupation. The depth at which the pupal cell is formed varies considerably at a depth of 2.5–17.5 cm (Jayaraj 1982) depending on hardness and wetness of the soil, presence of cracks and crevices and surface litter. Occasionally, pupation may also take place inside a tunnel in a maize cob (Reed 1965) or on the soil or leaf surface. The prepupal stage lasts for 1–4 days.

The adults of *Helicoverpa* spp. are stout-bodied moths. The males are uniformly pale cream and forewings generally tinged with green, whereas females are darker, and the forewings are without a green tinge. The longevity of adults depends upon the availability of food such as nectar, pupal weight, temperature and adult behaviour such as flight activity. In the absence of suitable food source, depletion of the fat bodies is rapid and death occurs in a few days (Armes 1989). In captivity, longevity varies from 1 to 23 days for males and 5 to 28 days for females. The importance of *Helicoverpa* is largely due to its well-developed survival strategies, diapause and dispersal abilities, which enable it to exploit food sources, separated by unfavourable times and distance, and it responds largely to local environmental cues and undertakes short- or long-distance flight in the direction largely governed by prevailing weather systems (Fitt 1989).

14.3 *Helicoverpa* spp. Control Methods

Pest management strategies for *Helicoverpa* spp. require integration of different control tactics to implement a threshold based on the relationship between population density and the economic loss.

14.3.1 *ETL and Population Monitoring Through Pheromone*

Adult moths are monitored with light and pheromone traps. Field monitoring of pest populations is necessary to determine whether the threshold has been exceeded, and control measures should be undertaken. Location-based economic threshold level (ETL) for *H. armigera* in cotton is reported by Simwat (1994) as 25 adults/trap/night or 1 larva/10 plants or 20 balls at Bapatla and Coimbatore, 10 % incidence in reproductive parts at Bhatinda and 2–7 larval unit/10 plants at Surat (India).

14.3.2 Cultural Control

Early or timely planting of crops can help avoid periods of peak abundance of *H. armigera*. The planting times are often decided by the rainfall pattern and availability of soil moisture. Late planted crops often suffer greater pest damage because of the build-up of pest populations over the cropping season. Short-season cultivars are used to minimise bollworm damage in cotton in the USA and plant growth regulators to shorten the crop maturity (Bradley et al. 1986). Early termination of flowering and fruiting can also check the population carry-over from one season to another or reduce the number of generations. Intercropping chickpea with wheat, mustard or safflower; pigeon pea with cowpea and sorghum; groundnut and coriander in chickpea; and tomato with radish result in reduced damage by *H. armigera*. Intercropping can also be used as a means of encouraging the activity of natural enemies. Trap crops and diversionary hosts have been widely used in the past to reduce the damage by *H. armigera*, but have seldom been successful (Fitt 1989). Sunflower, marigold, sesame and carrot are used as trap crop for *H. armigera* control. Ploughing destroys pupae of *H. armigera* in soil, and flooding with water affects *Helicoverpa* pupal survival and moth emergence (Fitt and Cotter 2005). Irrigation or flooding of cotton fields at the time of pupation reduces pupal survival and leads to decreased population densities in the following generation or season. However, chickpea grown under rainfed conditions is not amenable to flooding due to water scarcity (Sharma et al. 2008).

14.3.3 Host–Plant Resistance

The development of crop cultivars resistant or tolerant to *H. armigera* has a major potential for IPM (Lal et al. 1986; Fitt 1989; Sharma et al. 1999), particularly under subsistence farming in the developing countries. The development of crop varieties that are resistant or tolerant to *H. armigera* has received the major attention particularly for cotton, pigeon pea and chickpea. Many crop species possess some genetic potential, which can be exploited by breeders to produce varieties that are less susceptible to *H. armigera* damage (Sharma 2001). Cultivars with low to moderate level of resistance to insect pests have been identified in pigeon pea, chickpea, cowpea, black gram, green gram and field pea (Sharma et al. 2008).

14.3.4 Chemical Control

There is substantial literature on comparative efficacy of different insecticides against *Helicoverpa* on cotton, pulses, vegetables and other crops. Information on chemical control has been reviewed extensively by Zalucki et al. (1986) and

Matthews (1989). The early history of chemical control of corn earworms is given by Hardwick (1965), while COPR (1983) includes a list of 29 insecticides effective at the recommended rates. Jackson (2014) describes newer molecules of insecticides for the control of *Helicoverpa* infesting cotton, soybeans, corn, sorghum and peanuts. Insecticide resistance management strategies have been developed in several countries to prevent the development of resistance or to contain it (Sawicki and Denholm 1987; Whalon et al. 2007).

14.3.5 Biological Control

Information on natural enemies and biological control of *Helicoverpa* spp. is reviewed by several workers (King and Coleman 1989; Romeis and Shanower 1996). In India, 75 arthropod parasitoids and 33 predators are reported to occur on *H. armigera* (Manjunath et al. 1989). Sharma (2001) published the recent update on arthropod natural enemies of *H. armigera* in India including spiders, birds and microorganisms. In most areas, species of *Telenomus*, *Trichogramma* and *Trichogrammatoidea* are important egg parasitoids, whereas at least one species each of Ichneumonidae (e.g. *Campoletis chlorideae*) and Tachinidae (e.g. *Carcelia illota*) is larval parasitoids. The most common predators of *Helicoverpa* include species of *Chrysopa*, *Chrysoperla*, *Nabis*, *Geocoris*, *Orius* and *Polistes* and species belonging to Pentatomidae, Reduviidae, Coccinellidae, Carabidae, Formicidae and Araneidae (Zalucki et al. 1986; Romeis and Shanower 1996; Sharma 2001). Although effective in large numbers, high cost of large-scale production precludes its economic use in biocontrol of *H. armigera* (King et al. 1986).

There has been considerable interest in the inundative biological control using Bt and HearSNPV including entomopathogenic fungi and EPNs. *H. armigera* is highly susceptible to HearSNPV (Rabindra and Jayaraj 1988) and can be successfully controlled if the application coincides with the occurrence of early stages of the larvae; however, late-stage larvae are more tolerant to the virus, although to date, these tactics have not provided a viable alternative to insecticides. Granular formulations of Bt (based on wheat meal and yeast extract as phagostimulant, protect Bt against environmental degradation and formulation) have been found to be more effective than Bt sprays (Navon et al. 1997). The entomopathogenic fungus *Nomuraea rileyi* at a concentration of 1.0×10^6 spores/ml resulted in 90–100 % larval mortality, while *Beauveria bassiana* at a concentration of 2.7×10^7 spores/ml resulted in 6 % damage on chickpea compared to 16 % damage in untreated plots (Saxena and Ahmad 1997).

14.3.6 IPM Strategies

In view of the need to exploit the existing spectra of natural enemies and to reduce excessive dependence on chemical control, particularly where there is resistance to insecticides, various IPM programmes have been suggested in which different control tactics are combined to suppress pest numbers below threshold level (Sharma 2001, 2005; Deguine et al. 2008; Sharma et al. 2008; Fathipour and Sedaratian 2013). These vary from judicious use of insecticides, based on economic thresholds and regular scouting to ascertain pest population levels, to sophisticated population models to assess the need, optimum timing and selection of insecticides for sprays. Traps (light, pheromone, suction or wind traps) can be used to monitor pest populations to develop pest-forecasting models. The model, HEAPS, incorporates modules based on adult movement, oviposition, development, survival and host phenology (Dillon and Fitt 1990). Decision support systems for managing insect pests in cotton in Australia have been updated as SIRATAC and CottonLOGIC (Hearn and Bange 2002).

Genetic transformation with the Bt genes has been developed; however, the deployment of transgenic crops for pest management is raising concerns and may take time to be fully integrated in cultivation. Therefore, host–plant resistance, cultural practices, biological control with natural enemies and microbial pathogens remain the backbone of pest management systems favourable to most agroecosystems (Sharma 2005).

14.4 EPNs as Biological Control Agent of *Helicoverpa* spp.

EPNs along with their symbiotic bacteria are lethal obligatory parasites of insects (Gaugler and Kaya 1990; Kaya and Gaugler 1993; Dillman et al. 2012; Poinar and Grewal 2012; Stock and Goodrich-Blair 2012) (refer to Chap. 13 of this volume for the detailed account of bacterial symbiosis with EPNs). These beneficial nematodes can be considered good candidates for IPM and sustainable agriculture due to a variety of attributes such as rapid insect kill, host-finding ability, recycling persistence in the environment and safety for all vertebrates and nontarget invertebrates (Lacey and Georgis 2012). However, they exhibit differences in host range, infectivity, environmental tolerance and suitability for commercial production and formulation. These nematodes have been tested against a large number of insect pest species with results varying from no effect to excellent control (Begley 1990; Bedding et al. 1993; Grewal et al. 2005; Georgis et al. 2006; Kaya et al. 2006). The target insects include those from foliar, soil surface, cryptic and subterranean habitats (Arthurs et al. 2004; Georgis et al. 2006; Lacey and Georgis 2012).

From the producer's point of view, marketing EPNs is a success; however, from the global point of view, the revenues and the market size are limited. According to CPL Business Consultants report in 2008 and BBC Research report in 2009, revenue generated from chemical pesticide at the distributor level was US$37,315 million (97.92 %) compared with biopesticides in total (US$814.2 million, 2.08 %) (Lacey and Georgis 2012). Within the market share occupied by biopesticides, the revenue from essential/industrial oil, plant extract and others was maximum (66.2 %) followed by Bt (18.9 %), *B. subtilis* (7.8 %), *Trichoderma* spp. (2.8 %) and *Beauveria* spp. (2.7 %), whereas EPN's share was minimum (1.7 %). Factors such as cost, shelf life, handling, mixing, coverage, new caution signal-based pesticides, compatibility and profit margins to manufacturers and distributors have prevented nematode-based biopesticides to gain significant market share (Lacey and Georgis 2012).

14.4.1 Susceptibility of Helicoverpa *spp. at Different Life Stages*

Of the four different life stages of *Helicoverpa* spp., larvae and pupae are most susceptible to EPNs (Cabanillas and Raulston 1994; Raulston et al. 2001; Hussain and Ahmad 2011). A literature search on the infectivity, bioefficacy or susceptibility of *Helicoverpa* spp. to EPNs retrieves numerous references; describing all of them seems unnecessary and even not required; therefore, only some of the relevant works are discussed to highlight the issue. The pathogenicity of EPNs to *Helicoverpa* species has been demonstrated previously (Tanada and Reiner 1962; Samsook and Sikora 1981; Bong and Sikorowski 1983). Several other authors carried out the susceptibility studies and reported *Steinernema abbasi*, *S. carpocapsae*, *S. feltiae*, *S. glaseri*, *S. masoodi* (*Species inquirenda* by Nguyen and Hunt 2007), *S. riobrave*, *S. siamkayai*, *Heterorhabditis bacteriophora* and *H. indica* as effective in killing *Helicoverpa* larvae under laboratory conditions (Glazer and Navon 1989; Karunakar et al. 1999; Navon et al. 2002; Jothi and Mehta 2006; Shoeb et al. 2006; Hussain and Ahmad 2011).

The dose-mortality response indicates that *H. armigera* is highly susceptible to infective juveniles (IJs) with respect to different larval stages. Laboratory bioassays showed 100 % mortality of fourth instar larvae of *H. armigera* at 72 h post-exposure of 75 IJs of *H. indica* (Meerut strain) per larva (Fig. 14.1). In filter paper bioassay, mean larval mortality of different stages of *H. armigera* ranged from 43 to 75 % after 72 h post-exposure of various concentrations ranging from 25 to 150 IJs of *S. masoodi* per larva (Fig. 14.2). Third and fourth instar larvae (74 and 75 % mortality, respectively) were more susceptible than fifth (66 %) and second (43 %) instars. At the dose of 100 IJs of *S. masoodi*/larva, 100 % mortality of fourth instar larvae was recorded at 72 h post-exposure, whereas lowest mortality (23 %) of second-stage larva was obtained at nematode concentration of 25 IJs (Hussain and Ahmad 2011). The variation in virulence of nematodes to different life stages of insect is reported by several workers as well (Glazer and Navon 1989; Karunakar et al. 1999; Banu et al. 2007; Stock and Goodrich-Blair 2012). Jothi and

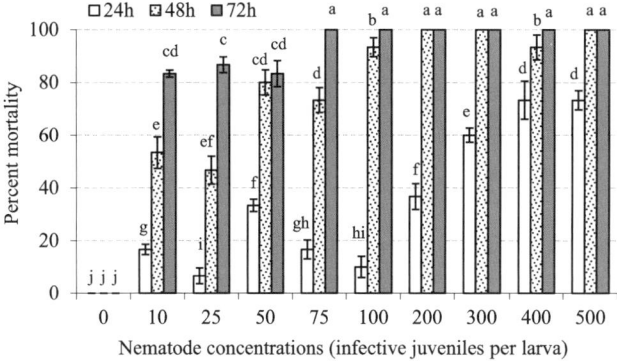

Fig. 14.1 The percent mortality of fourth instar larvae of *Helicoverpa armigera* observed at 24, 48 and 72 h post-exposure of various concentrations of *Heterorhabditis indica* (Meerut strain) infective juveniles at 30 ± 1 °C and 92 % RH. *Bars* (means, arcsine transformed values) indicated with the same letter are not significantly different according to LSD test at $P < 0.05$. Error bar = standard error of means (reproduced from Prasad et al. 2012)

Mehta (2006) obtained 83 % mortality of *H. armigera* (irrespective of larval stages) at 72 h after inoculation of *S. glaseri*, and 100 % mortality of third to fifth instar larvae was obtained at a dose of 40 IJs. Shoeb et al. (2006) recorded 93 % mortality of fourth instar larvae of *H. armigera* by *S. abbasi* and *H. bacteriophora* at 72 h post-exposure. One hundred percent mortality of last instar larvae of *H. armigera* was achieved with 200 IJs of *S. feltiae* (All strain) (Glazer and Navon 1989) and by 40 IJs of *S. glaseri* or *S. feltiae* (Karunakar et al. 1999). Mortality of 100 % fourth instar larvae of *H. armigera* was caused by feeding them on 1,000 IJ/g of *S. carpocapsae* (All strain) in the calcium alginate gel for 24 h, whereas mortality of larvae exposed to 500 IJ/g of gel ranged from 45–55 % at 4 h to 90–95 % at 48 h (Fig. 14.3).

The developmental stage of the insect plays an important role in susceptibility to the nematode. There exists a negative relationship between larval age and susceptibility to the nematodes (Samsook and Sikora 1981; Glazer and Navon 1989). There are many reasons that could explain the differences, including the size, immune response and host behaviour of the host insect. The portals of entry for nematodes may be smaller in the younger instars (Jackson and Brooks 1995), and smaller instars may be less attractive in terms of host cues such as CO_2 or kairomones (Kaya 1985). However, in a grown-up insect larva, nematodes may get crushed by insect's mandibles (Gaugler and Molloy 1981), or frequent defecation may expel nematodes entering through anus (Dowds and Peters 2002). After invasion, older larvae may also become less susceptible if nematodes fail evasion of host defences and are not able to overcome the insect's immune system (Simões and Rosa 1996).

The susceptibility of heliothinid prepupae and pupae to nematodes is reported by several authors, for example, *H. armigera* and *H. zea* prepupae and/or pupae of *H. indica*, *S. carpocapsae*, *S. glaseri*, *S. riobrave* and *S. masoodi* (Kaya and Hara

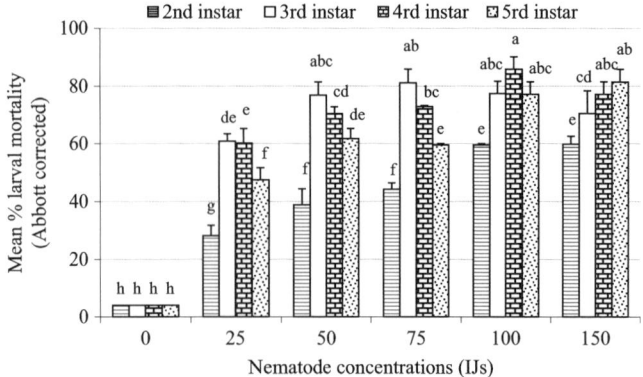

Fig. 14.2 The percent mortality of second to fifth instars of *Helicoverpa armigera* after 72 h exposure to various concentrations of *Steinernema masoodi* infective juveniles at 28 °C and 92 % RH. *Error bars* indicate the standard errors of the means. *Bars* (means of three replicates where each replicate comprised of 12 *H. armigera* larvae) indicated with the same letter are not significantly different according to Tukey's HSD test at $P < 0.05$ (arcsine transformed values) (reproduced from Hussain and Ahmad 2011)

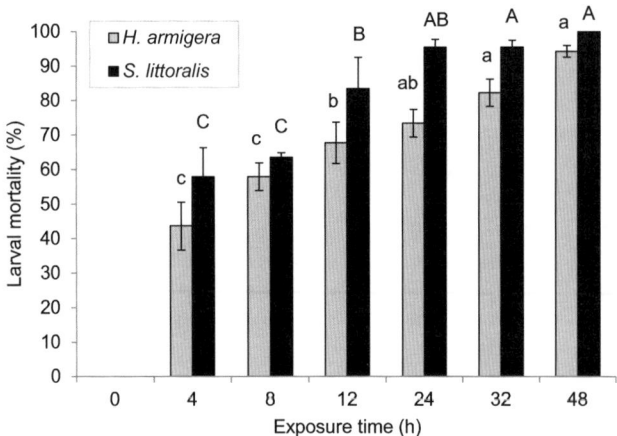

Fig. 14.3 Effect of larval exposure time to nematodes on the mortality of fourth instar larvae of *Helicoverpa armigera* and fifth instar *Spodoptera littoralis* feeding for 24 h on the nematode *S. carpocapsae* (All strain) in an alginate gel at 500 IJ/g. Means denoted by the same lower-case letter for *H. armigera* and capital letter for *S. littoralis* larvae are not significantly different ($P < 0.05$). Two-way ANOVA followed by a Student–Newman–Keuls (*SNK*) test (reproduced from Navon et al. 2002)

1981; Cabanillas and Raulston 1994; Raulston et al. 1992, 2001; Banu et al. 2007; Hussain and Ahmad 2011). The median lethal concentration of 13 IJs of *S. riobrave* per prepupa of *H. zea* was reported by Cabanillas and Raulston (1994), whereas Saravanapriya and Subramanian (2007) recorded LC_{50} values of 104 and 122 IJs of *H. indica* and *S. glaseri*, respectively, for *H. armigera* pupa. One hundred percent mortality of *H. zea* prepupae was achieved with the exposure to 100 IJs of *S. riobrave* (Cabanillas and Raulston 1994). However, lesser infectivity (59 %) to prepupae was obtained by introduction of 1,000 IJs of *S. masoodi* per earthen pot containing 500 g soil (Hussain and Ahmad 2011). The variation in percent kill could be due to the differences in nematode species, host insect species and/or testing method employed. In cage condition studies, conducted twice, Bell (1995) reported 57 and 66 % reduction in *H. virescens* adult emergence from soil under cotton plants treated with *S. riobrave* at dose 2.4×10^5 IJs/m^2. In a survey, the scouting of prepupae and pupae of *H. zea* over a period of 5 years from fruiting maize fields in Texas (USA) and northern Tamaulipas (Mexico) showed the natural parasitism of 34 % by *S. riobrave* (Raulston et al. 1992). These results strongly indicate the control potential of these nematodes in managing soil-dwelling stages of *Helicoverpa* species.

14.4.2   Field Efficacy of Nematodes for Helicoverpa *spp.*

Extensive research over the past three decades has demonstrated successes and failures of EPNs for the control of insect pests of crops, ornamental plants, trees, lawn and turf including soil insect pests (Georgis et al. 2006; Kaya et al. 2006; Lacey and Georgis 2012). Nonetheless, research highlights the potential of such bio-agents against above-ground pests under certain circumstances. Arthurs et al. (2004) analysed data from published research articles of field trials in which these nematodes were applied for control of insect pests in above-ground habitats. The highest efficacy was found for cryptic habitats compared with exposed foliage habitat.

14.4.2.1   Foliar Application

The status of the foliar application of nematodes has been reviewed to control insect pests feeding on above-ground parts (Arthurs et al. 2004; Georgis et al. 2006; Shapiro-Ilan et al. 2006; Lacey and Georgis 2012). Poor to moderate levels of suppression were achieved when nematodes were applied to foliage to control *H. armigera* and *H. zea* (Bong and Sikorowski 1983; Glazer and Navon 1989; Richter and Fuxa 1990; Vyas et al. 2003; Prabhuraj et al. 2008; Hussain and Ahmad 2011).

The limited viability of nematodes on chickpea foliage is reported by Ahmad et al. (2009) and Prabhuraj et al. (2005). When *S. masoodi*-sprayed chickpea leaves and pods were offered to *H. armigera* larvae soon after spraying, 85 % larval mortality was recorded after 72 h, whereas larval mortality declined gradually from 95 to 75 %, 55 % and 35 % when nematode-treated foliage at 1, 2 and 3 h post-spray was offered, respectively (Fig. 14.4). No larval mortality was recorded when leaves and pods (nematode sprayed at 16 and 24 h post-sprays) were fed to larvae. The

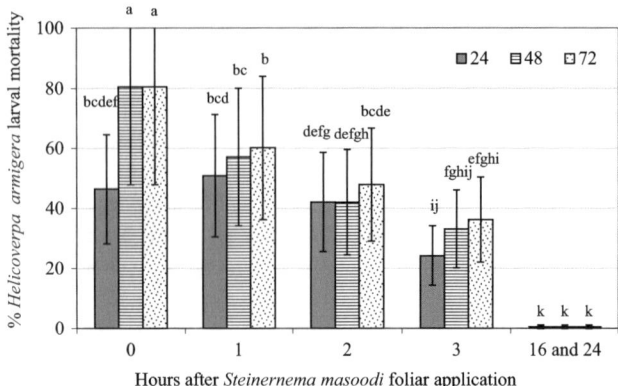

Hours after *Steinernema masoodi* foliar application

Fig. 14.4 Efficacy of *Steinernema masoodi* in killing *Helicoverpa armigera* fed on *S. masoodi*-sprayed chickpea foliage at dose 3×10^9 IJs/ha. Legends (24, 48 and 72) show the hours after *S. masoodi*-treated chickpea twigs fed to *H. armigera* larvae. *Bars* ($n = 6$) indicated with the same letters are not significantly different according to LSD test at $P < 0.05$. *Error bars* indicate the standard errors of the means (reproduced from Hussain and Ahmad 2011)

results indicate that *S. masoodi* treatments were effective up to initial 3 h only. Thereafter, nematodes desiccated over the foliage hampering its viability, mobility and effectiveness in killing the target insect (Hussain and Ahmad 2011). In contrary to above results, Patel and Vyas (1995) obtained 24.6 % mortality of *H. armigera* 6 days after *S. glaseri* application at the rate of 200 IJs/ml on chickpea in pots, whereas Vyas et al. (2003) recorded 59 and 71 % larval mortality on the 4th and 6th day after application of *Heterorhabditis* spp. @ 2,000 IJs/pot, respectively, and reported increased yield of chickpea. Glazer et al. (1992) reported low (22 %) effectiveness of *S. carpocapsae* against *H. armigera* on cotton at a dose of 5,000 IJs/ml in distilled water; however, higher mortality of 85–95 % was observed when antidesiccants like Biosys 627, natural wax and Folicote were added in spray suspension.

In field trial conducted at Kanpur (India), significant results were obtained with respect to percent pod damage and grain yield. The lowest pod damage (12 %) was recorded by the foliar application of *S. carpocapsae* (Kanpur isolate) at a dose of 3×10^9 IJs/ha + antidesiccant + UV retardant, and chickpea yield of 26.85 q/ha was obtained resulting in 42 % increased yield over the untreated control (Hussain and Ahmad 2011). Yet in another field trial carried out at Meerut (India) during 2007–2008 by Hussain and Ahmad (2011), least pod damage (11 %) was achieved by the application of 5×10^9 IJs of *H. indica* (Meerut strain)/ha + glycerine + Teepol, and the highest yield of 21.5 q/ha was obtained compared with the control plots (14.6 q/ha; pod damage 29 %). Hussain and Ahmad (2011) did not obtain greater level of control as reported by Vyas et al. (2002) and Prabhuraj et al. (2008). In pigeon pea field trial, Vyas et al. (2002) reported reduced larval population of *H. armigera* by 17 and 28.5 % over the initial population when *Heterorhabditis* sp. was sprayed alone (at a dose of 1×10^9 IJs/ha) and with adjuvants (5 % starch + gum arabic), respectively. Similarly, by the foliar spray of *H. indica* RCR isolate at a dose of 3×10^5 IJs/12 m^2 plot (equivalent to 0.25×10^9 IJs/ha), Prabhuraj et al. (2008) recorded chickpea var. A-1 pod damage up to 15.4 and 16.2 % during *rabi* 2003–2004 and 2004–2005, respectively, and obtained seed yield of 1.48 kg/plot (equivalent to 12.3 q/ha) and 18.1 q/ha, respectively. This level of control is not acceptable by end users and still a higher level of control is desirable. Overall, these data show that EPNs are not really suitable for foliar application to control *H. armigera* if we calculate cost-benefit ratio. Newer technologies and/or novel molecules are warranted to improve the viability of EPNs in order to control insect pests feeding on above-ground habitat.

The use of EPNs to manage insect pests feeding on aerial parts poses a considerable challenge as above-ground conditions are detrimental to nematodes (Arthurs et al. 2004). Infective juveniles (IJs) get inactivated quickly and are sensitive to extremes of physical environment particularly rapid desiccation (Womersley 1990), high temperature (Glazer 2002; Grewal et al. 1994), lethal UV radiation (Gaugler et al. 1992; Nickle and Shapiro 1992) and difficulty in establishing attraction gradients (Glazer 1992). To prolong nematode survival on the leaf, several adjuvants have been evaluated (Glazer 1992; Baur et al. 1997; Mason et al. 1998), but further improvements are needed to increase the feasibility of nematode foliar application (Baur et al. 1997; Schroer et al. 2005; Lacey and Georgis 2012). Besides the infective nematodes itself, other additives and adjuncts are needed to be formulated into the compositions of nematodes such as humectants, phagostimulants, UV protestants, inert fillers and dispersants, etc. to increase nematode's viability and infectivity (Raulston et al. 2001).

14.4.2.2   EPNs Application in Epigeal (Soil Surface) Habitat

The young larvae of *Helicoverpa* feed on tender foliage, but later instars switch over to feed on harvestable part of the plant. When EPN is applied on foliage to control these larvae, abiotic factors reduce the viability of nematodes to few hours only (Begley 1990; Glazer 1992; Prabhuraj et al. 2005; Ahmad et al. 2009). As a consequence, *Helicoverpa* larvae escape nematode's pathogenic attack and continue damaging fruiting bodies of the crop. The other possibility to control this dreaded pest is in the upper soil profile, the very own habitat of EPNs, while last instar larvae crawl to the ground and enter crevices or loose soil for pupation. There is likelihood that pupating larvae may encounter infective juveniles present in soil and infect them prior to insect metamorphosis into the next developmental stage (Hussain et al. 2014).

The potential of two species of EPNs for control of *Helicoverpa* spp. needs mention here. *S. riobrave* was isolated from soil samples taken from corn plots after harvest at Lower Rio Grande Valley of Texas, USA (Cabanillas et al. 1994), whereas *S. masoodi* was isolated from sandy soil of pigeon pea field when temperature was soaring high (40–45 °C) at Kanpur, India (Ali et al. 2005). The thermal tolerance of both the nematodes suggests that they could play a potential role at

Table 14.2 Percent suppression of *Helicoverpa armigera* adult emergence by *Steinernema masoodi* at prepupal stage in soil under laboratory conditions

Nematode concentration (IJs/60 g soil/container)	Percent suppression of adult emergence[a] (SE)
Experiment 1	
0	0.8 (0) b
500	80.4 (8.8) a
1,000	76.3 (12.9) a
2,000	89.2 (0) a
3,000	89.2 (0) a
5,000	89.2 (0) a
Experiment 2	
0	0.6 (0) c
50	29.9 (3.3) bc
100	45.0 (11.8) ab
150	45.0 (5.8) ab
200	56.8 (0) ab
250	53.8 (3.0) ab
300	63.4 (0) a
350	57.1 (6.3) ab
400	50.9 (5.9) ab
450	64.2 (7.4) a
500	60.1 (3.3) ab

[a]Within a column, means (Abbott corrected and arcsine transformed values) followed by the same lower-case letter are not significantly different, ANOVA, Tukey's HSD test at $P < 0.05$. Experiments 1 and 2 were analysed separately. Means of three replicates where each replicate comprised of ten last instar larvae of *H. armigera* (reproduced from Hussain et al. 2014)

high temperature regimes in the management of pupating *H. armigera* (Gouge et al. 1999; Ali et al. 2007). Both the nematode species were evaluated for the suppression of *H. armigera* and *H. zea* at prepupal stage while undergoing pupation in soil. In filter paper bioassay, exposure of 10, 20, 40, 80 and 100 IJs of *S. riobrave* per prepupa resulted in mortalities of 40, 55, 85, 90 and 100 %, respectively (Cabanillas and Raulston 1994), whereas in laboratory, suppression of 76–89 % adult emergence was obtained with no significant differences at nematode concentrations 500, 1,000, 2,000, 3,000 and 5,000 IJs of *S. masoodi*/60 g soil/last instar larva (Hussain et al. 2014). However, at lower concentrations (0–500 IJs/60 g soil/larva with increment of 50 IJs in the treatments), varied percent suppression was recorded, highest (64 %) being at 450 IJs/60 g soil/larva and lowest (30 %) at 50 IJs/60 g soil/larva (Table 14.2).

Cabanillas et al. (2005) describe many factors that play crucial roles in the successful use of nematode in corn earworm, *H. zea* control. First, nematode applications should be matched with the most susceptible stage of the maize earworm. Cabanillas and Raulston (1995) obtained insect mortalities of 100 and 95 % in maize fields by applying *S. riobrave* to the soil when 50 % of the larvae were late instars (still in the maize ears) and when 10 % of the larvae had left the ears to pupate, respectively (Fig. 14.5). Second, irrigation method and timing, and nematode concentration should be optimum. In chickpea micro-plot, the suppression of 70 % adult emergence was obtained at the dose of 600,000 IJs/m^2 (Fig. 14.6), whereas *S. riobrave*, at the most effective nematode concentration of 200,000 nematodes/m^2, caused higher insect mortalities when it was applied via in-furrow irrigation (95 %) than when it was applied after irrigation (84 %) or before irrigation (56 %) (Fig. 14.7). Third, the nematode species and the application method should be matched with the target ecosystem. *S. riobrave* (TX strain), at the most effective concentration of 200,000 nematodes/m^2, caused 95 % maize earworm prepupae and pupae mortality, while *S. carpocapsae* (All strain) did not cause any insect mortality in maize fields (Cabanillas and Raulston 1996b). The superiority of *S. riobrave* was attributed to its greater tolerance of warm soil temperatures (>38 °C) compared with *S. carpocapsae* (Grewal et al. 1994; Gouge et al. 1999). Cabanillas and Raulston (1996a) found that subsurface nematode incorporation produced higher insect infections than soil surface applications in the fields that received nematodes before or after irrigation. Subsurface application provided greater nematode protection against desiccation and sunlight than soil surface application (Gaugler 1988). Similarly, Feaster and Steinkraus (1996) achieved excellent results, by applying *S. riobrave* to the soil in Arkansas maize to control maize earworm. Mean mortalities from *S. riobrave* infections were 79 and 91 % at nematode levels of 3.7×10^6 and 12.0×10^6 nematodes/m^2 of soil, respectively. Although similar results were obtained in irrigated and nonirrigated plots, higher infection occurred in the plots receiving flood irrigation.

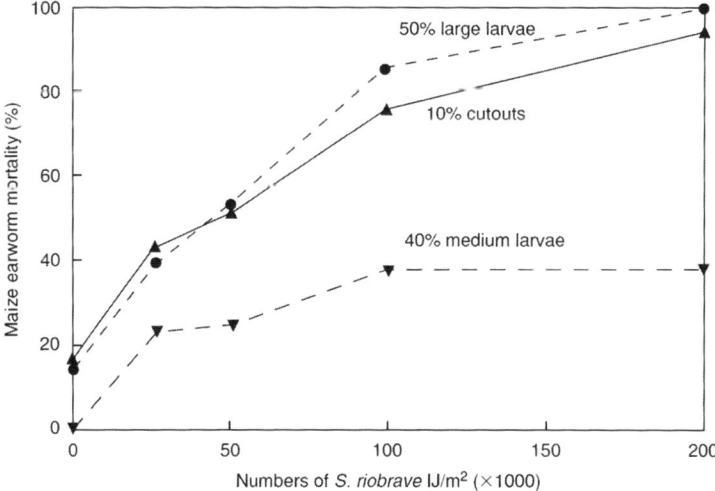

Fig. 14.5 Effect of *Steinernema riobrave* concentration and timing of soil application on parasitism of maize earworm *Helicoverpa zea* prepupae and pupae in maize (reproduced from Cabanillas and Raulston 1995)

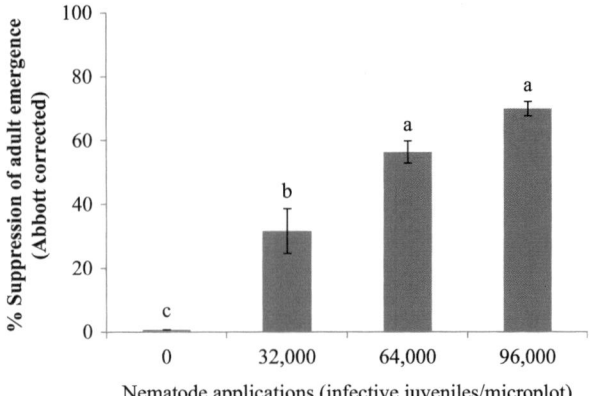

Fig. 14.6 Percent suppression of *Helicoverpa armigera* adult emergence by *Steinernema masoodi* at prepupal stage in chickpea micro-plot (40 cm × 40 cm). Infective juveniles were applied to soil surface, and *H. armigera* last instar larvae were released in netted micro-plot, and adult emergence was monitored up to 1 month. *Bars* (means, Abbott corrected and arcsine transformed values, of four replicates where each replicate comprised of 20 *H. armigera* larvae) indicated with the same letter are not significantly different according to Tukey's HSD test at $P < 0.05$. *Thin bars* are standard errors of means (reproduced from Hussain et al. 2014)

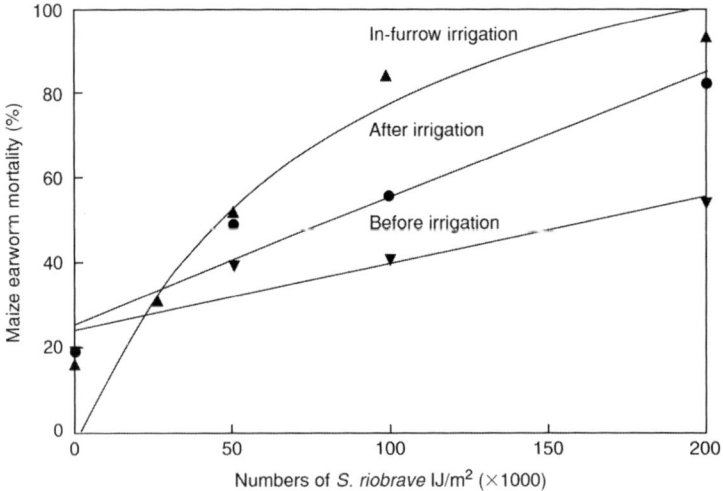

Fig. 14.7 Effects of irrigation timing and concentration of *Steinernema riobrave* on mortality of maize earworm *Helicoverpa zea* prepupae buried within 6 days after nematode application in soil in a maize field (reproduced from Cabanillas and Raulston 1996a)

They demonstrated that this nematode has potential as an inundative biocontrol agent for this pest.

In semiarid tropics, chickpea is grown under rainfed cool-weather crop or dry-climate crop and on residual moisture (Smithson et al. 1985), but no irrigation is needed to raise the crop. If desired, only one-time irrigation is given during flowering and/or fruiting period, which is the most appropriate time of nematode application as high incidence of *H. armigera* is expected during this period. Thus, control option at soil-dwelling stages could play a critical role in the management of *H. armigera* at the time when larva is finding a hiding place in cracks, crevices or loose soil in the field to undergo pupation (Hussain et al. 2014). These results substantiate a better approach to suppress the population of forthcoming generations of *Helicoverpa* spp., thus preventing adult emergence and subsequent migration and causing damage to cotton, corn, chickpea, pigeon pea or any other host crops. However, the apprehension of Bergvinson (2005) is that this approach may have application for commercial crop protection, but the larvae must complete their development before being controlled, thereby only reducing insect pest pressure for the next cropping cycle—an approach that is unlikely to be economically feasible for farmers.

14.4.3   Factors Affecting Success/Failure of Nematodes

In the laboratory, most EPN species infect a variety of insects where host contact is certain, environmental conditions are optimal and no ecological or behavioural barriers to infection exist. Foliage-feeding lepidopteran larvae are highly susceptible to infection in petri dishes, but are seldom impacted in the field, where nematodes tend to be quickly inactivated by the environmental factors (i.e. desiccation, radiation, temperature) characteristics of exposed foliage (Kaya and Gaugler 1993). However, in the field, EPNs attack a significantly narrower host range than in the laboratory (Georgis et al. 2006). Several factors related to the nematode's biology are critical for successful application; foremost is the matching of appropriate nematode species with the target pest. Proper match of the nematode to the host includes virulence, host finding and environmental tolerance (Shapiro-Ilan et al. 2006). If a nematode does not possess a high level of virulence towards the target pest, there is little hope of success.

Matching the appropriate nematode host-seeking strategy with the pest is also essential. These nematodes employ different foraging strategies to locate an insect host, which range from one extreme of sit-and-wait (ambush) to the other of widely foraging strategy (cruise) (Lewis 2002). Fenton et al. (2001) emphasised that different species and strains of EPNs are not the same. For instance, *S. glaseri* and *H. bacteriophora* exhibit strong dispersal (cruising) tendencies and are characterised by high motility and are distributed throughout the soil profile. Cruisers orient to volatile host cues and switch to a localised search after host contact and are well adapted to infect deep soil-dwelling sedentary hosts such as scarab and lepidopteran prepupae and pupae, whereas the sedentary nematodes, e.g. *S. carpocapsae* adopt sit-and-wait strategies or ambushers and are characterised by low motility and a tendency to stay near the soil surface. Ambushers tend not to respond to volatile and contact host cues unless presented in an appropriate sequence and are more effective against surface-dwelling pests such as the codling moth, cutworms and mole crickets near the soil surface (Georgis and Gaugler 1991; Lewis et al. 1993; Campbell et al. 1995).

Fenton et al. (2001) analyses suggest that pre-emptive application may be the optimum strategy and applying before pest invasion can result in greater control than applying afterwards. Simulations showed that, to prevent any damage occurring, it may be preferable to apply nematodes before pest invasion, possibly during the time window between adult arrival and larval emergence. However, the success of this approach relies on low levels of nematode mortality. A number of field trials have shown that survival can be enhanced if plots are heavily irrigated both before and after application, thereby increasing the period for which pre-emptive control is viable (Georgis and Gaugler 1991; Downing 1994). The results of trials carried out by Feaster and Steinkraus (1996) on the use of *S. riobrave* as a control agent of the corn earworm *H. zea* agree with Fenton et al. (2001) analyses; nematodes applied 24 h before pest invasion resulted in greater suppression than a larger dose of nematodes applied 24 h after pest invasion. Additionally, irrigated plots produced greater rates of suppression than nonirrigated, through enhanced nematode survival.

14.5 Future Perspective of *Helicoverpa* spp. Control with Nematodes

Since the release of first transgenic crops in 1996, the option for the management of *Helicoverpa* spp. by EPNs was taken a back seat and given the least priority as it is a costly business compared with the deployment of Bt crops. The major concern of Bt cotton has been widespread development of insect resistance to the Bt toxins, but this has so far proven to be minimal. The future of transgenic crops appears stable for some time to come (James 2012). Shapiro-Ilan et al. (2002) support the current management tactics through the use of transgenic crop that is far more economical than nematodes and, unlike nematodes, will provide control of several major cotton pests simultaneously, including pink bollworm and tobacco budworm. For example, the seed cost of transgenic cotton is US$116.54 per ha in the USA (Finger et al. 2011), whereas nematode costs US$1,895 at the recommended dose of 2.5×10^9 infective nematodes per ha in a single application.

The bollworm appears to be a prime example of an insect that can be managed using EPNs, but for practical reasons, nematodes will remain unused in conventional cotton (Shapiro-Ilan et al. 2002). That's the reason why we don't see literature after 1996 on large-scale field trials with nematodes for the control of *Helicoverpa* spp. in cotton, corn, soybeans, tomato or any other high-value cash crops wherever Bt transgenic crops are cultivated. However, the interest of using nematodes in refuge is renewed as to delay the development of resistance in insects. The experimental and modelling results suggest that these nematodes could slow the evolution of pest resistance to Bt crops, but only under some conditions

(Gassmann et al. 2008, 2012). In combination with the high dose-refuge strategy, other strategies such as stacking two or more insecticidal genes in the same plant; tissue-, time- or signal-dependent expression of transgene; rotation of insecticidal toxins with different modes of action; and deploying different toxins in different crops in a production system can be used to avoid or delay the development of resistance in insects to transgenic plants (Gopala Swamy et al. 2009). For example, transgenic *Arabidopsis thaliana* plant expressing the 283-kDa protein, toxin A, encoded by tcdA gene of EPN's symbiotic bacteria (*Photorhabdus luminescens*) conferred resistance to corn rootworm (*Diabrotica undecimpunctata*) and had a strong growth-inhibitory effect on the insect (Liu et al. 2003).

The phenomenon of anhydrobiosis inherited in several groups of plant parasitic nematodes [such as stem and bulb nematode (*Ditylenchus dipsaci*), seed gall nematode (*Anguina tritici*) and white tip nematode (*Aphelenchoides besseyi*) remain viable for several years in quiescent stage (Duncan and Moens 2006; Wright and Perry 2006)] is always baffling us why these nematodes show true anhydrobiotic behaviour whereas EPNs do not. EPNs exhibit only quiescent anhydrobiosis (Womersley 1990). Is there anything that can be done in this direction of research to bring the trait of anhydrobiosis in EPNs by utilising molecular and biotechnological means? Grewal et al. (2006) and Perry et al. (2012) reviewed the research work on the physiology, genetics and molecular biology of anhydrobiosis in the infective juveniles. If this is really possible, then this innovation will undoubtedly contribute to the expansion of EPNs as biocontrol agent of most of the susceptible insect pests encompassing above- as well as below-ground habitats. The advantages of incorporated trait of anhydrobiosis will keep the EPNs and their symbiotic bacteria in inactivated form not only in formulation and prolonged storage but also on foliage after above-ground application and resume their activity upon getting favourable condition. If this is not possible, then there must be something else such as novel molecules in nature or may be invented in near future, which can be admixed to encapsulate these nematodes and safeguard them from detrimental environmental conditions upon application, but recommence their activities soon after getting entered actively or passively by a susceptible host insect.

14.6   Conclusions

The crop production has been severely threatened by increasing difficulties in controlling *Helicoverpa* spp. as it has developed high levels of resistance to commonly used insecticides. Use of bio-intensive IPM strategies can reduce existing overdependence on insecticides and their negative effects on the environment. Manipulation of cultural practices, host–plant resistance including the deployment of transgenic crops and biological control could play a crucial role in reducing the ravages caused by this pest. Among microbial control agents, EPNs (*Steinernema* and *Heterorhabditis*) have been used to control insect pests. A

number of studies support the effectiveness of these nematodes in controlling
H. armigera in laboratory as well as field conditions. However, the outcome of
foliar applications indicates that nematode alone is not effective on foliage; as a
consequence, *Helicoverpa* larvae escape nematode's pathogenic attack and continue damaging fruiting bodies. However, charging nematode in irrigation water
could substantiate an approach to suppress the forthcoming generations of
Helicoverpa spp., thus preventing adult emergence, subsequent migration and
damage to succeeding host crops.

Acknowledgements We are grateful to anonymous colleagues for reviewing the manuscript and
for providing useful comments and suggestions.

References

Acharjee S, Sarmah BK (2013) Transgenic *Bacillus thuringiensis* (Bt) chickpea: India's most
    wanted genetically modified (GM) pulse crop. Afr J Biotechnol 12:5709–5713
Ahmad R, Hussain MA, Ali SS, Pervez R (2009) Survival of *Steinernema masoodi* and
    S. carpocapsae (Rhabditida: Steinernematidae) on pigeonpea and chickpea after foliar appli-
    cation. Arch Phytopathol Plant Protect 42:112–117
Ali SS, Shaheen A, Pervez R, Hussain MA (2005) *Steinernema masoodi* sp. n. and *S. seemae*
    sp. n. (Nematoda: Rhabditida: Steinernematidae) from India. Int J Nematol 15:89–99
Ali SS, Pervez R, Hussain MA, Ahmad R (2007) Effect of temperature on survival of *Steinernema
    seemae*, *S. masoodi* and *S. carpocapsae* (Rhabditida: Steinernematidae) and their subsequent
    infectivity to prepupa of *Helicoverpa armigera* (Hübner). Arch Phytopathol Plant Protect
    40:183–187
Alvi AHK, Sayyed AH, Naeem M, Ali M (2012) Field evolved resistance in *Helicoverpa armigera*
    (Lepidoptera: Noctuidae) to *Bacillus thuringiensis* Toxin Cry1Ac in Pakistan. PLoS One 7
    (10):e47309
Armes NJ (1989) Summary of life-table experiments on *Heliothis armigera* (Hübner) (Lepidop
    tera: Noctuidae). Natural Resources Institute, Chetham, p 55
Armes NJ, Jadhav DR, DeSouza KR (1996) A survey of insecticide resistance in *Helicoverpa
    armigera* in the Indian subcontinent. Bull Entomol Res 86:499–514
Arthurs S, Heinz KM, Prasifka JR (2004) An analysis of using entomopathogenic nematodes
    against above-ground pests. Bull Entomol Res 94:297–306
Banu JG, Jothi BD, Narkhedkar NG (2007) Susceptibility of different stages of cotton bollworm,
    Helicoverpa armigera (Lepidoptera: Noctuidae) to entomopathogenic nematodes. Int J
    Nematol 17:41–45
Barwale RB, Godwal VR, Usha Z, Zehr B (2004) Prospects for Bt cotton technology in India.
    AgBioForum 7:23–26
Baur ME, Kaya HK, Gaugler R, Tabashnik B (1997) Effects of adjuvants on entomopathogenic
    nematode persistence and efficacy against *Plutella xylostella*. Biocontrol Sci Technol
    7:513–525
Bedding R, Akhurst R, Kaya HK (eds) (1993) Nematodes and the biological control of insects.
    CSRIO, East Melbourne
Begley JW (1990) Efficacy against insects in habitats other than soil. In: Gaugler R, Kaya HK
    (eds) Entomopathogenic nematodes in biological control. CRC Press, Boca Raton, pp 215–231
Bell MR (1995) Effects of an entomopathogenic nematode and a nuclear polyhedrosis virus on
    emergence of *Heliothis virescens* (Lepidoptera: Noctuidae). J Entomol Sci 30:243–250

Bergvinson DJ (2005) *Heliothis/Helicoverpa* problem in the Americas: biology and management. In: Sharma HC (ed) *Heliothis/Helicoverpa* management. Oxford and IBH Publishing, New Delhi, pp 7–37

Bong CFJ, Sikorowski PP (1983) Use of the DD 136 strain of *Neoaplectana carpocapsae* Weiser (Rhabditida: Steinernematidae) for control of corn earworm (Lepidoptera: Noctuidae). J Econ Entomol 76:590–593

Bradley JR, Herzog GA, Roach SA, Stinner RE, Terry LI (1986) Cultural control in southeastern US cropping systems. In: Johnson SJ, King EG, Bradley JR (eds) Theory and tactics of *Heliothis* population management: 1. cultural and biological control, vol 316. Southern Cooperative Series Bull, pp 22–27

Cabanillas HE, Raulston JR (1994) Pathogenicity of *Steinernema riobravis* against corn earworm, *Helicoverpa zea* (Boddie). Fundam Appl Nematol 17:219–223

Cabanillas HE, Raulston JR (1995) Impact of *Steinernema riobravis* (Rhabditida: Steinernematidae) on the control of *Helicoverpa zea* (Lepidoptera: Noctuidae) in corn. J Econ Entomol 88:58–64

Cabanillas HE, Raulston JR (1996a) Effects of furrow irrigation on the distribution and infectivity of *Steinernema riobravis* against corn earworm in corn. Fundam Appl Nematol 19:273–281

Cabanillas HE, Raulston JR (1996b) Evaluation of *Steinernema riobravis*, *S. carpocapsae*, and irrigation timing for the control of corn earworm, *Helicoverpa zea*. J Nematol 28:75–82

Cabanillas HE, Poinar GO Jr, Raulston JR (1994) *Steinernema riobravis* sp. nov. (Rhabditida: Steinernematidae) from Texas. Fundam Appl Nematol 17:123–131

Cabanillas HE, Wright RJ, Vyas RV (2005) Cereal, fibre, oilseed and medicinal crop applications. In: Grewal PS, Ehlers R-U, Shapiro-Ilan DI (eds) Nematodes as biocontrol agents. CABI Publishing, Wallingford, pp 265–279

Campbell JF, Lewis E, Yoder F, Gaugler R (1995) Entomopathogenic nematode (Heterorhabditidae and Steinernematidae) seasonal population dynamics and impact on insect populations in turfgrass. Biol Control 5:598–606

Cook SM, Khan ZR, Pickett JA (2007) The use of push-pull strategies in integrated pest management. Annu Rev Entomol 52:375–400

COPR (1983) Pest control in tropical tomatoes. COPR, London

Deguine JP, Ferron P, Russell D (2008) Sustainable pest management for cotton production: a review. Agron Sustain Dev 28:113–137

Dillman AR, Chaston JM, Adams BJ, Ciche TA, Goodrich-Blair H, Stock SP, Sternberg PW (2012) An entomopathogenic nematode by any other name. PLoS Pathog 8:e1002527

Dillon ML, Fitt GP (1990) HEAPS: a regional model of *Heliothis* population dynamics. In: Proceedings of the 5th Australian cotton conference, 8–9 August 1990, Broadbeach, Queensland. Australian Cotton Growers' Research Association, Brisbane, Queensland, Australia, pp 337–344

Dowds BCA, Peters A (2002) Virulence mechanisms. In: Gaugler R (ed) Entomopathogenic nematology. CABI Publishing, Wallingford, pp 79–98

Downes S, Parker T, Mahon R (2010) Incipient resistance of *Helicoverpa punctigera* to the Cry2Ab Bt toxin in Bollgard II® cotton. PLoS One 5:e12567

Downing AS (1994) Effect of irrigation and spray volume on efficacy of entomopathogenic nematodes (Rhabditida: Heterorhabditidae) against white grubs (Coleoptera: Scarabaeidae). J Econ Entomol 87:643–646

Duncan LW, Moens M (2006) Migratory endoparasitic nematodes. In: Perry RN, Moens M (eds) Plant nematology. CAB International, Wallingford, pp 123–152

Fathipour Y, Sedaratian A (2013) Integrated management of *Helicoverpa armigera* in soybean cropping systems. In: El-Shemy HA (ed) Soybean – pest resistance. InTech, Rijeka, pp 231–280. http://dx.doi.org/10.5772/54522

Feaster MA, Steinkraus DC (1996) Inundative biological control of *Helicoverpa zea* (Lepidoptera: Noctuidae) with the entomopathogenic nematode *Steinernema riobravis* (Rhabditida: Steinernematidae). Biol Control 7:38–43

Fenton A, Norman R, Fairbairn JP, Hudson PJ (2001) Evaluating the efficacy of entomopathogenic nematodes for the biological control of crop pests: a nonequilibrium approach. Am Nat 158:408–425

Finger R, Benni NE, Kaphengst T, Evans C, Herbert S, Lehmann B, Morse S, Stupak N (2011) A meta analysis on farm-level costs and benefits of GM crops. Sustainability 3:743–762

Fitt GP (1989) The ecology of *Heliothis* species in relation to agroecosystems. Annu Rev Entomol 34:17–52

Fitt GP, Cotter SC (2005) The *Helicoverpa* problem in Australia: biology and management. In: Sharma HC (ed) *Heliothis/Helicoverpa* management: emerging trends and strategies for future research. Oxford and IBH Publishing, New Delhi, pp 45–61

Gassmann AJ, Stock SP, Sisterson MS, Carrière Y, Tabashnik BE (2008) Synergism between entomopathogenic nematodes and *Bacillus thuringiensis* crops: integrating biological control and resistance management. J Appl Ecol 45:957–966

Gassmann AJ, Hannon ER, Sisterson MS, Stock SP, Carrière Y, Tabashnik BE (2012) Effects of entomopathogenic nematodes on evolution of pink bollworm resistance to *Bacillus thuringiensis* toxin Cry1Ac. J Econ Entomol 105:994–1005

Gaugler R (1988) Ecological considerations in the biological control of soil-inhabiting insects with entomopathogenic nematodes. Agric Ecosyst Environ 24:351–360

Gaugler R, Kaya HK (eds) (1990) Entomopathogenic nematodes in biological control. CRC Press, Boca Raton, p 365

Gaugler R, Molloy D (1981) Instar susceptibility to *Simulium vittatum* (Diptera: Simuliidae) to the entomogenous nematode *Neoaplectana carpocapsae*. J Nematol 13:1–5

Gaugler R, Bednarek A, Campbell JF (1992) Ultraviolet inactivation of heterorhabditid and steinernematid nematodes. J Invertebr Pathol 59:155–160

Georgis R, Gaugler R (1991) Predictability in biological control using entomopathogenic nematodes. J Econ Entomol 84:713–720

Georgis R, Koppenhöfer AM, Lacey LA, Bélair G, Duncan LW, Grewal PS, Samish M, Tan L, Torr P, van Tol RWHM (2006) Successes and failures in the use of parasitic nematodes for pest control. Biol Control 38:103–123

Glazer I (1992) Survival and efficacy of *Steinernema carpocapsae* in an exposed environment. Biocontrol Sci Technol 2:101–107

Glazer I (2002) Survival biology. In: Gaugler R (ed) Entomopathogenic nematology. CABI Publishing, Wallingford, pp 169–187

Glazer I, Navon A (1989) Activity and persistence of entomoparasitic nematodes tested against *Heliothis armigera* (Lepidoptera, Noctuidae). J Econ Entomol 83:1795–1800

Glazer I, Klein M, Novan A, Nakache Y (1992) Comparison of efficacy of entomopathogenic nematodes combined with antidesiccants applied by canopy sprays against three cotton pests (Lepidoptera: Noctuidae). J Econ Entomol 85:1636–1641

Gopala Swamy SVS, Prasad NVVSD, Rao NHP (2009) Transgenic Bt crops: a major component of integrated pest management – an overview. Indian J Crop Sci 4:1–10

Gouge DH, Lee LL, Henneberry TJ (1999) Effect of temperature and lepidopteran host species on entomopathogenic nematode (Nematoda: Steinernematidae, Heterorhabditidae) infection. Environ Entomol 28:876–883

Grewal PS, Selvan S, Gaugler R (1994) Thermal adaptation of entomopathogenic nematodes: niche breadth for infection, establishment, and reproduction. J Thermal Biol 19:245–253

Grewal PS, Ehlers R-U, Shapiro-Ilan DI (eds) (2005) Nematodes as biocontrol agents. CABI Publishing, Wallingford, p 505

Grewal PS, Bornstein-Forst S, Burnell AM, Glazer I, Jagdale GB (2006) Physiological, genetic, and molecular mechanisms of chemoreception, thermobiosis, and anhydrobiosis in entomopathogenic nematodes. Biol Control 38:54–65

Hardwick DF (1965) The corn earworm complex. Memoirs Entomol Soc Canada 40:1–247

Hearn AB, Bange MP (2002) SIRATAC and CottonLOGIC: persevering with DSSs in the Australian cotton industry. Agric Syst 74:27–56

Hussain MA, Ahmad W (2011) Management of *Helicoverpa armigera* by entomopathogenic nematodes. Lambert Academic Publishing GmbH & Co. KG, Saarbrücken, p 168

Hussain MA, Ahmad R, Ahmad W (2014) Evaluation of *Steinernema masoodi* (Rhabditida: Steinernematidae) against soil-dwelling life stage of *Helicoverpa armigera* (Lepidoptera: Noctuidae) in laboratory and microplot study. Can J Plant Protect 2:4–8

Jackson GB (2014) Insect control guide for agronomic crops. Mississippi State University Extension Service, Publication No. 2471, p 104

Jackson JJ, Brooks MA (1995) Parasitism of western corn rootworm larvae and pupae by *Steinernema carpocapsae*. J Nematol 27:15–20

Jallow MFA, Cunningham JP, Zalucki MP (2004) Intra-specific variation for host plant use in *Helicoverpa armigera* (Hübner) (Lepidoptera: Noctuidae): implications for management. Crop Prot 23:955–964

James C (2012) Global status of commercialized biotech/GM crops: 2012. ISAAA Brief No. 44. ISAAA, Ithaca

Jayaraj S (1982) Biological and ecological studies of *Heliothis*. In: Reed W, Kumble V (eds) Proceedings of the international workshop on *Heliothis* management, 15–20 November 1981. ICRISAT, Patancheru, pp 17–28

Jothi BD, Mehta UK (2006) Pathogenicity of three species of EPN against cotton bollworm *Helicoverpa armigera* Hüb. Entomon 31:259–266

Karunakar G, Easwaramoorthy S, David H (1999) Susceptibility of nine lepidopteran insects to *Steinernema glaseri*, *S. feltiae* and *Heterorhabditis indicus* infection. Int J Nematol 9:68–71

Kaya HK (1985) Susceptibility of early larval stages of *Pseudaletia unipuncta* and *Spodoptera exigua* (Lepidoptera: Noctuidae) to the entomogenous nematode *Steinernema feltiae* (Rhabditida: Steinernematidae). J Invertebr Pathol 46:58–62

Kaya HK, Gaugler R (1993) Entomopathogenic nematodes. Annu Rev Entomol 38:181–206

Kaya HK, Hara AH (1981) Susceptibility of various species of lepidopterous pupae to the entomogenous nematode *Neoaplectana carpocapsae*. J Nematol 13:291–294

Kaya HK, Aguillera MM, Alumai A, Choo HY, de la Torre M, Fodor A, Ganguly S, Hazir S, Lakatos T, Pye A, Wilson M, Yamanaka S, Yang H, Ehlers R-U (2006) Status of entomopathogenic nematodes and their symbiotic bacteria from selected countries or regions of the world. Biol Control 38:134–155

King EG, Coleman RJ (1989) Potential for biological control of *Heliothis* species. Annu Rev Entomol 34:53–75

King EG, Baumhover A, Bouse LF, Greany P, Hartstack AW, Hooper KR, Knipling EF, Morrison RK, Nettles WC, Powell JE (1986) Augmentation of entomophagous arthropods. In: Johnson SJ, King EG, Bradley JR (eds) Theory and tactics of Heliothis. Population management. 1. Cultural and biological control, vol 316. Southern Cooperative Series Bull, pp 116–131

Kranthi KR, Jadhav DR, Kranthi S, Wanjari RR, Ali SS, Russel DA (2002) Insecticide resistance in five major insect pests of cotton in India. Crop Prot 21:449–460

Kumari A, Reddy DJ, Sharma HC (2010) Stability of resistance to pod borer, *Helicoverpa armigera* in pigeonpea. Indian J Plant Protect 38:6–12

Lacey LA, Georgis R (2012) Entomopathogenic nematodes for control of insect pests above and below ground with comments on commercial production. J Nematol 44:218–225

Lal SS, Yadava CP, Sachan JN (1986) Strategies for the development of an integrated approach to control grain pod borer, *Heliothis armigera* infesting chickpea. Pesticides 20:39–51

Lewis EE (2002) Behavioural ecology. In: Gaugler R (ed) Entomopathogenic nematology. CABI Publishing, Wallingford, pp 205–223

Lewis EE, Gaugler R, Harrison R (1993) Response of cruiser and ambusher entomopathogenic nematodes (Steinernematidae) to host volatile cues. Can J Zool 71:765–769

Liao C, Heckel DG, Akhursta R (2002) Toxicity of *Bacillus thuringiensis* insecticidal proteins for *Helicoverpa armigera* and *Helicoverpa punctigera* (Lepidoptera: Noctuidae), major pests of cotton. J Invertebr Pathol 80:55–63

Liu D, Burton S, Glancy T, Li Z-S, Hampton R, Meade T, Merlo DJ (2003) Insect resistance conferred by 283-kDa *Photorhabdus luminescens* protein TcdA in *Arabidopsis thaliana*. Nat Biotechnol 21:1222–1228

Manjunath TM, Bhatnagar VS, Pawar CS, Sithanantham S (1989) Economic importance of *Heliothis* spp. in India and an assessment of their natural enemies and host plants. In: King EG, Jackson RD (eds) Proceedings of the workshop on biological control of *Heliothis*: increasing the effectiveness of natural enemies, 15–20 November 1985. Far Eastern Regional Research Office, USDA, New Delhi, pp 197–228

Mason JM, Matthews GA, Wright DJ (1998) Screening and selection of adjuvants for the spray application of entomopathogenic nematodes against a foliar pest. Crop Prot 17:463–470

Matthews GA (1989) Cotton insect pests and their management. Longmans, Harlow, p 199

Matthews M (1999) Heliothine moths of Australia: a guide to pest bollworms and related noctuid groups, vol 7, Monograph on Australian Lepidoptera. CSIRO Publishing, Callingford, p 320

Mendelsohn M, Kough J, Vaituzis Z, Matthews K (2003) Are Bt crops safe? Nat Biotechnol 21:1003–1009

Mihm JA (ed) (1997) Insect resistant maize: recent advances and utilization. In: Proceedings of an international symposium held at the International Maize and Wheat Improvement Center (CIMMYT), 27 November–3 December, 1994. CIMMYT, Mexico, p 304

Naseri B, Fathipour Y, Moharramipour S, Hosseininaveh V, Gatehouse AM (2010) Digestive proteolytic and amylolytic activities of *Helicoverpa armigera* in response to feeding on different soybean cultivars. Pest Manag Sci 66:1316–1323

Navon A (2000) *Bacillus thuringiensis* insecticides in crop protection – reality and prospects. Crop Prot 19:669–676

Navon A, Keren S, Levski S, Grinstein A, Riven Y (1997) Granular feeding baits based on *Bacillus thuringiensis* products for the control of lepidopterous pests. Phytoparasitica 25:101–110

Navon A, Nagalakshmi VK, Levski S, Salame L, Glazer I (2002) Effectiveness of entomopathogenic nematodes in an alginate gel formulation against lepidopterous pests. Biocontrol Sci Technol 12:737–746

Nguyen KB, Hunt DJ (eds) (2007) Entomopathogenic nematodes: systematics, phylogeny and bacterial symbionts, vol V. Brill, Leiden-Boston, p 816

Nickle WR, Shapiro M (1992) Use of a stilbene brightener, Tinopal LPW, as a radiation protectant for *Steinernema carpocapsae*. J Nematol 24:371–373

Patel MC, Vyas RV (1995) Efficacy of *Steinernema glaseri* against *Helicoverpa armigera* on chickpea in pots. Int Chickpea Pigeonpea Newsl 2:39–40

Perry RN, Ehlers R-U, Glazer I (2012) A realistic appraisal of methods to enhance desiccation tolerance of entomopathogenic nematodes. J Nematol 44:185–190

Poinar GO Jr, Grewal PS (2012) History of entomopathogenic nematology. J Nematol 44:153–161

Prabhuraj A, Girish KS, Shivaleela (2005) Persistence of *Heterorhabditis indica* on chickpea foliage. Indian J Nematol 35:24–27

Prabhuraj A, Girish KS, Shivaleela PBV (2008) Integration of *Heterorhabditis indica* with other biorationals for managing chickpea pod borer, *Helicoverpa armigera* (Hüb.). J Biol Control 22:433–448

Prasad CS, Hussain MA, Pal R, Prasad M (2012) Virulence of nematode *Heterorhabditis indica* (Meerut strain) against lepidopteran and coleopteran pests. Vegetos 25:343–351

Rabindra RJ, Jayaraj S (1988) Efficacy of NPV with adjuvants as high volume and ultra low volume application against *Heliothis armigera* (Hbn.) on chickpea. Trop Pest Manag 34:441–444

Raulston JR, Pair SD, Loera J, Cabanillas HE (1992) Prepupal and pupal parasitism of *Helicoverpa zea* and *Spodoptera frugiperda* (Lepidoptera: Noctuidae) by *Steinernema* sp. in cornfields in the Lower Rio Grande Valley. J Econ Entomol 85:1666–1670

Raulston JR, Pair SD, Enrique C (2001) *Steinernema* sp. nematode for suppression of *Helicoverpa zea* and *Spodoptera frugiperda*. US Patent US 6184434 B1

Reddy GVP, Manjunatha M (2000) Laboratory and field studies on the integrated pest management of *Helicoverpa armigera* (Hübner) in cotton, based on pheromone trap catch threshold level. J Appl Entomol 124:213–221

Reed W (1965) *Heliothis armigera* (Hub.) (Noctuidae) in Western Tanganyika. Biology with special reference to pupal stage. Bull Entomol Res 56:117–125

Reed W, Pawar CS (1982) *Heliothis*: a global problem. In: Reed W, Kumble V (eds) Proceedings of the international workshop on *Heliothis* management, 15–20 November 1981. International Crops Research Institute for the Semi-Arid Tropics (ICRISAT) Center, Patancheru, pp 9–14

Richter AR, Fuxa JR (1990) Effect of *Steinernema feltiae* on *Spodoptera frugiperda* and *Heliothis zea* (Lepidoptera: Noctuidae) in corn. J Econ Entomol 83:1286–1291

Romeis J, Shanower TG (1996) Arthropod natural enemies of *Helicoverpa armigera* (Hübner) (Lepidoptera: Noctuidae) in India. Biocontrol Sci Technol 6:481–508

Sachan JN, Lal SS (1997) Integrated pest management of pod borer complex of chickpea and pigeonpea in India. In: Asthana AN, Ali M (eds) Recent advances in pulses research. Indian Society of Pulses Research and Development, Indian Institute of Pulses Research, Kanpur, pp 349–376

Samsook V, Sikora RA (1981) Influence of parasitic density, host developmental stage and leaf-wetness duration on *Neoaplectana carpocapsae* parasitism of *Heliothis virescens*. Meded Fac Landbouww Gent 46:685–693

Sanahuja G, Banakar R, Twyman RM, Capell T, Christou P (2011) *Bacillus thuringiensis*: a century of research, development and commercial applications. Plant Biotechnol J 9:283–300

Saravanapriya B, Subramanian S (2007) Pathogenicity of EPN to certain foliar insect pests. Ann Plant Prot Sci 15:219–222

Sawicki RM, Denholm I (1987) Management of resistance to pesticides in cotton. Trop Pest Manag 33:262–272

Saxena H, Ahmad R (1997) Field evaluation of *Beauveria bassiana* (Balsamo) Vuillemin against *Helicoverpa armigera* (Hübner) infecting chickpea. J Biol Control 11:93–96

Schroer S, Ziermann D, Ehlers R-U (2005) Mode of action of a surfactant-polymer formulation to support performance of the entomopathogenic nematode *Steinernema carpocapsae* for control of diamondback moth larvae (*Plutella xylostella*). Biocontrol Sci Technol 15:601–613

Shanower TG, Kelley TG, Cowgill SE (1998) Development of effective and environmentally sound strategies to control *Helicoverpa armigera* in pigeonpea and chickpea production systems. In: Saini RK (ed) Proceeding of the 3rd international conference on tropical entomology. International Center for Insect Physiology and Ecology Science Press, Nairobi, pp 239–260

Shapiro-Ilan DI, Gouge DH, Koppenhöfer AM (2002) Factors affecting commercial success: case studies in cotton, turf and citrus. In: Gaugler R (ed) Entomopathogenic nematology. CABI Publishing, Wallingford, pp 333–356

Shapiro-Ilan DI, Gouge DH, Piggott SJ, Fife JP (2006) Application technology and environmental considerations for use of entomopathogenic nematodes in biological control. Biol Control 38:124–133

Sharma HC (2001) Cotton bollworm/legume pod borer, *Helicoverpa armigera* (Hübner) (Noctuidae: Lepidoptera): biology and management. Crop Protection Compendium, CABI, Wallingford, p 72

Sharma HC (ed) (2005) *Heliothis/Helicoverpa* management. Oxford and IBH Publishing, New Delhi, p 469

Sharma HC (2007) Host plant resistance to insects: modern approaches and limitations. Indian J Plant Prot 35:170–184

Sharma HC, Singh BU, Hariprasad KV, Bramel-Cox PJ (1999) Host-plant resistance to insects in integrated pest management for a safer environment. Proc Acad Environ Biol 8:113–136

Sharma HC, Ahmad R, Ujagir R, Yadav RP, Singh R, Ridsdill-Smith TJ (2005) Host plant resistance to cotton bollworm/legume pod borer, *Heliothis/Helicoverpa*. In: Sharma HC (ed) *Heliothis/Helicoverpa* management: emerging trends and strategies for future research. Oxford and IBH Publishers, New Delhi, pp 167–208

Sharma HC, Clement SL, Ridsdill-Smith TJ, Ranga Rao GV, El Bouhssini M, Ujagir R, Srivastava CP, Miles M (2008) Insect pest management in food legumes: the future strategies. In: Kharkwal MC (ed) Food legumes for nutritional security and sustainable agriculture. Proceedings of the IVth international food legumes research conference, vol 1. Indian Society of Genetics and Plant Breeding, New Delhi, pp 522–544

Shoeb MA, Atalla FA, Matar AM (2006) Pathogenicity of the entomopathogenic nematodes, *Steinernema abbasi* and *Heterorhabditis bacteriophora* to certain economic insect pests. Egypt J Biol Pest Control 16:99–102

Simões N, Rosa JS (1996) Pathogenicity and host specificity of entomopathogenic nematodes. Biocontrol Sci Technol 6:403–412

Simwat GS (1994) Modern concepts in insect pest management in cotton. In: Dhaliwal GS, Arora R (eds) Trends in agricultural insect pest management. Commonwealth Publishers, New Delhi, pp 186–237

Singh H, Singh G (1975) Biological studies of *Heliothis armigera* in Punjab. Indian J Entomol 37:154–164

Smithson JB, Thompson JA, Summerfield RJ (1985) Chickpea (*Cicer arietinum* L.). In: Summerfield RJ, Roberts EH (eds) Grain legume crops. Collins, London, pp 312–390

Soleimannejad S, Fathipour Y, Moharramipour S, Zalucki MP (2010) Evaluation of potential resistance in seeds of different soybean cultivars to *Helicoverpa armigera* (Lepidoptera: Noctuidae) using demographic parameters and nutritional indices. J Econ Entomol 103:1420–1430

Stock SP, Goodrich-Blair H (2012) Nematode parasites, pathogens and associates of insects and invertebrates of economic importance. In: Lacey LA (ed) Manual of techniques in invertebrate pathology, 2nd edn. Academic, San Diego, pp 373–426

Tanada Y, Reiner C (1962) The use of pathogens in the control of the corn earworm, *Heliothis zea* (Boddie). J Insect Pathol 4:139–154

Tay WT, Soria MF, Walsh T, Thomazoni D, Silvie P, Behere GT, Craig Anderson C, Downes S (2013) A brave new world for an old world pest: *Helicoverpa armigera* (Lepidoptera: Noctuidae) in Brazil. PLoS One 8:e80134

van Emden HF, Ball SL, Rao MT (1988) Pest and disease problems in pea, lentil, faba bean, and chickpea. In: Summerfield RJ (ed) World crops: cool season food legumes. Kluwer Academic Publishers, Dordrecht, pp 519–534

Vyas RV, Patel NB, Patel P, Patel DJ (2002) Efficacy of entomopathogenic nematode against *Helicoverpa armigera* on pigeonpea. Int Chickpea Pigeonpea Newsl 9:43–44

Vyas RV, Patel NB, Yadav P, Ghelani YH, Patel DJ (2003) Performance of entomopathogenic nematodes for management of gram pod borer, *Helicoverpa armigera*. Ann Plant Prot Sci 11:107–109

Whalon ME, Mota-Sanchez D, Hollingsworth RM, Duynslayer L (2007) IRAC arthropod pesticide resistance database. Michigan State University, East Lansing, MI. http://www.pesticideresistance.org/

Womersley CZ (1990) Dehydration survival and anhydrobiotic potential. In: Gaugler R, Kaya HK (eds) Entomopathogenic nematodes in biological control. CRC Press, Boca Raton, pp 117–137

Wright DJ, Perry RN (2006) Reproduction, physiology and biochemistry. In: Perry RN, Moens M (eds) Plant nematology. CAB International, Wallingford, pp 187–209

Zalucki MP, Dalglish G, Firempong S, Twine P (1986) The biology and ecology of *Heliothis armigera* (Hübner) and *H. punctigera* Wallengren (Lepidoptera: Noctuidae) in Australia: what do we know? Aust J Ecol 34:779–814

Zhang H, Yin W, Zhao J, Jin L, Yang Y, Wu S, Tabashnik BE, Wu Y (2011) Early warning of cotton bollworm resistance associated with intensive planting of Bt cotton in China. PLoS One 6(8):e22874

Chapter 15
Sustainability of Entomopathogenic Nematodes Against Crop Pests

S. Sivaramakrishnan and M. Razia

15.1 Introduction

Entomopathogenic nematodes (EPNs) are represented as new arsenal for management of crop pests above and below ground and are growing steadily at the expense of conventional pesticides in many countries. Use of biopesticides is expanding, drawing on developments in biotechnology. In general, EPNs are beneficial to plant community by regulating the root-invading pests. Few nematode species, tetradomatids and mermithids, have been used for minor pest control, but they are host specific, poorly suited to exploit their pest control properties, and difficult to mass produce. Later in the 1970s, there had been tremendous research and commercial interest in insect-parasitic nematodes. This interest was sparked due to the lack of adequate tools to control soil-inhabiting insect pests in an effective, environmentally acceptable manner and the possibility to produce these nematodes monoxenically in vitro which was followed by a scale-up of commercial production levels. More than 100 laboratories and around 70 countries are working on EPNs. EPNs are small roundworms, about 1 mm long, soft bodied, and non-segmented. They are obligate or sometimes facultative parasites of insects. Recognition of new strains and novel species has increased rapidly with the development and improvement of isolation and characterization techniques (Kanga et al. 2012). As of early 2012, 78 valid EPN species have been assigned to Steinernematidae and 15 Heterorhabditidae.

S. Sivaramakrishnan (✉)
Department of Biotechnology, Bharathidasan University, Tiruchirappalli 620024, Tamil Nadu, India
e-mail: sivaramakrishnan123@gmail.com

M. Razia
Department of Biotechnology, Mother Teresa Women's University, Kodaikanal 624102, Tamil Nadu, India

© Springer International Publishing Switzerland 2015
K.S. Sree, A. Varma (eds.), *Biocontrol of Lepidopteran Pests*, Soil Biology 43, DOI 10.1007/978-3-319-14499-3_15

15.2 Life Cycle of EPNs

In general, the life cycle of nematode includes the egg, four larval stages and the adult. Nematodes are infective to insects only in their third juvenile stage and size range from 0.4 to 1.5 mm in length and observed under microscope; they do not require food and can survive without host in the soil. The infective juveniles actively locate attack and infect target pests. They enter the host through openings on the insect body such as mouth and anus. Once inside the insect host, nematodes release bacteria from their gut. The bacteria multiply and release insecticidal toxins, which kills the insect within 24–48 h. These bacteria break down the host tissues which are then taken up by the nematodes as a food. After the host dies, the nematodes feed on insect tissues and bacteria and reproduced within host. EPN come out only third stage of juveniles to seek out new hosts. Development of nematode under optimal conditions (soil temperatures 77–82 °F) takes approximately 3–7 days for one life cycle inside a host from egg to egg.

15.2.1 Steinernematids

In the nematode *Steinernema*, mode of reproduction is amphimictic. The third stage of IJs matures to become either a male or a female. The life cycle of *Steinernema* consists of an egg stage, four juvenile stages, and an adult stage (male and female). The cycle proceeds from IJs (third-stage infective juveniles) to IJs. Inside the host the third-stage nematodes, the IJs, develop to adult (males and females) of the first generation if the nutrients are sufficient. Adult female hatches the eggs and the juveniles develop to become adult males or females of the second generation. The completion of life cycle takes 8–10 days. Another route of reproduction exists when the nutrient supply is insufficient inside the host due to overcrowding of nematodes. In this case, the IJs develop into adults (males and females) of the first generation, and eggs produced by the females develop directly into IJs. This cycle completes in 6–7 days. *S. scapterisci* are less tolerant to lower temperatures and more tolerant to higher temperatures when compared with other species of the same genus. Sex ratio in nematodes is influenced by temperature. At 15 and 24 °C, females constitute 54 and 60 % of the population, respectively, but at 30 °C females constitute 47 % of the population (Nguyen and Smart 1990).

15.2.2 Heterorhabditids

During the life cycle of *Heterorhabditis*, the IJs mature to give first-generation hermaphrodite females, and the females give rise to a second generation of amphimictic females and IJs. The males and females of *Heterorhabditis* are environmentally determined. Under in vivo conditions, three adult generations were produced. *Heterorhabditis* hermaphrodites lay 1,000 eggs which develop into second-generation males and females, and the first-generation hermaphrodites hold about 500 eggs which develop into IJs. Second-generation females lay 6–10 eggs which develop into another generation of adults, and the retained 30 eggs within the nematode body develop into IJs. Both first and second generations produced IJs through *endotokia matricida*. The third-generation females do not oviposit and their eggs develop into IJs. These nonfeeding IJs come out from insect cadaver and present in the soil (Wang and Bedding 1996).

15.2.3   Symbiotic Bacteria

Xenorhabdus (Thomas and Poinar 1979) and *Photorhabdus* (Boemare et al. 1993), members of the family *Enterobacteriaceae*, engage in a mutualistic association with *Steinernema* and *Heterorhabditis* spp., respectively. In *Steinernema* spp. the bacterial symbionts are carried monoxenically in a special vesicle in the intestine of IJs, whereas in *Heterorhabditis* spp. they are present in the entire intestine and pharynx of IJs. Nematodes provide protection and transport facility to their bacterial symbionts. The bacteria are carried by IJs as a vector and infect the insect host by penetrating through natural openings such as the mouth, anus, or spiracles. These bacteria are released from the nematode into the insect's hemocoel which kill the host within 48 h. The nematodes consequently reproduce for several generations in the hemocoel feeding on both the bacteria and the nutrients derived from insect sources. With the depletion of nutrient supplies, the nematodes develop into IJ stage acquiring its bacterial partner from the hemocoel before emerging from the insect cadaver into the soil in order to search for new insect hosts (Poinar 1990; Endo and Nickle 1991).

Symbiotic bacteria tend to produce two colony forms, a primary form and a secondary form. The unstable primary form is preferentially taken up by the IJs and is often converted into the secondary form when cultured in vitro. The mechanism by which these variants arise is still unidentified. Several genes coding for different characteristics are switched on and off spontaneously at the same time (Gerritsen et al. 1992).

This complex life cycle that involves mutualistic as well as pathogenic interactions makes *Photorhabdus* and *Xenorhabdus* spp. ideal models to study symbiosis and pathogenesis. During the life cycle, the bacteria not only have to kill the insect host using several protein toxins but also have to defeat several other microbes that are direct food competitors. Saprophytic microbes from the soil as well as bacteria adhering to the insect gut or cuticle of the nematode represent other potential sources of competitors that can grow within the insect cadaver.

15.3   EPNs as Biocontrol Agents

EPNs occur naturally in soil environments and locate their host in response to carbon dioxide, vibrations, and volatile compounds released from insect-infected plant roots (Lewis et al. 2006). EPNs are robust components of integrated pest management because they are nontoxic to humans and are relatively specific to their target pests.

15.3.1   Distribution of EPNs

EPNs have global distribution and numerous surveys have documented their occurrence in Australia, Europe, America, South Africa, Sri Lanka, Malaysia, India, China, and Japan. EPNs are more frequently detected in reduced tillage regimes (Hummel et al. 2002). In addition, many countries are concerned about the introduction of exotic EPNs, since they might have a harmful impact on nontarget organisms. Surveys have been conducted in many parts of the world providing a clue of which species are indigenous for a given area (Kaya et al. 2006).

EPNs applied for insect control are dependent on the motility and persistence of the applied IJs. Active IJ dispersal is rather limited with up to 90 cm in 30 days in both horizontal and vertical dispersals to seek suitable host. On the other hand, inactive nematode dispersal by water, wind, infected hosts, human activity, etc. can cover much greater distances and increases the chance of prevalent distribution. Many species and strains of EPNs illustrate a variety of adaptations to extremes in soil and plant environments. Developmental dormancy and diapause are important for seasonal survival and long-term longevity. EPNs undergo temporary quiescence in response to environmental stress and enter into anhydrobiosis or other extreme states which allow long-term survival in strangely stressful environments. EPNs infect over 200 insect hosts (Shapiro-Ilan et al. 2002). In agroecosystems, habitat complexity can be created through the planting of diverse carpet of vegetation within or adjacent to crop areas that remain undisturbed during the field season.

Abiotic and biotic factors that influence the persistence, infectivity, and motility of individual IJs also influence nematode recycling. Some of these factors may even be more crucial for recycling than for persistence and infection.

15.3.2   Behavior of EPNs

Based on the search strategies used, EPNs could be ambushers or cruisers. Ambushers such as *S. carpocapsae* have an energy-conserving approach and sit and wait to attack mobile insects like codling moth, cutworms, and mole crickets in the upper soil. Cruisers like *S. glaseri* and *H. bacteriophora* are highly active and generally subterranean, moving significant distances using volatile cues to find their host which are less mobile such as white grubs in the underground soil. A few nematode species such as *S. feltiae* and *S. riobrave* use an intermediate foraging strategy (combination of ambusher and cruiser type) to find their hosts such as prepupal insects, fungus gnats, or weevil larvae.

15.3.3 Factors Influencing EPNs

Steinernematidae and Heterorhabditidae are obligate pathogens in nature. They have been recovered from soils throughout the world and their distribution may be limited by the availability of susceptible hosts. Habitat density may affect soil biota positively or negatively due to plant type and diversity (John et al. 2006). The environmental factors control EPN's potential in terms of infectivity, motility, and distribution in the soil environment, the most common being temperature, moisture, aeration, soil type (texture), and biotic factors (soil fauna) (Shapiro-Ilan et al. 2012).

15.3.3.1 Soil and Moisture

Soil physical properties, such as those characteristic of sandy soils (porous and aerated), facilitate nematodes when compared to denser soils, such as clay. EPN species were recovered from rainforest habitats with sandy or sandy loam soil types and a pH that ranged from slightly acidic to slightly alkaline (Uribe-Lorio et al. 2005).

Moisture is the most important factor for nematode performance. In soil, IJs move through the water film that coats the interstitial spaces. In dry soil, the film becomes too thin and in saturated soil, the interspaces are completely filled with water; in both cases the nematode movement is restricted. IJs can survive low-moisture conditions by lowering their rate of metabolism. Gradual water removal from the IJs gives them time to adapt to the desiccating conditions (Koppenhfer et al. 1995).

15.3.3.2 Temperature

Temperature is one of the key factors affecting the infectivity of nematodes; however, the effect of temperature on nematode survival and infectivity varies with nematode species and strains and their habitats. EPN populations indicate the temperature and moisture changes due to seasonal variations (Puza and Mracek 2007). *Heterorhabditis* species are able to cause infection at temperature as low as 5–7 °C. Heat tolerance of the IJs under field conditions is one of the important factors restricting their application on exposed surfaces like foliage and warm ecosystems. The species *S. feltiae* can be infective from 2 to 30 °C, whereas some *Heterorhabditis* can infect host insects from 7 to 35 °C, and *S. carpocapsae* is nearly inactive at 10 °C. More information is required on the relationship between abiotic factors and nematodes' persistence and infectivity. A lack of genetic diversity in the lab-reared populations might be overcome by surviving natural populations for desired traits. Native isolates have the ability to tolerate the extreme temperature.

15.3.3.3 pH

Soil pH in most agroecosystems, having a range of 4–8, is not likely to have any significant effect on EPNs, but a pH of 10 or higher is likely to be detrimental (Kung et al. 1990). *Heterorhabditis* is well distributed in coastal sandy soils and is capable of tolerating a wide range of pH and salinity. The relationship between nematode populations and soil pH suggests that EPNs may not be as effective on pests in acidic soils as they are in soils of near-neutral pH. Recommendation on addition of lime to acidic soils may raise the pH, thus enhancing the effectiveness of nematodes in farms.

15.3.3.4 Ultra Violet

Emissions produced by UV source corresponding to the specific wavelengths in solar radiation (290–320 nm) were responsible for inactivation of nematodes. Consequently, even with adequate moisture, 60 min of direct exposure to sunlight turned most IJs of *Steinernema* uninfective. The virulence of nematodes was unaffected because of trehalose accumulation which may protect the biological and physiological functions of EPNs against UV stress (Gaugler et al. 1992). *Heterorhabditis* might retain significant pathogenicity up to 30 min of exposure to UV stress (Gaugler and Boush 1978). EPN appears to be the most vulnerable of all biological insecticides to UV irradiation and also to the influence of sunlight on persistence and virulence of IJs in exposed habitats (Grewal et al. 2002).

   When exposed to natural sunlight, UVA and UVB were responsible for total inactivation of IJs. In order to protect EPNs from UV irradiation, nematodes can be encapsulated using starch or other flours.

15.3.3.5 Insect Host

Biotic factors affect nematode survival. Among the natural enemies of nematodes, nematophagous fungi play a major role, for example, *Hirsutella rhossiliensis*. Further natural enemies of infective juveniles include collembolans, mites, tardigrades, and predatory nematodes, but their impact under field conditions is not well understood. Nematodes are unable to attack some insects due to the insect's morphology, i.e., structure of insect mouth, for example, IJs are blocked by oral filters of wireworms, the presence of sucking or piercing or chewing mouthparts, the anus constricted by muscles, or the spiracles covered with septa or sieve plates as in scarab grubs (Eidt and Thurston 1995).

15.3.3.6   Tillage

EPN populations vary with respect to cropping and tillage practices in an agroecosystem. No-till and conservation tillage practices potentially conserve and thereby enhance native nematodes (Millar and Barbercheck 2002). The nematode survival and pathogenicity is affected by tillage, EPN population can be preserved in agroecosystems by recommending reduced tillage practices.

15.3.4   Pathogenicity: Mode of Action of EPN-Associated Symbiotic Bacteria

Gupta (2002) stated that in insect cellular and humoral immune responses, hemocytes are involved in phagocytosis of microbes and microaggregation and encapsulation of large pathogens. Antimicrobial peptides, such as cecropins and lysozyme, hemolymph clotting system, and activation of the prophenoloxidase cascade leading to melanization are among the other host responses to the attack of organisms like bacteria, fungi, or nematodes.

EPN–bacterium complexes are able to kill a wide range of insect species by overcoming and evading the defense mechanisms of insect host that involves interactions with the humoral and cellular factors as reported under laboratory conditions (Gupta 1991). Exo- and endotoxins are produced by EPN-symbiotic bacteria that are toxic to the hemocytes of *Galleria mellonella* (Ffrench-Constant and Bowen 2000). The purified insecticidal toxin from symbiotic bacteria when given through feeding method killed the model insect *Manduca sexta*.

The genomes of *X. nematophila* and *X. bovienii* have been sequenced and revealed several biosynthesis gene clusters involved in the biosynthesis of secondary metabolites, namely, benzylidene acetone, iodinine, phenethylamides, indole derivatives, and complex compounds like xenocoumacins (XCNs) derived from hybrid polyketide synthase (PKS)/nonribosomal peptide synthetase (NRPS) systems, xenorhabdins, and xenorxides. Xenematide and xenortide insecticidal compounds are produced by *X. nematophila* (Lang et al. 2008).

The genome of *P. luminescens* sp. *laumondii* strain TT01 has been completely sequenced. *Photorhabdus* spp. produce various metabolites having antibacterial, antifungal, cytotoxic, and nematicidal properties, which are apparently of mixed peptide–polyketide origin. *P. luminescens* TT01 was analyzed to assign the functional relationship with PKS. Many hypothetical proteins which are presented in this organism have shown homology to PKS family. The mechanism involved in biosynthesis and regulation of potential metabolites with insecticidal properties is a vital process in nematode–bacteria (Razia et al. 2010).

Use of *Photorhabdus* and *Xenorhabdus* genes encoding entomotoxins to create transgenic plants for crop protection was proposed by Ffrench-Constant and Bowen (1999). Members of both of these genera produce insecticidal proteins that are toxic to a wide variety of lepidopteran insects. These toxic proteins provide new insights for production of transgenic crop plants like that of transgenic Bt crops. As with other bacterial toxins, the mutation rate of bacteria in a population may result in the variation of the sequence of toxin genes. Preferably, the toxins are active against Lepidoptera, Coleoptera, and Diptera (Hofte and Whiteley 1989).

15.3.5   *Efficacy of EPNs as Biopesticides*

Out of the one million known species of insects, about 15,000 of them are considered as pests. Some of the lepidopteran, coleopteran, dipteran, orthopteran, and homopteran species have become serious pests and are currently a major threat to several crops. Due to the unavailability of successful biological control agents to manage these pests, farmers mainly depend on chemical pesticides. However, the most effective and persistent insecticides used to control soil dwelling insects are banned. Exploration and field testing of the biological control agents of these pests is one of the major thrust areas of research in agriculture, which can lead to the development of the commercial biological control agents needed at present.

The use of nematodes for biological pest control began in the 1930s. Nematode appears promising because it is effective against a large number of insects, including Japanese beetles, root and vine weevils, fire ants, mole crickets, cutworms, and potato beetles. Unfortunately, field evaluations have not been very effective so far due to the lack of understanding of the conditions required which facilitate infestation, including abiotic factors such as soil moisture, temperature, and solar radiation, as well as biotic factors such as nematode strain, host stage and its defenses, and insect's behavioral challenge events. The proper identification and characterization of nematode–bacteria are fundamental importance in an EPN-based biocontrol perspective (Sicard et al. 2003).

EPNs are applied with substrates that are regularly treated with many other agents, including chemical or biorational pesticides, soil amendments, and fertilizers. Depending on the agents, application timing, and physicochemical characters, the nematodes may or may not interact with these other agents. EPNs appear to be compatible with many, but not all, herbicides, fungicides, acaricides, insecticides, nematicides, azadirachtin, *Bacillus thuringiensis* products, and pesticidal soaps. Inorganic fertilizer may be compatible with nematodes for short-term inundative pest control. *Heterorhabditis* tend to be more sensitive to physical challenges, including pesticides, than *Steinernema* (Grewal et al. 2001). Various studies were carried out where plant-boring insects have been controlled by injecting nematode suspensions into the borer holes with sponges soaked with nematode suspensions

(Yang et al. 1993). For the control of banana weevil, a nematode suspension can be placed into insect-attracting cuts in residual rhizomes of bananas.

15.4 Mass Production of Biopesticides

For the past few decades, agriculture is dependent on synthetic and/or organic insecticides for crop protection. Today awareness has been directed toward natural enemies such as predators, parasites, and pathogens. Yet none of the predators or parasites can be mass produced and stored for long periods of time, like that of synthetic insecticides because it must be produced under in vivo conditions. Technology is being developed to achieve high rates of production, to increase the efficacy of a biological agent making it highly toxic to the target organism, to enhance their shelf life, and to deliver it to the farmer in an effective way. Entomopathogens like fungi, viruses, and bacteria have been suggested as controlling agents of insect pests for over a century. More attention needs to be given to the selection of broad-spectrum biopesticides and improvement in their production, formulation, and application technologies.

15.4.1 Methods of Production of EPNs

EPNs are currently produced by in vivo or in vitro (solid and liquid culture) techniques. Each method has its own advantages and disadvantages relative to production cost, technical know-how required, economy of scale, and product quality. In vivo EPN production is the appropriate method for field testing and laboratory use at a small scale. It requires small amounts of resources like labor, insect culture, and technical expertise. For commercial use in vitro liquid culture is usually considered relative to other production methods, and it demands greater capital investment and a higher level of technical expertise.

15.4.1.1 In Vitro Production of EPNs

Method to produce EPNs initiates axenic nematode eggs placed on a pure culture of the symbionts. Fermenters like airlift fermenter are used for liquid culture into which symbiotic bacteria are first introduced followed by the nematodes. Various ingredients for liquid culture media have been used, viz., soy flour, yeast extract, canola oil, corn oil, thistle oil, egg yolk, casein peptone, milk powder, liver extract, and cholesterol. Incubation time varies with the media and species being used and may be as long as three weeks, although many species can reach maximum IJ production in two weeks or less. Factors affecting yield of in vitro liquid cultures in case of both *Steinernematids* and *Heterorhabditids* include the lack of nutritional factors, aeration, CO_2, lipid content, and temperature. The maximum yield of nematodes in liquid culture fluctuates based on their life cycle and reproductive biology.

15.4.1.2 In Vivo Production of EPNs

The general approach to in vivo culture is a two-dimensional system that relies on production in trays and shelves (Ehlers and Shapiro-Ilan 2005). In vivo production yields vary greatly among insect hosts and nematode species. The wax moth, *G. mellonella*, is the insect of choice for in vivo production. It is produced in many parts of the world as source of fish bait and bird food (Costa et al. 2007). The silkworm is highly susceptible to EPNs and is a potential host for multiplication of nematodes (Goldsmith et al. 2004). The most common insect hosts used for laboratory and commercial EPN culture are Wax Moth (*G. mellonella*) and the yellow mealworm (*Tenebrio molitor*), whereas other hosts which have been studied include the navel orangeworm (*Amyelois transitella*), tobacco budworm (*Heliothis virescens*), cabbage looper (*Trichoplusia ni*), pink bollworm (*Pectinophora gossypiella*), beet armyworm (*Spodoptera exigua*), corn earworm (*Helicoverpa zea*), gypsy moth (*Lymantria dispar*), house cricket (*Acheta domesticus*), etc. In vivo production yields vary greatly among insect hosts and nematode species used, and the most efficient and cost-effective host would be promoted for use by farmers.

Factors affecting yield of in vivo culture especially are nematode dosage and host density. Crowding of hosts can lead to oxygen deficiency or increase of ammonia. Thus, optimization of host density and inoculation rate of IJs for maximum yield is recommended. Infection efficiency and yield of IJs can be affected by inoculation method. Approximately 25–200 IJs per insect are sufficient (depending on nematode species and method of inoculation), whereas higher rates are generally needed for the yellow mealworm, *T. molitor* (100–600 IJs per insect). The methods of inoculation of nematodes include pipetting or spraying of nematodes onto a substrate, immersion of insects in a nematode suspension, or applying the nematodes to the insect's food. Environmental factors including temperature, aeration, and moisture can affect the in vivo yield.

15.4.2 Application

Nematodes can be stored and formulated in different ways including the use of polyurethane sponges, water-dispersible granules, vermiculite, alginate gels, and baits. Formulated EPNs can be stored for 2–5 months depending on the nematode species and storage media and conditions. Unlike other microbial control agents (fungi, bacteria, and viruses), EPNs do not have a fully dormant resting stage, and they use their limited energy during storage. The quality of the nematode product can be determined by nematode virulence and viability assays, age, and the ratio of viable to nonviable nematodes.

EPNs can be applied using nearly all agronomic or horticultural ground equipment including pressurized sprayers, mist blowers, and electrostatic sprayers or aerial sprayers (Shapiro-Ilan et al. 2006). The application equipment used depends on the cropping system, and in each case there are a variety of handling considerations including volume, agitation, nozzle type, and pressure. It is important to ensure adequate agitation during application. For small-plot applications, water cans or backpack sprayers can be used. When nematodes are applied to larger plots, a suitable spraying apparatus such as a boom sprayer should be considered. Conceivably, other methods such as through Microjet Irrigation Systems, subsurface injection, or baits could also be used. Enhanced efficacy of EPN applications can be facilitated through cadaver application. Bait formulations and insect host cadavers can enhance EPN persistence and reduce the quantity of nematodes required when compared to liquid medium (Shapiro-Ilan et al. 2012). An overview of the commercial use of EPNs is tabulated in Table 15.1.

Direct exposure to sunlight (UV light) can be minimized by applying infective juveniles early in the morning or in the evening. EPNs should be prepared for field application no earlier than 1 h ahead of application time. If nematodes are in a liquid suspension, shake them well before use; if they are on a sponge, soak the sponge in water; and if nematodes are in vermiculite, add the vermiculite–nematode mixture directly to water allowing their dispersion. Mass production of these nematodes in liquid media has become a major challenge for commercialization. A better understanding of both the nematode and its bacterial symbiont in liquid culture might increase the chance of success.

Table 15.1 Commercial use of entomopathogenic nematodes (EPNs) *Steinernema* and *Heterorhabditis* and insect pest

EPN species	Major insect pest
S. glaseri	White grubs (scarabs, Japanese beetle, *Popillia* sp.)
S. kraussei	Black vine weevil, *Otiorhynchus sulcatus*
S. carpocapsae	Cutworms, armyworms, black vine weevil, peach tree borer
S. feltiae	Fungus gnats(*Bradysia* spp.), western flower thrips
S. scapterisci	Mole crickets
S. riobrave	Citrus root weevils
H. bacteriophora	White grubs, cutworms, black vine weevil, flea beetles, corn rootworm
H. marelatus	White grubs (scarabs), cutworms, black vine weevil
H. indica	Fungus gnats, root mealybug, grubs
H. megidis	Weevils

15.5   Conclusion

Research and development in the area of production of biological insecticides magnetize extremely modest financial support compared that of chemical pesticides. In addition to scientific studies, regulatory policies of the government should aim at supporting further introduction of EPN-based products into biological control practices, and the government should take measures to make the farmers aware of EPNs as natural insect killers. EPNs are well adapted to their native habitat and hosts through natural selection; however, in the near future, additional useful traits could be established in their genome to make them even more competent against insect hosts in different environments. With advancements, EPNs might serve to reduce chemical insecticide inputs and add to environmentally sustainable increase of crop yields.

References

Boemare NE, Akhurst RI, Mourant KG (1993) DNA relatedness between *Xenorhabdus* spp. (Enterobacteriaceae), symbiotic bacteria of entomopathogenic nematodes, and a proposal to transfer *Xenorhabdus luminescens* to a new genus, *Photorhabdus* gen. nov. Int J Syst Bacteriol 43:249–255

Costa JR, Dias RJP, Morenze MJF (2007) Determining the adaptation potential of multiplication of a *Heterorhabditis* sp and *Steinernema carpocapsae* (Rhabditidae::Heterorhabditidae and Steinernematidae) in larvae of *Alphitobius diaperinus* (Coleoptera: Tenebrionidae) and *Galleria mellonella* (Lepidoptera: Pyralidae). J Parasitol 102:139–144

Eidt DC, Thurston GS (1995) Physical deterrents to infection by entomopathogenic nematodes in wire worms (Coleoptera: Elateridae) and other soil insects. Can Entomol 127:423–429

Endo BY, Nickle WR (1991) Ultrastructure of the intestinal epithelium, lumen and associated bacteria in *Heterorhabditis bacteriophora*. Proc Helminthol Soc Wash 58:202–212

Ehlers RU, Shapiro-Ilan DI (2005) Mass production. In: Grewal PS, Ehlers RU, Shapiro-Ilan DI (eds) Mass production. CABI Publishing, Wallingford, pp 65–78

Ffrench-Constant RH, Bowen D (1999) *Photorhabdus* toxins: novel biological insecticides. Microbiology 2:284–288

Ffrench-Constant RH, Bowen DJ (2000) Insecticidal toxin complex proteins from *Xenorhabdus nematophilus*: structure and pore formation. Cell Mol Life Sci 57:828–833

Gaugler R, Boush GM (1978) Effects of ultraviolet radiation and sunlight on the nematode *Neoaplectanu carpocapsae*. J Invertebr Pathol 32:291–296

Gaugler R, Bednarek A, Campbell JF (1992) Ultraviolet inactivation of Heterorhabditid and Steinernematid nematodes. J Invertebr Pathol 59:155–160

Gerritsen LJM, DeRaay G, Smits PH (1992) Characterization of form variants of *Xenorhabdus luminescens*. Appl Environ Microbiol 58:1975–1979

Grewal PS, Nardo DE, Marineide ABE, Aguillera M (2001) Entomopathogenic nematodes: potential for exploration and use in South America. Neotrop Entomol 30:191–195

Grewal PS, Wang X, Taylor RAJ (2002) Dauer juvenile longevity and stress tolerance in natural populations of entomopathogenic nematodes: is there a relationship? Int J Parasitol 32:717–725

Goldsmith M, Toru S, Hiroaki A (2004) The genetics and genomes of silkworm *Bombyx mori*. Annu Rev Entomol 50:71–100

Gupta AP (1991) Insect immunocytes and other hemocytes: roles in cellular and humoral immunity. In: Gupta AP (ed) Immunology of insects and other arthropods. CRC, Boca Raton, pp 19–18

Gupta AP (2002) Immunology of invertebrates: cellular. In: Encyclopedia of live sciences. Wiley/ Nature Publishing Group, London, p 9

Hofte H, Whiteley HR (1989) Insecticidal crystal proteins of *Bacillus thuringiensis*. Microbiol Rev 53:242–255

Hummel RL, Walgenbach JF, Barbercheck ME, Kennedy GG, Hoyt GD, Arellano C (2002) Effects of production practices on soil-borne entomopathogens in Western North Carolina vegetable systems. Environ Entomol 31:84–91

John MG, Wall DH, Behan-Pelletier VM (2006) Does plant species co-occurrence influence soil mite diversity? Ecology 87:625–633

Kanga FN, Waeyenberge L, Hauser S, Moens M (2012) Distribution of entomopathogenic nematodes in Southern Cameroon. J Invertebr Pathol 109:41–51

Kaya HK, Aguiller MM, Alumai A, Choo HY, De la Torre M, Fodor A, Ganguly S, Hazır S, Lakatos T, Pye A, Wilson M, Yamanaka S, Yang H, Ehlers R-U (2006) Status of entomopathogenic nematodes and their symbiotic bacteria from selected countries or regions of the world. Biol Control 38:134–155

Koppenhfer AM, Kaya HK, Taormino S (1995) Infectivity of entomopathogenic nematodes (Rhabditida: Steinernematidae) at different soil depths and moistures. J Invertebr Pathol 65:193–199

Kung S, Gaugler R, Kaya HK (1990) Influence of soil, pH and oxygen on persistence of *Steinernema* spp. J Nematol 22:440–445

Lang G, Kalvelage T, Peters A, Wiese J, Imhoff JF (2008) Linear and cyclic peptides from the entomopathogenic bacterium *Xenorhabdus nematophilus*. J Nat Prod 71:1074–1077

Lewis EE, Campbell J, Griffin C, Kaya H, Peters A (2006) Behavioral ecology of entomopathogenic nematodes. Biol Control 38:66–79

Millar LC, Barbercheck ME (2002) Effects of tillage practices on entomopathogenic nematodes in a corn agroecosystem. Biol Control 25:1–11

Nguyen KB, Smart GC (1990) *Steinernema scapterisci* n. sp. (Steinernematidae: Nematoda). J Nematol 22:187–199

Poinar GO (1990) Taxonomy and biology of Steinernematidae and Heterorhabditidae. In: Gaugler R, Kaya H (eds) Entomopathogenic nematodes in biological control. CRC, Boca Raton, pp 23–61

Puza V, Mracek Z (2007) Natural population dynamics of entomopathogenic nematode *Steinernema affine* (Steinernematidae) under dry conditions: possible nematode persistence within host cadavers? J Invertebr Pathol 96:89–92

Razia M, Raja KR, Padmanaban K, Sivaramakrishnan S, Chellapandi P (2010) A phylogenetic approach for assigning function of hypothetical proteins in *Photorhabdus luminescens* subsp. *laumondii* TT01 genome. J Comp Sci Syst Biol 3:21–29

Sicard M, Le Brun N, Pages S, Godelle B, Boemare N, Moulia C (2003) Effect of native *Xenorhabdus* on the fitness of their *Steinernema* host: contrasting types of interactions. Parasitol Res 91:520–524

Shapiro-Ilan DI, Bruck DJ, Lacey LA (2012) Principles of epizootiology and microbial control. In: Vega FE, Kaya HK (eds) Insect pathology, 2nd edn. Academic, San Diego, pp 29–72

Shapiro-Ilan DI, Gouge DH, Piggott SJ, Patterson FJ (2006) Application technology and environmental considerations for use of entomopathogenic nematodes in biological control. Biol Control 38:124–133

Shapiro-Ilan DI, Gouge D, Koppenhofer A (2002) Factors affecting commercial success: case studies in cotton, turf and citrus. In: Gaugler R (ed) Entomopathogenic nematology. CBI Publishing, Oxon, pp 333–356

Thomas GM, Poinar GO (1979) *Xenorhabdus* gen. nov., a genus of entomopathogenic nematophilic bacteria of the family Enterobacteriaceae. Int J Syst Bacteriol 29:352–360

Uribe-Lorio L, Mora M, Stock SP (2005) First record of entomopathogenic nematodes (Steinernematidae and Heterorhabditidae) in Costa Rica. J Invertebr Pathol 88:226–231

Wang J, Bedding RA (1996) Population development of *Heterorhabditis bacteriophora* and *Steinernema carpocapsae* in the larvae of *Galleria mellonella*. Fund Appl Nematol 19:363–367

Yang H, Zhang G, Zhang S, Jian H (1993) Biological control of tree borers (Lepidoptera: Cossidae) in China with the nematode *Steinernema carpocapsae*. In: Bedding R, Akhurst R, Kaya H (eds) Nematodes and the biological control of insect pests. CSIRO Publications, East Melbourne, pp 33–40

Chapter 16
An Overview of Some Culture Collections of Entomopathogenic Microorganisms in the World

K. Sowjanya Sree and Ajit Varma

16.1 Introduction

Pure cultures of microorganisms are becoming increasingly important in diverse fields such as agriculture, medicine and industrial biotechnology. Culture collection centres play an important role in making these pure cultures available for scientific and industrial use. One of the functions of the culture collection facility is to act as a repository. Receipt, verification, preservation, maintenance and accession of cultures are the main activities undertaken by such a facility. Standard protocols are followed to conserve microbial strains under optimal conditions. Most commonly microbial stock cultures are preserved in glycerol, mineral oil, sterile distilled water or in lyophilised form. In addition, cryopreservation of authentic strains is also carried out in liquid nitrogen. Information about the cultures maintained at the culture collection centre is maintained in the form of inventories and as online databases which could be of use also to the ones who seek services from them. Apart from these activities, the collection centres also provide various services to the academic and industrial communities. These include morphological and molecular identification of microbial cultures, deposition and accession of microbial cultures and supply of authentic microbial strains to academia, research institutions and industries. The culture collection facilities are also centres of active research. A few of the culture collection centres for microbes in general and specifically entomopathogenic microorganisms have been detailed below. The main aim of the present chapter is to introduce a few of the entomopathogen culture collection centres around the world to the scientists who want to start with research on entomopathogens, enabling them to acquire pure and authentic cultures of entomopathogens.

K.S. Sree (✉) • A. Varma
Amity Institute of Microbial Technology, Amity University Uttar Pradesh, Noida 201303, Uttar Pradesh, India
e-mail: ksowsree@gmail.com

© Springer International Publishing Switzerland 2015   329
K.S. Sree, A. Varma (eds.), *Biocontrol of Lepidopteran Pests*, Soil Biology 43,
DOI 10.1007/978-3-319-14499-3_16

16.2   American Type Culture Collection

American Type Culture Collection (ATCC) is a non-profit biological resource centre (BRC) and is a private research organisation. The centre aims at the acquisition, authentication, production, preservation, development and distribution of reference standards for microorganisms, cell lines and other research materials in the field of life sciences. A group of scientists realised the necessity of a collection of microorganisms with a purpose to serve the scientists worldwide, and this resulted in the establishment of ATCC in 1925.

ATCC possesses a broad collection of biological research materials, which includes microorganisms, cell lines, molecular genomics tools and bioproducts. More than 3,400 human, animal and plant cell lines are available at ATCC. The collection of molecular genomics includes eight million cloned genes from human, mouse, soybean, rat, monkey, zebrafish and several disease vectors. In addition, collections of protozoans, yeasts and fungi having more than 49,000 yeast and fungal strains and 2,000 strains of protists are maintained at ATCC. The collection of microorganisms includes over 18,000 strains of bacteria, 2,000 different animal viruses and 1,000 plant viruses. Entomopathogenic viruses, bacteria and fungi form a part of the collection of microorganisms at ATCC (cf. ATCC website).

16.3   World Federation for Culture Collections

The World Federation for Culture Collections (WFCC) is a multidisciplinary commission of the International Union of Biological Sciences (IUBS) and a federation within the International Union of Microbiological Societies (IUMS). The centre is involved in the collection, authentication, maintenance and distribution of microbial and cell cultures. It aims at promoting and supporting the setting up of new culture collections and their related services, providing liaison and establishing an information network between the collection centres and their users, organising workshops and conferences, bringing out publications and newsletters and also ensuring the long-term growth of important collection centres.

The WFCC is a pioneer in the development of a database, WFCC World Data Centre for Microorganisms (WDCM), on culture collection centres around the world for international use which is an important resource for information on all microbiological activities. WDCM also acts as a centre for data activities among the members of WFCC. The National Institute of Genetics (NIG) at Japan maintains this database and has close to 476 culture collections from 62 countries as records. The data on the organisation, management, services and scientific interests of the collections are stored in this database.

The Culture Collections Information Worldwide which includes CCINFO and STRAIN is a database management system for culture collections in the world. STRAIN is a database which is composed of the list of holdings present in the culture collection centres around the world, and CCINFO maintains a directory of all the culture collections of the world which are registered. The information on 671 culture collections located in 70 countries and regions is provided in the database, including 2,421,998 microorganisms (bacteria, 1,034,647; fungi, 706,083; virus, 37,666) (cf. WFCC website).

16.4 Microbial Type Culture Collection

The Microbial Type Culture Collection (MTCC) and Gene Bank is a national facility at the Institute of Microbial Technology (IMTECH) located in Chandigarh, India. It was established in 1986 as a joint effort of the Department of Biotechnology and the Council of Scientific and Industrial Research, Government of India. It is a registered member of the WFCC and WDCM. Similar to other culture collection centres, it aims at serving as a depository, at supplying authentic microbial strains and at delivering appropriate services to the scientific community affiliated to universities, research institutions and industries.

The MTCC, on 4 October 2002, was recognised as an International Depositary Authority (IDA) by the World Intellectual Property Organization (WIPO) located at Geneva, Switzerland. It is the first IDA established in India. Under the Budapest Treaty, the deposit of microorganisms is recognised in 55 member countries so as to fulfil the requirements of patent procedures in all these countries.

Currently, the MTCC houses five sections. These sections include the cultures of actinomycetes, bacteria, fungi, yeasts and plasmids. Collectively they hold more than 9,000 cultures. In general, the cultures are preserved under liquid nitrogen and by freeze-drying. The fungal strains are also preserved in mineral oil. In order that the cultures continue to represent the strains which were originally deposited, viability of the strain and some other key characteristics of the organisms are assessed periodically, although some specific properties like secondary metabolite production, degradation of specific compounds and so on are not assessed on a normal basis. It also maintains a database which provides relevant information on the microbial strains available at the MTCC (cf. MTCC website).

16.5 National Fungal Culture Collection of India

India hosts a vast diversity of fungal resources. Moreover, India has one of the largest fungal biodiversity gene pools in the world. Since decades, mycologists have been researching on fungi and have reported and described new and interesting fungi from India. Long-term preservation of fungal strains in the form of pure cultures allows the availability of these fungal strains for scientific and industrial research.

The Agharkar Research Institute at Pune maintains the National Fungal Culture Collection of India (NFCCI) which is a unique national facility established by the Department of Science and Technology, Government of India. It is also a registered with the WFCC and the WDCM. This facility aims at conserving fungal strains as genetic resource pools for use in future scientific research and to provide services to academia and industry. Over 2,800 fungal strains belonging to different groups are included in the NFCCI repository (cf. NFCCI website).

16.6   Agricultural Research Station Collection of Entomopathogenic Fungal Cultures

The Agricultural Research Station Collection of Entomopathogenic Fungal Cultures (ARSEF), established in the early 1970s, is one of the culture collections worldwide which is concentrated mainly on entomopathogens. This culture collection centre aims at providing fundamental support for research on the fungal pathogens of invertebrates in both basic and applied fields.

The ARSEF culture collection centre is run by the Biological Integrated Pest Management Research Unit of the USDA Agricultural Research Service. ARSEF is one of the largest germplasm collections of fungal pathogens of invertebrates globally. ARSEF is also registered with the WFCC and WDCM. The ARSEF cultures are preserved as cryogenic stocks in liquid nitrogen. Lyophilised forms of the cultures which can tolerate lyophilisation are also available.

The culture collection holds fungal strains which are isolated from insects, other arthropods and nematodes. This facility renders general research resources for the isolation, collection, preservation and distribution of these fungal strains. The collections served to distribute the fungal strains to researchers worldwide on request. It also seeks to acquire fungal strains. Fungal systematics, fungal cytology, methodologies for long-term preservation of fungal germplasm and pathobiology are the areas of basic research which are directly associated with this culture collection. The slide collection and the collection of herbarium specimens at ARSEF extend support to researchers in the field of fungal taxonomy and also aid in the diagnoses and identification of fungal pathogens associated with invertebrates. ARSEF also provides free-of-charge services for the identification of fungal cultures and follows up-to-date fungal taxonomic classifications, and the curator, Richard A. Humber, of the culture collection centre is also keen at receiving any information about nomenclatural or taxonomic changes or possible misidentifications involving any ARSEF strains. All these efforts are made with an intention to provide accurately identified and pure fungal cultures to the researchers in this field (cf. Humber, ARSEF website).

16.7   University of California Davis Nematode Collection

The University of California Davis Nematode Collection (UCDNC) is one of the largest collections of nematodes worldwide. It is a very comprehensive collection of various types of nematodes including freshwater, marine, free-living soil and plant-parasitic nematodes. The collection includes samples in the form of slides and wet specimens from more than 90 countries including 47 of the USA, representing a wide geographical area. The nematode collection largely conserves original specimens of the newly described species, research-associated voucher specimens and specimens which were collected from different parts of the world. The slide collection of the type specimens, ranging over 11,000 specimens, includes 3,454 primary types representing 930 species and 14 of the 18 recognised orders. More than 53,000 specimens representing 2,221 species belonging to 16 orders are included in the general slide collection. These specimens are of utmost use to researchers in the field of comparative morphology, taxonomy, ecology, evolution, biological control and agriculture. The UCDNC, having a huge reference source of identified specimens, is extremely useful for the precise identification of nematode specimens, in turn helping the nematology-related research. The huge catalogued wet collection of the UCDNC is highly appreciable. This includes 5,717 samples from 96 localities, out of which 926 are from the USA and 4,791 are from the other parts of the world. In collaboration with the laboratory of Dr. Harry Kaya, UCDNC maintains live cultures of entomopathogenic nematodes. This collection is currently represented by 21 species and 30 isolates from the USA and other locations worldwide (cf. UCDNC website).

16.8   Conclusions and Future Prospects

With the ever-increasing number of microbes being identified and with the advancements in molecular and genetic tools, the role of culture collection centres in maintaining, identification and supply of microbial cultures for academic and industrial use is gaining importance. Establishment of more specific culture collection centres with the financial support of the government or non-profit private organisations is being looked forward to.

Acknowledgements   KSS is grateful to the Science and Engineering Research Board, Government of India, for support through the fast-track Young Scientist project.

References

ARSEF website: http://www.ars.usda.gov/News/docs.htm?docid=12125. Date last accessed: 12.01.2015

ATCC   website: http://www.atcc.org/en/Products/Collections/Microbiology_Collections.aspx.
    Date last accessed: 12.01.2015
Humber RA: http://vivo.cornell.edu/display/individual10325. Date last accessed: 12.01.2015
MTCC website: http://mtcc.imtech.res.in/. Date last accessed: 12.01.2015
NFCCI website: http://nfcci.aripune.org/. Date last accessed: 12.01.2015
UCDNC website: http://nematology.ucdavis.edu/about/facility/collection.php. Date last accessed:
    12.01.2015
WFCC website: http://www.wfcc.info/. Date last accessed: 12.01.2015

Index

© Springer International Publishing Switzerland 2015

K.S. Sree, A. Varma (eds.), *Biocontrol of Lepidopteran Pests*, Soil Biology 43,

DOI 10.1007/978-3-319-14499-3

Printed by Printforce, the Netherlands